Springer Series in Molecular Biology

Series Editor: Alexander Rich

Springer Series in Molecular Biology

Series Editor: Alexander Rich

J. F. T. Spencer, Dorothy M. Spencer, A. R. W. Smith (Eds.)
Yeast Genetics · Fundamental and Applied Aspects

Eugene Rosenberg (Ed.)
Myxobacteria · Development and Cell Interactions

Aharon Razin, Howard Cedar, Arthur D. Riggs (Eds.)
DNA Methylation · Biochemistry and Biological Significance

Terrell L. Hill
Cooperativity Theory in Biochemistry · Steady-State and Equilibrium Systems

Roger L. P. Adams, Roy H. Burdon
Molecular Biology of DNA Methylation

Boyd Hardesty, Gisela Kramer (Eds.)
Structure, Function and Genetics of Ribosomes

Arnold W. Strauss, Irving Boime, Günther Kreil (Eds.)
Protein Compartmentalization

Theodor Wieland
Peptides of Poisonous Amanita Mushrooms

Akira Wake, Herbert R. Morgan
Host-Parasite Relationships and the Yersinia Model

Akira Wake · Herbert R. Morgan

Host-Parasite Relationships and the Yersinia Model

With 86 Figures

Springer-Verlag New York Berlin Heidelberg
London Paris Tokyo

Dr. Akira Wake
Department of Bacteriology
National Institute of Health
10–35, Kamiosaki 2-chome
Shinagawa-ku, Tokyo 141
Japan

Dr. Herbert R. Morgan
Emeritus Prof. of Microbiology
University of Rochester
School of Medicine and Dentistry
Rochester, NY
USA

Series Editor:
Alexander Rich
Department of Biology
Massachusetts Institute of Technology
Cambridge, Massachusetts 02139, USA

Library of Congress Cataloging in Publication Data.
Wake, Akira, 1927– . Host-parasite relationships and the Yersinia model.
(Springer series in molecular biology) Bibliography: p. . Includes index.
1. Host-parasite relationships. 2. Host-parasite relationships – Genetic aspects.
3. Yersinia infections. 4. Virulence (Microbiology). 5. Infection. 6. Immune response. 7. Molecular biology. I. Morgan, Herbert R. II. Title. III. Series.
RB153.W334 1986 616.9′047 86-15584

Softcover reprint of the hardcover 1st edition 1986

2131/3130-543210

ISBN-13: 978-3-642-71346-0 e-ISBN-13: 978-3-642-71344-6
DOI: 10.1007/978-3-642-71344-6

Series Preface

During the past few decades we have witnessed an era of remarkable growth in the field of molecular biology. In 1950 very little was known of the chemical constitution of biological systems, the manner in which information was transmitted from one organism to another, or the extent to which the chemical basis of life is unified. The picture today is dramatically different. We have an almost bewildering variety of information detailing many different aspects of life at the molecular level. These great advances have brought with them some breath-taking insights into the molecular mechanisms used by nature for replicating, distributing and modifying biological information. We have learned a great deal about the chemical and physical nature of the macromolecular nucleic acids and proteins, and the manner in which carbohydrates, lipids and smaller molecules work together to provide the molecular setting of living systems. It might be said that these few decades have replaced a near vacuum of information with a very large surplus.

It is in the context of this flood of information that this series of monographs on molecular biology has been organized. The idea is to bring together in one place, between the covers of one book, a concise assessment of the state of the subject in a well-defined field. This will enable the reader to get a sense of historical perspective – what is known about the field today – and a description of the frontiers of research where our knowledge is increasing steadily. These monographs are designed to educate, perhaps to entertain, certainly to provide perspective on the growth and development of a field of science which has now come to occupy a central place in all biological studies.

The information in this series has value in several perspectives. It provides for a growth in our fundamental understanding of nature and the manner in which living processes utilize chemical materials to carry out a variety of activities. This information is also used in more applied areas. It promises to have a significant impact in the biomedical field where an understanding of disease processes at the molecular level may be the capstone which ultimately holds together the arch of clinical research and medical therapy. More recently in the field of biotechnology, there is another type of growth in which this science can be used with many practical consequences and benefit in a variety of fields ranging from agriculture and chemical manufacture to the production of scarce biological compounds for a variety of applications.

This field of science is young in years, but it has already become a mature discipline. These monographs are meant to clarify segments of this field for the readers.

Cambridge, Massachusetts

ALEXANDER RICH
Series Editor

Preface

As science progresses, its pattern changes. In 1953, Watson and Crick reported the double helix structure of DNA. I joined research life in 1954 and was encouraged when I read *The clonal selection theory of acquired immunity* published in 1957 by F. M. Burnet. At that time (1960), as a beginner in this specialized field I witnessed the first presentation by Professor Akiba of the discovery of the drug-resistant (R-) factor to the Kanto Branch Meeting of Japanese Society for Bateriology. Since then, many excellent Japanese bacteriologists have joined in this research on the R-plasmids, with the result that the traditional Japanese trend to research on host-parasite relationships, founded independently of Theobald Smith by Rokuzoh Kobayashi in the 1930's has lost its successors.

One of Kobayashi's pupils, Dr. Masami Kurokawa, knowing that I had begun plague research, asked me "What is the mechanism of septicemia?" and said that although this question has remained as enigma, plague study would offer the key to its resolution. So I continued my isolated work on plague, a disease extinct in Japan since 1929, until I was promoted to the chairmanship of our institute, the National Institute of Health, Tokyo.

I regret that my bench work had to be interrupted before more effective tools of research, e.g., Southern, Western, and Northern blotting methods and modern technology using bacterial genetics, were introduced and could result in the more sophisticated studies undertaken by many researchers in the U.S.A. and Europe. It now appears to me that "catechin", of which the peptide structures and the DNA sequences encoding its biosynthesis have been elucidated as released from reticuloendothelial cells triggered by *Yersinia pestis* organisms, might well be the mechanism of the septicemic form of plague (see Nature 320: 184, 1986).

However, I do not regret having been able to work without the restriction of the Biohazard Regulation Rules, which were only recognized as important at the end of the 1970's in Japan. Thus, I believe it was possible for me to seize the "essence" of the concept: "Host-parasite relationships" during my years of research.

Dr. Konrad F. Springer kindly suggested the possibility of this publication when I visited Heidelberg in 1980. My thanks go to staff of Springer-Verlag, Dr. Philip C. Manor in particular. Mrs. Akemi Matsumoto, a friend since 1949, has devoted her time and efforts to preparing the typewritten drafts of my manuscript which were corrected and rearranged repeatedly thanks to the cooperative suggestions of my co-author, Herbert R. Morgan. The latter lives far from Tokyo in Maryland, U.S.A. although we have seen each other several times on the occasion of international congresses. This book is thus the result of a real international friendship.

AKIRA WAKE, M.D.

Acknowledgments. We gratefully acknowledge Dr. Emil Weiss, Uniformed Services Medical School, Bethesda, Maryland 20014, US.A. for providing us with the manuscript of summarized description of *Rickettsiae*, which is used in Part I of this book. We also thank Dr. Willard T. Charnetzky, Department of Biological Sciences, Florida Atlantic University, Boca Raton, Florida 33431, U.S.A. for helping us by reviewing and giving valuable advice for the improvement of Part III of this book.

Contents

Part III The *Yersinia*-Host Relationship

Part I
Virulence Factors and Their Genetic Control

The purpose of this book is to illustrate the dynamic state of *infectious disease* as new scientific knowledge is acquired. The fundamental concept is to analyze disease as a product of the ecology of microorganisms within a particular environment which in this case is a living macroorganism, the host, with both interacting organisms subject to genetic control. The objective of this study is the analysis of these two interacting, genetically controlled systems, i.e., host-parasite relationships.

This analysis is presented in three parts. Part I is devoted to the consideration of microorganismic factors which make them pathogenic, while Part II considers host factors which play defensive roles against microorganisms; Part III describes the interactions between both classes of factors which are expressed in a particular field of host-parasite relationships, i.e., *Yersinia* infection.

Pathogenicity is the ability to produce disease, which is operationally divided into three categories: (1) superficial infections, (2) exudative (partially invasive) infections, and (3) invasive (penetrating) infections. Due to the recent progress of the research in this field, substances responsible for producing each category of disease have been identified and in some cases even isolated which makes it possible to delineate the molecular mechanisms of certain pathogenic processes.

An erroneous concept exists that all of these pathogenic factors are toxins or of similar nature; however, many complex mechanisms of pathogenesis that are nontoxic per se play roles in the production of disease.

Disease virulence is the degree of pathogenicity of microorganisms as indicated by case fatality rates in the corresponding disease and/or its ability to invade the tissues of the host-macroorganism. It is measured experimentally by determining the median lethal dose (LD_{50}) or median effective dose (ED_{50}).

The elucidation of the genetic controls of the factors responsible for the virulence of microorganisms is a remarkable contribution of modern microbial genetics to our understanding of infection which has also opened new avenues for future research. It has become clear that a number of microorganisms cannot be virulent if they do not harbor small genetic components, i.e., plasmids or phages. In other words, certain nucleotide base sequences found in plasmids or phage nucleic acids contain the information that transforms the otherwise innocuous microorganisms into pathogenic parasites.

Chapter 1 Microorganisms

The biological sphere of the earth is a complex system of ecological chains of animals, plants, and microorganisms. In considering this complex system, this book focuses on the association between mammalian forms, especially man, and disease-producing microorganisms. The definition of a microorganism has been considered to require the presence of a nucleic acid until recently when Prusiner (1982) reported that novel proteinaceous infectious particles called "prions" cause the disease "scrapie" in sheep, and probably Creutzfeld-Jacob syndrome in man. These agents appear to lack many of the properties of classical microorganisms, most significantly nucleic acid. Aside from this newly proposed "microorganism", living forms multiply, differentiate, and mature according to their own unique patterns directed by codes preserved in their genetic material which may have DNA or RNA as the information source. These genetic codes, consisting of nucleotide triplets, are found universally in all living forms on the earth. The sequences, length, and structures of these codes are unique to each species of organism, and differences can be found even within a single species.

The hereditary continuity of genetic information is preserved within the nucleic acids (DNA or RNA), because these molecules themselves are self-replicative. The nucleic acids alone may serve as the simplest of infectious agents as viroids (naked nucleic acids) which can replicate within the cells of other living organisms.

Viruses consist of nucleic acids either DNA or RNA with proteins and sometimes lipids which are assembled in their specific host cells under the direction of the viral nucleic acids before their release from the cell as nonliving extracellular organisms. As these viruses are assembled they may incorporate host-cell components as in the envelopes of viruses that bud from host-cell membranes – nuclear or cytoplasmic. On the other hand, certain of the intracellular viral DNA's can be incorporated into host-cell chromosomes and remain there and replicate. In the case of viral RNA, this occurs after its translation into proviral DNA which is then integrated into host-cell DNA.

Viruses also infect unicellular organisms, e.g., bacteria as their "host" and are called bacteriophages or phages. Other unicellular organisms, e.g., actinomyces, fungi, or mycoplasmas, are infected by their own parasitic "phages". Usually this infection occurs by the attachment of the tail-end of the phage envelope to the receptor of the microorganismic cell surface. Then the phage nucleic acid (DNA or RNA) is injected into the host cell. Phage DNA can also be incorporated into the chromosome of the bacterial or other unicellular host microorganisms, after infection. Genetic information of a host-bacterial cell can translocate itself into another cell via phages by transduction after being attached to the phage DNA. This is carried out by lysogenic phages (i.e., non-lethal phages), which translocate certain DNA fragments from one bacteria to another. On the other hand, lysogenic conversion means the change of some bacterial genetic character by the presence of the phage DNA itself.

In addition to this "host-parasite relationship at the cellular level", viruses relate to their multicellular, differentiated host in more general ways, i.e., route of infection, spread to organs, etc. These complex relationships are so varied that independent monographs are needed to discuss "viruses and host-parasite relationships" and they will not be included in this presentation.

However, it is noteworthy that recent studies of microorganisms have shown the importance of viruses and viruslike particles (phages or plasmids) for the development of the pathogenic processes caused by unicellular pathogens, e.g., bacteria. Examples of their action will follow throughout Part I, beginning with the role of β-phage in the pathogenesis of diphtheria, and Part III, ending with the role of the plasmid family in the immunogenetics of plague.

This monograph will review host-parasite relationships from the viewpoint of "triple biology" – mammalian hosts parasitized with microorganisms that harbor organized, foreign genetic particles.

Other major groups of microorganisms, bacteria, actinomyces, yeast, fungi, and mycoplasmas have more complicated body structures than viruses and multiply independent of other cells. In contrast, rickettsia and chlamydia are intracellular parasites and while having a considerable degree of independent metabolism still require the internal environment of living cells for growth.

For an understanding of the heterogeneous world of microorganisms, their biological classification must be briefly considered here.

Organisms are classified as eukaryotic or prokaryotic. Fungi, yeasts, algae, protozoa, and some others belong to the former, because their genetic materials are covered by membranes, forming a nucleus which is not found in the latter. Furthermore, they have internal organelles such as microtubules, microfilaments, mitochondria, endoplasmic reticulum, Golgi apparatus, lysosomes, microbodies, and chloroplasts, although some of these may be absent depending on the position of the organism in the classification system. All of these organelles are composed of highly developed membranous systems, unique to eukaryotic cells which are the characteristic cells of multicellular organisms, including mammals. Figure I.1 shows an ultra-thin section electron micrograph (EM) of a unicellular eukaryotic microorganism, a yeast cell which shows many of these structures. In prokaryotic organisms, the structures of the genetic materials are quite different. They are naked because of the lack of nuclear membranes, but are concentrated in one place within the cytoplasm. In spite of such a rather diffuse morphology, they are called a chro-

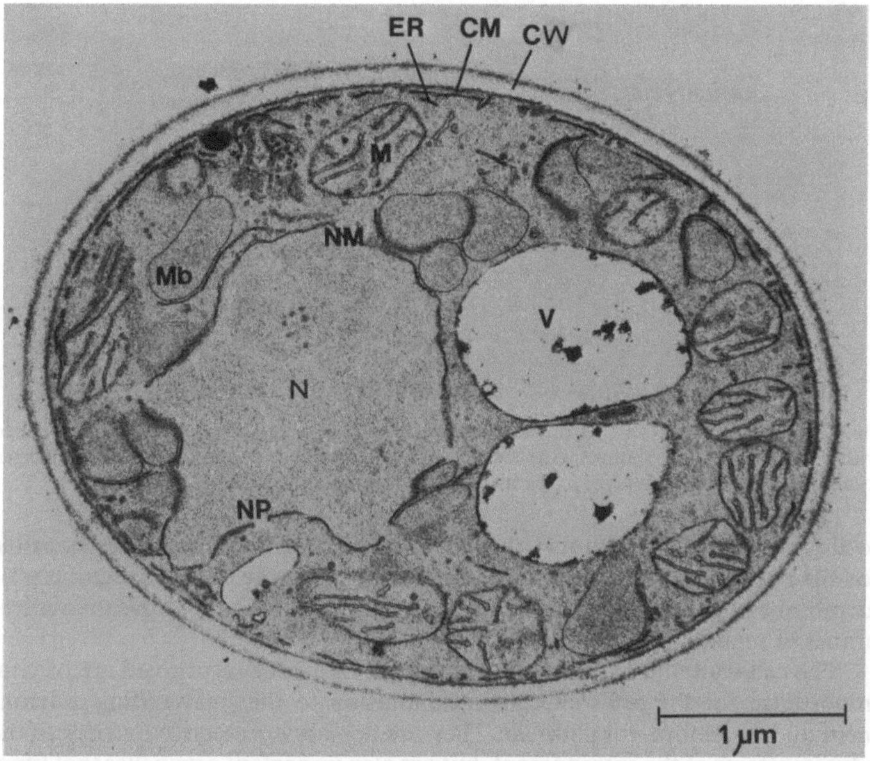

Fig. I.1. Ultra-thin section electron micrograph (EM) of a *Candida tropicalis* cell (×30,000). Fixed with glutaraldehydekaliumpermanganate. Abbreviations: *CM*, cell membrane (protoplasmic membrane); *CW*, cell wall; *ER*, endoplasmic reticulum; *M*, mitochondria; *Mb*, microbody; *N*, nucleus; *NM*, nuclear membrane; *NP*, nucleomembranous pore; *V*, vacuole. [S. Mizushima and K. Miura (1979)]

mosome(s) just as are the genetic materials of eukaryotic organisms which are enclosed within the nuclear membrane. Prokaryotic membranous systems are much simpler than those of the eukaryote as the ultra-thin section EM of *Bacillus subtilis* (Fig. I.2) shows. However, the surface structure of prokaryotic organisms is more or less sophisticated depending on the classification of the microorganisms. Mammalian cells, like mycoplasma microorganisms, have no cell wall. Their cell membranes consist of bilayers of phospholipids, with the hydrophobic lipid chains facing each other internally and the hydrophilic phosphate proteins exposed. These membranes have various cell receptors that play important roles in the cooperative functions of cells characteristic of multicellular organisms.

Outside the cell wall of prokaryotic organisms various substances are frequently found, i.e., capsular or slime layers which are encoded by the genetic elements of the cell. The nature of these substances can be unique, not only

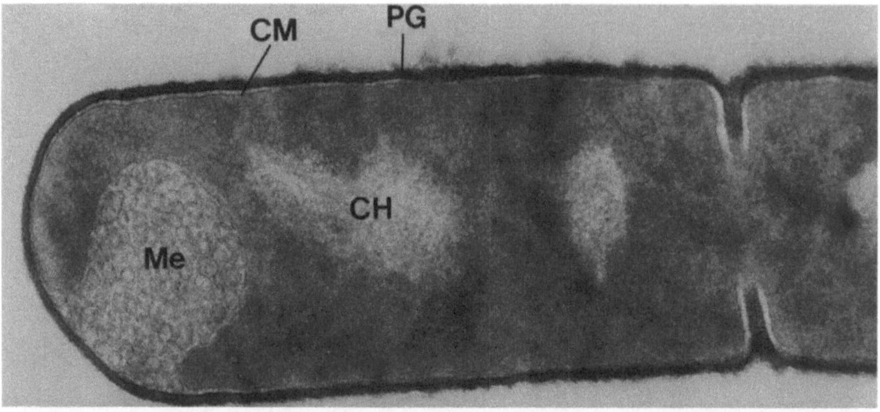

Fig. I.2. Ultra-thin section EM of *Bacillus subtilis*. Gram-positive bacteria, × 67,000; fixed with osmic acid. Abbreviations: *CM*, cytoplasmic membrane; *Me*, mesosome; *CH*, chromosome; *PG*, peptidoglycan layer (cell wall). [S. Mizushima and K. Miura (1979)]

to the species, but also to its subspecies and variants, and can be responsible for the antigenic specificity of the organism. Inside the cell wall, cytoplasmic membranes of the same structure as the cell membrane (protoplasmic membrane) of mammalian cells are found.

The cell wall and cytoplasmic membrane of a prokaryotic cell are of vital importance for the cell's successful adaptation to the surrounding environment and, therefore, for survival. They are not only protective barriers to the adverse effects of the environment, but are also important organelles that function for nutrition, excretion, and interaction with other cells. In the host-parasite relationship, every interaction of a parasite with the host starts with these microbial surface structures. Cellular functions involving enzymatic proteins functioning for energy production, energy exchange, transportation of ions and other substances, and protein biosynthesis are related to the inner cell membrane. Some of these enzymes locate themselves within periplasmic space between the inner membrane and the peptidoglycan layer of the cell wall.

Peptidoglycan is the basal structure of the bacterial cell wall. It is principally composed of parallel threadlike chains of N-acetylmuramic acid bridged with short peptide chains between the muramic acid molecules. A high percentage of D-amino acids are incorporated into these peptides and this makes the peptidoglycan resistant to physicochemical and enzymatic degradation (Fig. I.3).

Before the elucidation of the chemical structure of peptidoglycan, bacteria were empirically divided into Gram-positive and -negative groups. In the former group, a thick consolidated peptidoglycan layer associated with acidic polysaccharides like teichoic acids makes decoloration treatment in the Gram's stain ineffective. The latter Gram-negative group has a similar chemical structure of peptidoglycan, but a smaller percentage (20–30%) of N-acetylmuramic acid-N-acetylglucosamine chains which are bridged with two tetra-

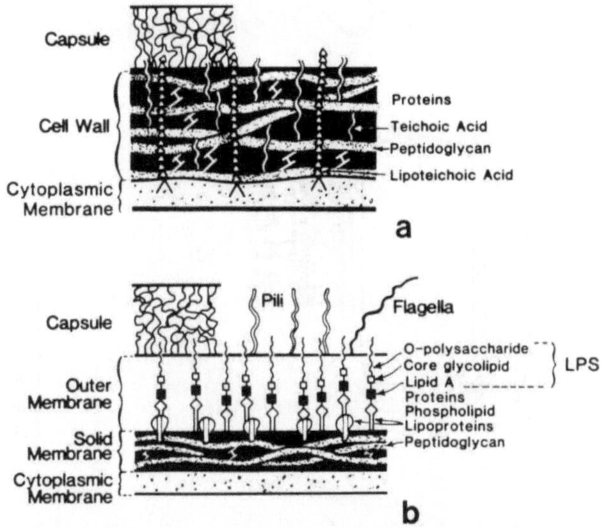

Fig. I.3. Cell-surface structures of Gram-positive (**a**) and Gram-negative (**b**) bacteria. (Courtesy of Prof. P. G. Quie)

peptides containing meso-2,6-diaminopimelic acids. This peptidoglycan layer is thinner and thus permits decoloration in the procedure of Gram's stain. Thus, the classification based on the Gram's stain is not only a powerful tool for taxonomy, but also indicates structural differences in the biochemical constituents of the peptidoglycan layer.

Further studies have revealed an "outer membrane" of Gram-negative bacteria composed of proteins, lipopolysaccharides, and phospholipids which covers the peptidoglycan layer. A lipopolysaccharide (LPS) of this membrane, is called endotoxin because of its pharmacological effects. It is composed of O side chains, core polysaccharide (KDO) and lipid A as is shown in Fig. I.4. This LPS plays an important role in the "host-parasite relationship" as will be discussed in Part III.

Many prokaryotes show active motility which is an important marker in taxonomy. The organelles responsible for this motility are flagella shown in Fig. I.4. The number of flagella varies depending on the species. The importance of motility for virulence will be discussed in Chap. 4. Morphologically similar to flagella, but much smaller, are bacterial pili which are also attached to the bacterial surface. Some of them relate to bacterial virulence, and others to the sexuality of bacteria, i.e., the transfer of genetic traits from a male individual cell equipped with pili to a female cell lacking them (Jacob and Wallmann 1961).

However, in general, the organelles of prokaryotes have simpler structures than those of eukaryotes. For example, the mesosomes of prokaryotic *Bacillus subtilis* recognizable in Fig. I.2 appear to be formed by internalization of the cytoplasmic membranes, while the endoplasmic reticulum, mitochondria, microbodies, nucleus, nuclear membrane, nuclear pores, and vesicles of the eukaryotic yeast cell (Fig. I.1) have complex organizations.

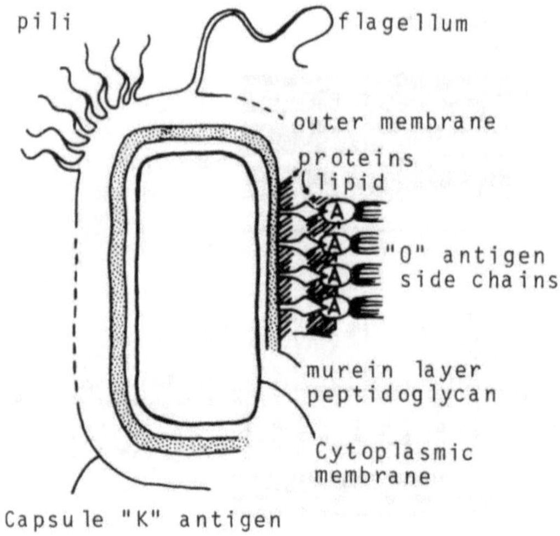

Fig. I.4. Major cell-wall antigens of *Escherichia coli* (core polysaccharide, i.e., KDO is localized between Lipid A and "O" antigen side chains. (Courtesy of Prof. P. G. Quie)

Other criteria for the classification of microorganisms are: obligatory or facultative anaerobiosis or aerobiosis, spore formation under certain physicochemical conditions, various metabolic pathways, and antigenic characteristics. Detailed discussion of these will not be included here.

Chapter 2 Host-Parasite Relationship

Any circumstance that provides the physicochemical conditions and nutritional sources suitable for the replication of microorganisms provides their ecological niche for growth. Multicellular organisms, including man and other mammals, also occupy their niche in the natural environment and, in turn, they provide on their surfaces living space for many microorganisms, so that all hosts have a natural microbial flora. It is only by using very artificial techniques that it is possible to produce and maintain "germ-free animals" which have been very valuable in medical research, e.g., to study the roles of normal flora.

A great number of microbial species live in the alimentary, respiratory, or urogenital canals of animals and man, and on their epithelial, mucous-secreting surfaces. These canals can be regarded as an extension of the external surfaces of these hosts which together form the boundary between the exterior and interior world of the host. However, some species of microorganisms live within the interior world as well.

Even a "healthy" human or other mammal with its normal flora is in itself a complex ecosystem, consisting of "host" and "parasites". Mutual interactions exist among the microbial components of the normal flora and in turn with the host itself. These interactions of host and microbe can be studied individually in germ-free animals which are subsequently monocontaminated to elucidate host relationships with a single microorganism. However, natural hosts live "healthy" lives in coexistence, with their normal flora, i.e., "in balance" with these parasitic microorganisms.

A disturbance of the balance between the host and an indigenous parasite is one of the causes of disease. For example, a usually nonpathogenic fungus *Candida albicans,* living in the human gastrointestinal tract can cause prolonged diarrhea when the human host cannot secrete a protein, immunoglobulin A, from the mucous membrane of his gut. In this case, multiplication of this microorganism is released from the limiting action of the secretory immunoglobulin A and it becomes predominant in the gut, causing irritation and subsequent diarrhea. In other instances, many inhabitants within the host can be released from their host balance by the presence of host-genetic defects or environmental changes, including radiation, misuse of drugs and antibiotics, malnutrition, etc.

In special cases, a cell of a multicellular organism itself can lose its normal identity converting into an abnormal cell which is released from the normal controls of the body, to cause a neoplasm. The special field of oncology studies these phenomena in detail, and overlaps with the field of host-parasite relationships because some viruses trigger this conversion of a normal cell to a malignant state.

Uncontrolled multiplication of microorganisms confined within the host is called infection and "opportunistic infections" develop when a resident microorganism enters a state of uncontrolled replication, e.g., as in the case of *Candida* as discussed previously. Resident organisms producing such infections are called "opportunistic pathogens".

On the other hand, there are some species of microorganisms that *usually* cause disease after they gain entry to the host and multiply and cause an infection. This potential to cause disease is called pathogenicity, and the microorganisms are called pathogenic. However, not all infections with pathogenic microorganisms lead to disease of the host so that the "onset of disease" should be discriminated from infection. In other words, microbial infections lead, sometimes, to diseases or in other cases, to inapparent or latent infections. There are also acute, subacute, or chronic types of infection depending on the unique characteristics of microorganisms and the state of resistance of the host. Furthermore, sometimes after recovering from disease the host coexists with the particular pathogenic microorganism without symptoms in the "carrier state", and thus may become a dangerous source of new infection for the associates of the carrier host.

Not all microorganisms of each pathogenic species can infect and cause disease in every suitable host. As will be described in Part II, multicellular hosts have developed complicated, multifaceted defense mechanisms during their millions of years of biological evolution which prevent their infection by many microorganisms. Humans and other mammals that have appeared in the latest period of evolutional history, have at least three facets of defenses; first, the cutaneous or mucous membrane boundaries as physical barriers between the external and internal world of the individual host; secondly, humoral molecules which prevent infection by the pathogens; and thirdly, cellular components that act as unicellular elements in resistance mechanisms. Some of these cellular components, phagocytes, engulf and destroy the pathogenic microorganisms. Others recognize and destroy virus-infected cells before the production of virulent viral particles occurs. If all facets of these defense mechanisms work in perfect cooperation, infection will not occur.

Among the pathogenic microorganisms, there are individuals that possess special properties which are produced by the expression of particular genetic codes and which enables them to produce infection. This genetic information plays a role in overcoming one or more facets of the complex host-defense mechanisms. This equipment is called a *virulence factor,* and the genetic code responsible for its expression, a *virulence determinant*. For any pathogenic microorganism to establish infection, one or more facets of host-defense mechanisms must be overcome.

The defense mechanisms of each mammalian species have more or less unique characteristics and, therefore, a particular pathogenic microorganism has its special susceptible or resistant hosts. This can be explained by the interrelationships between the virulence factors of the microorganisms and the counterpart resistance mechanisms of the host. Taking this specificity into consideration, the ability of a particular microorganism to cause disease in its susceptible host is called "virulence", and the microbial population containing those individual microorganisms possessing virulence are called *"virulent* strains" for the particular host. Thus, there are virulent and avirulent strains even within a single pathogenic microbial species.

Since the golden period of the study of pathogenic microorganisms, beginning with the contributions by Pasteur and Koch, most investigators in medical microbiology have paid attention mainly to the microorganisms, in other words, the "parasite". A new approach to host-parasite relationships was proposed independently by Theobald Smith, an American, and by Rokuzo Kobayashi, a Japanese, in the early 1930's when they emphasized the role of the host. In addition to this concept, rapid developments in molecular biology and immunology provide other new viewpoints which reexamine host-parasite relationships. For example, the concept of the opportunistic pathogen may have to be altered to some degree since recent findings indicate that plasmid-controlled virulence factors, including those not responsible for the production of any toxic substance, may be introduced into nonpathogenic microorganisms and convert them to pathogens.

For an approach to host-parasite relationships, the following formula is proposed:

$$\text{Disease} = \frac{\text{Dose} \times \text{Virulence factors (e.g. toxins)}}{\text{Natural resistance} \times \text{Normal flora}}$$

$$\frac{\times \text{Pathogenicity}}{\times \text{Host cell factors (e.g. cytokines)}}$$

$$\frac{\times \text{Adverse immune responses (humoral and cellular)}}{\times \text{Beneficial immune responses (humoral and cellular)}}$$

Of this formula, Part I deals with dose × virulence factors × pathogenicity, whereas the remaining parts will be discussed in Part II.

When this formula is applied to each pathogenic microorganism, the disease produced can be classified into the following three categories:

1. Superficial infections with or without toxinogenesis.
2. Exudative (partially invasive) infections.
3. Invasive (penetrating) infections.

Each of these disease categories will be illustrated with a few examples, and yersiniosis will be described separately in Part III as a special model of host-parasite relationships.

Chapter 3 Localization of Virulence Determinants

With the development of molecular genetics, the localization and nature of the behavior of particular DNA segments that regulate various biological characteristics (e.g., drug resistance, resistance to heavy metals, sugar fermentation) of the microorganism have been elucidated in many instances. Virulence determinants, such as toxin production, also have been identified as belonging to particular DNA segments within the *parasitic* microorganisms.

With the elucidation of the molecular events in genetic control, the unit of DNA nucleotide base sequences which can replicate independently, has been termed a "replicon". On the other hand, groups of discrete segments of DNA, which are capable of transposition from site to site within a replicon or between different replicons have been found in maize – a higher plant, at first, and then in the other living organisms, including prokaryotes. At both ends of these DNA segments, several to several tens of nucleotide base pairs are attached mutually with the same (normal repeat) or inverted (inverted repeat) sequences of direction with regard to the molecular strand of DNA. Transposition is thought to depend on enzymes, termed transponases, which act on these sequences after being biosynthesized through coding of particular nucleotide base pairs located near these terminal repeats.

The DNA segments composed of inverted repeats with the genomes coding the transponases are termed "insertion sequences: IS", and a more complex DNA segment in which genomes coding the expression of various biological characteristics are inserted between two insertion sequences, is termed a "transposon: Tn". Thus, insertion sequences (IS) and transposon (Tn) are essentially the same, but the transposition of the latter is detected by readily selectable traits, i.e., by the biological characteristics expressed. The types of transposon-promoted events are shown diagrammatically in Fig. I.5.

Therefore, transposons cause mutation in genes into which they insert themselves, alter the expression of genes near the insertion site, promote genome rearrangements, and facilitate the flow of genes between different bacterial species. They are probably important determinants of the rate of evolution in natural populations of bacteria (C. M. Berg and D. E. Berg 1981). The replicons into which transposons are inserted, can be chromosomes, phages, or plasmids.

Phages are unique replicons because they are the viruses that infect microorganisms as aforementioned with regard to "phage transduction or conversions", while plasmids are genetic factors that multiply extrachromosomally

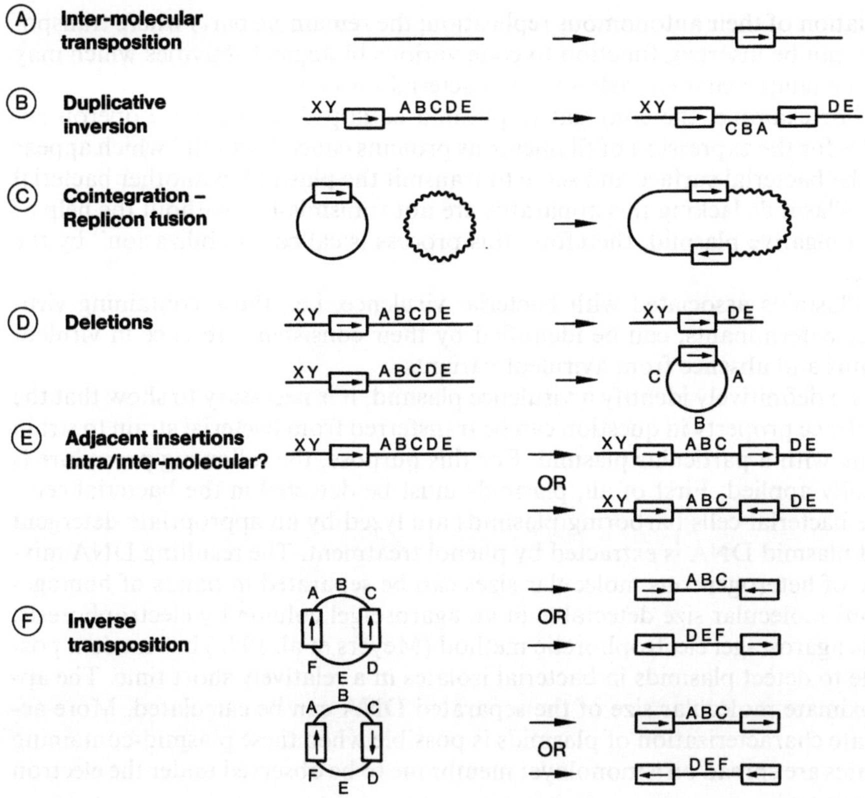

Fig. I.5. Types of transposon-promoted events. In these diagrams, *squares* indicate transposable elements, i.e., IS or transposon base pairs; *solid* or *wavy lines*, DNA base pairs on which genes detectable by the biological functions are located in sequence, e.g., *X, Y, A, B, C, D,* and *E.* Events promoted by single transposable elements are shown in *panels* **A–E.** Composite transposons can also promote intermolecular inverse transposition (**F**) and its intramolecular equivalent, an adjacent inversion-plus deletion event. **A** and **C** Insertion and transposition occur between two different base pairs. **B** Increase in the number of transposons or additional transposition within one molecule of the base pair. **C** One cointegrate is formed by the bindings of two base pairs being mediated by the duplicated transposon. **D** and **E** Inversion and deletion occur within the same base pair. This event of deletion accompanies the release of a partial base pair ring. **F** Two base pairs between two transposons can function as one large transposon as a whole. (From N. Klecker 1981)

within bacterial cells and are transferred to the cell progeny and maintained there stably by cell division. All known plasmids are parasitic factors and not essential to cellular life since bacteria can be "cured" from plasmids by certain treatments, e.g., acriflavine, acridine orange, ethidium bromide, sodium dodecyl sulfate (SDS), or cultivation at higher temperatures and survive.

Actually, plasmids exist as twisted forms of double-helixed DNA rings and the most essential part of their nucleotide base sequences is responsible for the initiation of their autonomous replication; the remaining parts where transposons can be inserted, function to code various biological activities which may induce unique characteristics in the bacterial host cell.

For example, the conjugative plasmid equipped with the *tra* operon encodes for the expression of filamentous proteins called "sex pili" which appear on the bacterial surface and serve to transmit the plasmid to another bacterial cell. Plasmids lacking this apparatus are not transmissible without the help of a conjugative plasmid; therefore, this process is called "mobilization" by the latter.

Plasmids associated with bacterial virulence, i.e., those containing virulence determinants, can be identified by their consistent presence in virulent strains and absence from avirulent variants.

To definitively identify a virulence plasmid, it is necessary to show that the virulence property in question can be transferred from bacterial strain to strain along with a particular plasmid. For this purpose, the following procedure is usually applied: First of all, plasmids must be detected in the bacterial cells. The bacterial cells harboring plasmids are lyzed by an appropriate detergent and plasmid DNA is extracted by phenol treatment. The resulting DNA mixture of heterogeneous molecular sizes can be separated in bands of homogeneous molecular size detectable in an agarose-gel column by electrophoresis. This agarose-gel electrophoretic method (Meyers et al. 1976) has made it possible to detect plasmids in bacterial isolates in a relatively short time. The approximate molecular size of the separated DNA can be calculated. More accurate characterization of plasmids is possible when these plasmid-containing lysates are spread on a monolayer membrane to be observed under the electron microscope.

In both cases, reference plasmids of known molecular sizes are simultaneously measured to estimate the molecular sizes of the plasmids under study in megadaltons (Mdal).

Molecular cloning (So et al. 1977) is a further advanced tool to characterize virulence plasmids. In this process, both the putative virulence plasmid and known plasmids, usually a derivative of Col El, are digested with a restriction endonuclease, e.g., Hind III to generate linear fragments with short cohesive ends. These ends can reassociate by reforming circular plasmid molecules.

In order to introduce the resulting hybrid plasmid into a new *E. coli* host, the cells are washed first with cold $MgCl_2$ solution followed by washing with 0.05 M $CaCl_2$ solution, and suspended in the latter solution under cold temperature. When DNA molecular rings, e.g., the hybrid plasmids are added to this bacterial suspension, they are taken up by the calcium-pretreated cells. This procedure is called "transformation". By the transformation procedure, closed-circular (twisted) forms as well as open-circular forms of DNA can enter the host cell.

The finding of "minicells" by Adler et al. (1967) provided another tool useful in a variety of biological studies, including the collection of large amounts of the hybrid plasmid containing the virulence determinant or its carrier plasmid, e.g., Col El derivative, free from chromosomal DNA. Minicells were unusually small anucleate cells that are regularly produced by a newly isolated mutant strain of *Escherichia coli* K12 during the logarithmic phase of growth. The minicell-producing strain was derived after treatment of a log-phase nutrient-broth culture with triethylenemelamine. Afterwards, minicell-producing mutant strains that divide abnormally at a location so extremely close to the cell wall that they produce tiny spherical "minicells" lacking DNA have also been isolated in *Salmonella* organisms and *Bacillus subtilis*, etc. Minicells do not divide, but they may be isolated from the normal, rod-shaped cells by density-gradient centrifugation. The purified minicells contain normal amounts of protein and RNA, but are deficient in chromosomal DNA and several enzymes that use chromosomal DNA as a template or substrate. However, the parent minicell-producing bacteria that harbor plasmids can deliver them into segregated minicells. Therefore, purified minicells can be used as a source of not only plasmids, but also of the proteins synthesized in minicells by the carrier plasmid (Col El derivative) and the hybrid plasmid containing the virulence determinant. By the comparison of the sodium dodecyl sulfate (SDS)-polyacrylamide gel electrophoretic patterns of these proteins, the specific polypeptides associated with plasmid-mediated virulence can be identified.

Thus, a given virulence determinant can be cloned into a suitable vector organism and the specific polypeptides associated with plasmid-mediated virulence can be identified by the applications of the above-mentioned techniques. For the studies of pathogenesis of divergent bacterial isolates or species with expressions of similar virulence, investigators often want to identify the presence of homologous nucleotide sequences that are the virulence determinants shared in common among these divergent bacteria. For this purpose, the following filter blot hybridization technique of Southern (1975) is effective.

After digestion of the probe plasmid DNA with a restriction endonuclease that cuts the DNA nucleotide base sequences at the site specific for it, the mixture of fragments is subjected to agarose-gel electrophoresis. The fragments are denatured into dissociated single-stranded DNA by heating in situ, and transferred (blotted) onto strips of cellulose nitrate. The cloned fragments carrying virulence determinant genes that had been labeled with radioactive compounds are hybridized with the fragments on the blots to detect the homologous one by autoradiography. Sites where the probe hybridized with fragments of the test DNA appear as spots.

By the application of these techniques, the nature of virulence has been disclosed in some cases as will be described in examples in Chap. 4.

Chapter 4 Superficial Infection With or Without Toxinogenesis

Pathogenic organisms causing superficial infections colonize certain local parts of the surface barriers of the host body and initiate multiplication. The localizing part is generally unique to each microorganismic species, e.g., either of skin, respiratory or digestive tract, etc., but infection is usually limited without spreading from that region. The reason pathogenic microorganisms cause diseases in contrast to organisms of the normal flora, is that they succeed in establishing dominant "colonization" and usually elaborate toxins or other bioactive substances which aid the invasiveness of these microorganisms into the adjacent epithelial stratums. However, the absence of additional virulence factor(s) required for penetration deeper into the tissue results in the characteristic process of exudative rather than invasive infection. Therefore, it is critical for extensive disease production that the organism elaborates toxins or other bioactive substances. Adhesive factors are also important because they make it possible for the organism to bind to the host-cell surfaces and thus to establish itself on these epithelial surfaces to initiate infection.

Example 1: Corynebacterium diphtheriae

The causative agent of diphtheria, *C. diphtheriae,* colonizes primarily the mucous membrane of the upper respiratory tract and releases an exotoxin which causes the development of local necrosis with characteristic patches of pseudomembrane which are formed as the result of plasma and leukocyte exudation, i.e., superficial inflammation of the nasopharyngeal region. This organism is selected for discussion because of its unique status among pathogenic bacteria in that the production of the main virulence factor, diphtheria toxin, is controlled by a chromosomal gene named *tox* which can be incorporated into the genome of a bacteriophage named β-phage (Chap. 1). *C. diphtheriae* organisms without the *tox* gene successfully colonize the local site, but are completely innocuous to the host. Lysogenization of such *C. diphtheriae* organisms with the β-phage results in toxin elaboration by the bacteria (lysogenic or phage conversion), and production of disease. This phenomenon which may serve as a typical example of phage-mediated transfer of a bacterial virulence determinant, was disclosed as early as 1951 by the following experiments.

When Freeman (1951) compared the toxicity of phage lysates of nontoxinogenic *C. diphtheriae* organisms using phages designated A and B, he found that only the phage B-lysate showed toxicity and the bacteria exposed to this lysate were converted from a nontoxinogenic *C. diphtheriae* strain to one secreting typical diphtherial toxin. It is now established that toxin synthesis is coded for by a phage-structural gene (tox^+) and that the gene serves no essential viral function. The tox^+ gene may be expressed either in the vegetative replication of phage, or when the phage is integrated as prophage, or lastly when the phage is within the bacterial cell as a nonreplicating and nonintegrated entity. The experimental results of Freeman that a phage lysate of nontoxinogenic *C. diphtheriae* organisms using phage B was toxic, can be explained by the fact that phage B was a lytic (virulent) mutant of phage β which can cause lysogenic conversion of the host bacteria. Returning to the pathogenesis of diphtheria, the biochemical mechanism of this toxic effect is unique, and it is the spread of toxin from the local site of bacterial colonization that injures the host and produces a high mortality. Therefore, the discussion of *C. diphtheriae* serves to illustrate superficial infection with associated systemic disease.

The role of diphtheria toxin as the main virulence factor in this disease was demonstrated in an unfortunate accident producing large-scale human intoxications involving 606 babies in Kyoto, Japan in 1948 which was due to diphtherial toxemia in the absence of infection. This was caused by the inadvertent immunization of these individuals with a "toxoid" preparation containing no formalin which was therefore active exotoxin. Among 68 deaths, 9 died of acute diphtherial paralysis (Kurokawa and Murata 1961). According to Barksdale's report on 54 cases published elsewhere, necrosis with ulceration at the site of inoculation of the toxoid was noted in 32 cases, local edema in 52, axillary lymphadenopathy in 7, pharyngeal paralysis as evidence by difficulty in swallowing in 34, and urinary failure in 9 (Barksdale 1960).

The chemical nature of diphtheria toxin has been investigated. Table I.1 outlines the data and shows that trypsin treatment of diphtheria toxin produces so-called nicked toxin at the initial step, by breaking one of two disulfide bonds interconnecting fragment A and fragment B of the toxin, the molecular weights of which are 24,000 and 38,000, respectively. The second step of cleavage results in the separation of these two subunits (Fig. I.6).

Table I.1. Activities of diphtheria toxin and related molecules

Preparation	Molecular weight	Enzyme activity (Aminoacyl transferase)	Toxicity
Intact	62,000	−	+
Nicked toxin	62,000	−	+
Fragment A	24,000	+	−
Fragment B	38,000	−	−
Reduced nicked toxin	62,000	+	+
crm_{197}, nicked[a]	62,000	−	−
crm_{45}, nicked[a]	45,000	+	−

[a] Antigenically similar, but nontoxic, proteins produced by *C. diphtheriae* being lysogenized by mutant phages derived from β, which has been found to be responsible for toxigenocity of *C. diphtheriae* by its lysogenization.

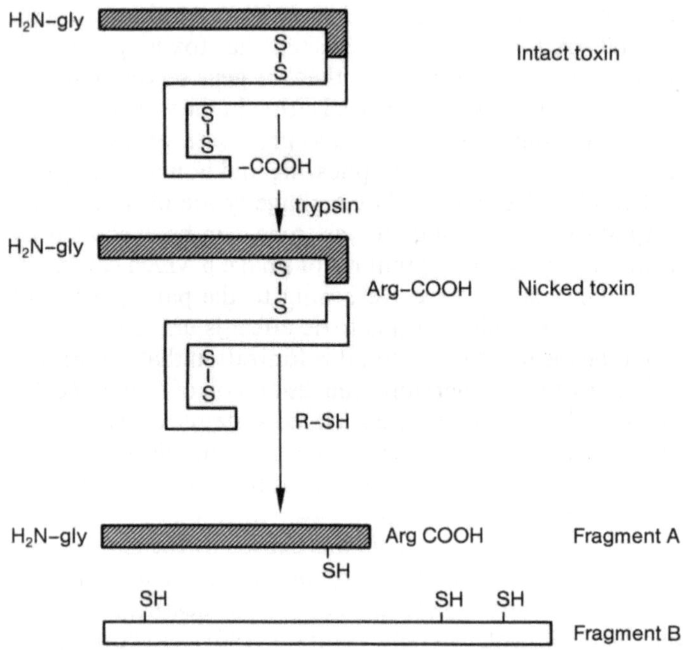

Fig. I.6. The structure of diphtherial toxin (T. Honjo 1972). Molecular characterization of each unit is given in Table I.1

The efficiency of using several phage mutants which code for antigenically similar, but nontoxic proteins (crm proteins) for the elucidation of toxic mechanisms of diphtherial toxin was demonstrated by Uchida et al. (1971).

One of these mutant phages produces a protein designated crm_{197} after lysogenization of nontoxinogenic *C. diphtheriae*. This protein has an intact B fragment and an altered A fragment, rendering it enzymatically nonfunctional (Table I.1). Due to the intact B fragment, crm_{197} absorbs to toxin-sensitive mammalian cells, but does not exert toxic activity, because its A fragment is altered to be nonfunctional. Furthermore, mammalian cells which absorb crm_{197} do not absorb added active toxin molecules as the result of competitive inhibition of the fragment B-receptor site by the crm_{197} protein.

On the other hand, another mutant phage lysogenized in nontoxinogenic *C. diphtheriae* results in the production of crm_{45} protein, which lacks a large region of the B moiety, but is equipped with a normal fragment A. This protein crm_{45} cannot act as a competitive inhibitor of toxin, nor is it itself toxic for mammalian cells (Table I.1).

On the basis of these experiments, the toxic mechanism of diphtheria toxin was finally elucidated as follows (Honjo et al. 1968, 1971).

The intact toxin molecule is absorbed to the plasma membrane of sensitive cells. But, separate A or B fragments are not toxic to mammalian cells (Table I.1). After absorption, the host cells accomplish the "nicking" and re-

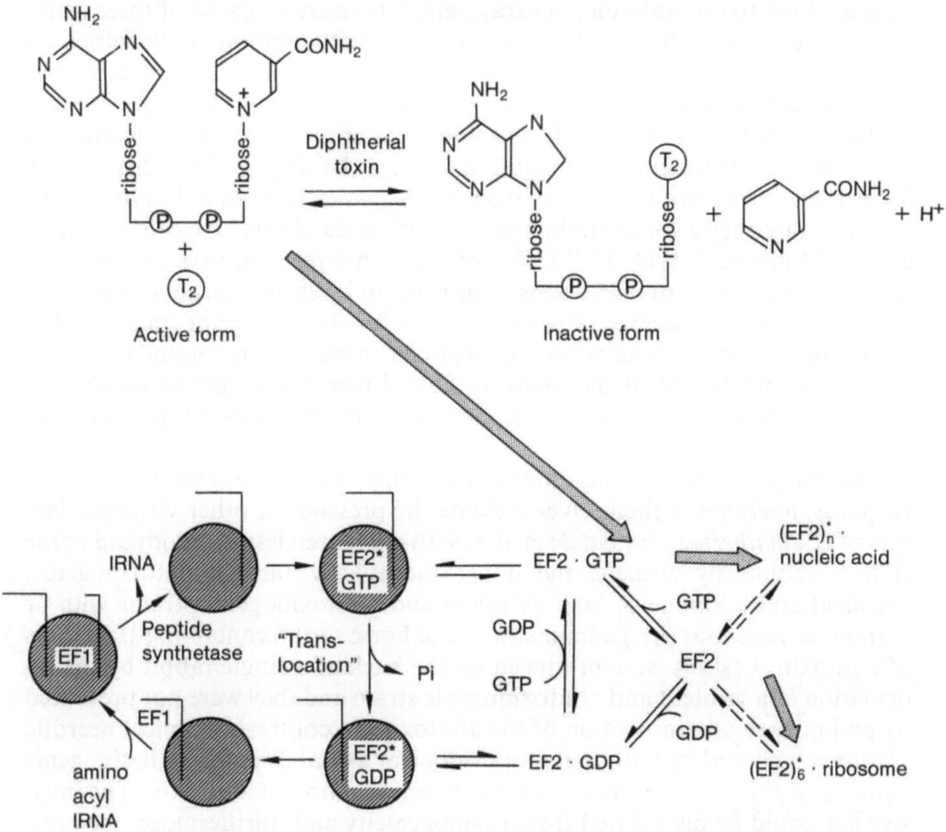

Fig. I.7. ADP-ribosylation reaction of aminoacyl transferase II by diphtheria toxin [T. Honjo, Taisha 9:206 (in Japanese)]. The aminoacyl transferase II activity inherent to fragment A of the intracytoplasmic toxin depletes one of the soluble enzymes inherent only to eukaryotic cells, i.e., the functional elongation factor (*EF2*). This depletion process mediates the linking of EF2 with the adenosine diphosphoribose moiety of nicotine adenine dinucleotide (NAD) to leave nicotinamide and H^+ ion. The probable interactions between EF2 and ribosomes, nucleotides, and nucleic acids are depicted *downwards*. A round of peptide chain elongation on a ribosome is shown diagrammatically on the *left*, the postulated EF2 cycle is shown in the *center*, and the binding observed in vitro of EF2 to free DNA, RNA, or to inappropriate sites on ribosomes is shown by the *dashed arrows* on the *right*. Such binding inhibits the diphtheria toxin, but the inhibition is lifted by adding GTP or GDP as if the EF2·GTP or EF2·GDP complexes cannot bind nonspecifically to nucleic acids, but only to a single (functional) ribosomal site. Under physiological conditions, insignificant amounts of EF2 will exist unbound to nucleotide and all of the factor is probably involved in the four-stage cycle depicted in the *center*. ADPR-EF2 binds to ribosomes in the presence of GTP, but is unable to catalyze the translocation of polypeptidyl tRNA on the ribosomes, although the associated hydrolysis of GTP continues. *EF2*:* Forms of EF2 that cannot act as substrates for diphtheria toxin

ducing of the toxin molecule, but fragment B is entirely devoid of toxic activity, while the toxic effect is a kind of enzyme activity residing totally within an active site on fragment A. After traversing the host-cell plasma membranes, the aminoacyl transferase II activity inherent to fragment A of the intracytoplasmic toxin depletes one of the soluble enzymes inherent only to eukaryotic cells, i.e., the functional elongation factor 2 (EF-2). EF-2 is indispensable for peptide formation using messenger RNA as templates by eukaryotic cells. In short, diphtheria toxin inhibits protein synthesis of sensitive cells (Figs. I.6 and I.7, Honjo et al. 1968, 1971). The difference between sensitive cells, organs, or species and resistant ones seems to be resident in the membranes of the cells which differ in their degree of permissiveness for the passage of fragment A.

These observations have been interpreted to mean that fragment B is required for attachment of the toxin to the cell membrane, and in its absence fragment A is unable to traverse the membrane and reach its cytoplasmic substrate.

Virulence discriminations among subgroups of *C. diphtheriae*, such as *gravis, mitis, intermedius* themselves indicate the presence of other virulence factors in *C. diphtheriae*. Barksdale et al. (1960) compared lesions produced in the skin of rabbits by virulent and nontoxinogenic, virulent and toxinogenic, avirulent and toxinogenic, and avirulent and nontoxinogenic strains with or without intravenous (i.v.) administration of horse serum containing 1000 units of antitoxin. Lesions were produced on the back of a single rabbit by the inoculation of a virulent and nontoxinogenic strain and they were not prevented by preliminary administration of the antitoxin. In contrast, a typical necrotic plaque was caused by diphtheria toxin alone or as elaborated by a toxinogenic strain, and this was prevented by prior administration of antitoxin. The invasiveness could be dissociated from toxinogenicity and, furthermore, the presence of a completely innocuous diphtheria bacillus was demonstrated because no lesion was produced in the rabbit where the avirulent and nontoxinogenic strain was inoculated.

Furthermore, two human cases of laboratory infections caused by virulent or invasive, but nontoxinogenic, strains demonstrated the validity of these animal experiments. Despite the fact that these individuals were immune to "diphtheria" as indicated by negative Schick tests, each developed a sore throat 48 h after the accidental swallowing or inhaling of perhaps a million organisms. About 70 h after exposure to the bacilli, a small amount of tenacious, grayish-white membranous exudate was seen on the tonsils, especially around the edges of the crypts. This probably resulted from the slight exudative reaction of the host against the invasions of the nontoxinogenic strain.

The *C. diphtheriae mitis* isolated from these patients was shown to be nontoxinogenic by the intracutaneous test in guinea pigs. Although both men were asymptomatic 36 h following the institution of antibiotic therapy, the same strain could be isolated from their throat 1 yr. after the onset of these mild infections; one of them was still carrying the organism 2 y. later (Barksdale et al. 1960). Therefore, these cases serve as evidence for the long-term persistence of a nontoxinogenic, slightly invasive *C. diphtheriae* strain which had established itself in the host-parasite relationship of carrier state.

Example 2: Vibrio cholerae and Enterotoxinogenic Escherichia coli

These two bacterial species both induce diarrheal symptoms following superficial infection of the intestinal tract. One of the virulence factors of *V. cholerae,* i.e., heat-labile enterotoxin is essentially identical with that produced by some enterotoxinogenic strains of *E. coli.*

Vibrio cholerae is a unique organism which has been selected as an example of superficial intestinal infection because it causes extraordinarily severe dehydration due to diarrhea without any inflammatory lesions of the gut epithelium. However, this phase of dehydration is preceded by the successful contact of vibrios with the gut epithelial surface and colonization after the organism overcomes the defensive barriers of the host, such as the competing normal microbial flora and the mucous layer secreted from the Goblet's cells which covers all the surface of the gut epithelium. This mucus contains immunoglobulin A antibodies with specific inhibiting activities directed against microorganisms which the host has met in the past. In addition, the mucus contains lactoferrin with so high an affinity for free iron that microorganisms without special virulence factors to take over the iron bound by the lactoferrin cannot multiply (refer to Part II).

Since superficial colonization and the production of the main virulence factors, enterotoxin, are exclusively responsible for the dehydration symptoms, the *V. cholerae* organism is not required to resist the internal defensive barriers of the host, i.e., complement, humoral antibodies, and phagocytes and it is, therefore, a prototype of superficial pathogens and has been selected for extensive discussion.

However, some diarrheal diseases caused by certain so-called NAG vibrios (*V. cholerae* except 0:1 serovar, i.e., nonagglutinable with 0:1 antiserum) show inflammatory lesions at the epithelial surface of the gut, indicating that invasion occurs with these vibrios. In fact, clinical microbiologists have been confronted with a similar difficulty in that a single bacterial species belonging to the *Enterobacteriaceae* may cause either dehydrating diarrheal symptoms or inflammatory lesions of the gut due to bacterial invasion of the epithelial tissue. These facts indicate that a better classification of disease categories could be made with regard to symptoms and pathological changes rather than one based solely on the taxonomy of the bacteria.

It is clear that the identification of virulence factor(s) causing the respective pathological changes and more fundamentally, the analysis of the virulence determinants, i.e., genes controlling these determinants, are the best approaches to the understanding of pathogenesis.

This approach has been applied in the most efficient way for the analysis of the disease produced by enteropathogenic *E. coli* which belongs to *Enterobacteriaceae*. Another distinct characteristic of *E. coli* is that its specific animal-host range is diverse, in contrast with *V. cholerae* which causes disease only in man as the natural infection. For example, there are human, bovine, and porcine enteropathogenic *E. coli*, which can be differentiated. In addition, the heat-labile enterotoxin (LT) of *E. coli* which is similar to cholera enterotoxin (CT) and with which it will be compared, is not the sole virulence factor causing diarrhea. Some enteropathogenic *E. coli* strains produce a heat-stable enterotoxin (ST) which is distinct from LT. Thus, there exist LT-only producing, and ST-only producing enterotoxinogenic *E. coli* in addition to those producing both. In addition to these enterotoxins, an enteropathogenic *E. coli* has other virulence factors that are responsible for adhesion to and colonization of the gut surface of a specific host. Furthermore, clinical studies have provided evidence for the existence of an *E. coli* strain which causes a partially invasive infection similar to that seen in shigellosis. Thus, a single species can cause a variety of diarrheal diseases.

The first major advance in the understanding of the mechanisms of diarrheal disease were the studies of Finkelstein et al. (1966) and Finkelstein and Lo Spalluto (1969) which demonstrated the role of CT as the main virulence factor in cholera. CT was purified from the sterile semisynthetic medium (Syncase) culture filtrate of a classical *V. cholerae* (serovar Inaba) strain 569B and injected into gut lumen of infantile rabbits (model of Sanarèry and Dutta); intraluminal fluid accumulation and lethal diarrhea typical of human cholera was produced. This was a serendipitous finding, because 569B was later found to be the highest enterotoxin-producing mutant among all of the 0:1 *V. cholerae* strains studied. The investigators designated this toxin "choleragen".

It was found that the complete form of the single CT molecule being excreted from the vibrio into medium consisted of two antigenically distinct polypeptide subunits designated A and B. Subunit A itself consists of A_1 with a molecular weight of 23,000 and A_2 of 11,000. Five or six subunits of B each with a molecular weight of 11,000 surround one A subunit being covalently bound to A_2.

In parallel with studies of cholera, investigators have observed the frequent occurrence of noncholera diarrheal disease caused by particular pathogenic strains of *E. coli* in developing countries where cholera is prevalent. Enterotoxins (ET) elaborated by these enterotoxinogenic *E. coli* organisms have been described by investigators who used the ligated rabbit ileal loop model as a test system (De's test or its modification; Gorbach et al. 1971; Sack et al. 1971; Kasai and Burrows 1966; Dupont et al. 1971) or by other investigators who employed the intragastric administration of millipore-filtered sterile culture supernatants of enterotoxinogenic *E. coli* strains into 7-day-old infant rabbits after overnight starvation (Gorbach and Khurana 1972). These ET have been characterized only in the most general terms and can be classified into a heat-stable enterotoxin (ST) and a heat-labile one (LT).

Intragastric administration of ST-producing *E. coli* into 3–4-day-old suckling mice produces intraluminal fluid accumulation, but this method is insensitive to the action of LT (Dean et al. 1972). By the use of this method for dis-

Table I.2. Physiochemical and biological properties of purified *E. coli* heat-labile toxin (LT) and cholera enterotoxin (CT) preparations

Properties and biological activities	*E. coli* LT	*V. cholerae* CT
Molecular weight measured by ultracentrifugation	102,000	82,000
Sedimentation constant S_{20}, w × 10^{13} (s)	5.4	5.54
Molecular weight, measured by SDS gel electrophoresis	102,000	56,000
Isoelectric point	6.90	6.60
Lipid	1%	1%
Heat stability	Inactivated at 65° C, 30 min	Inactivated at 65° C, 30 min
Acid stability	Inactivated at pH 4.0	Inactivated at pH 6.0
Antigenicity	+	+
Protection by immunization	+	+
ED_{50} dose by the use of:		
Rabbit ileal loop	4.75 μg	0.2 μg
Rabbit-skin capillary permeability	0.1 μg	0.00006–0.01 μg
Cardial muscle cell adenylate cyclase	0.65 μg	ND
Rat epididymis fat cells	ND	0.014 μg
Rounding of mouse adrenal Yl cells	10 μg	0.005 μg

Adapted from M. Ohashi and H. Zen-yoji (1977) *Escherichia coli* enterotoxins. Jpn. J. Bacteriol. 32:455–468.

crimination of ST from LT, an enterotoxinogenic *E. coli* which produces ST without producing LT has been isolated from various districts of the world (Sack et al. 1975; Ryder et al. 1976; Kudoh et al. 1977).

ST is a nonantigenic, dialyzable glycoprotein being heat-stable at 60 °C for 30 min, but inactivated at 100 °C for 30 min. It has a low molecular weight of 1,000–10,000. In contrast to LT which will be discussed as to mechanism, its biochemical mechanisms in causing diarrhea have remained unclear. Despite that, So et al. (1976) disclosed that the ST molecule is biosynthesized by encoding of the specific nucleotide sequences of structural genes having at least 1 Mdal length that are contained in the 69 Mdal plasmid designated Ent 41. The response of the small intestine to ST is characterized by an immediate accumulation of fluid in the lumen and a short-lived duration of action for the toxin. Furthermore, it has been established that the mechanisms of ST responsible for fluid accumulation in the small intestines of the infant mouse are not related to the activation of the adenyl cyclase of the intestinal epithelial cell membrane (see Table I.2). Therefore, ST is quite a distinct toxin from LT.

The difference between ST and LT was demonstrated by Gyles et al. (1969) who showed that LT was inactivated by the administration of specific antiserum or by heating the sample to 60 °C for 30 min.

Smith and Gyles (1970) differentiated two enteropathogenic strains of wild *porcine* isolates of *E. coli* into "LT-ST producers" and "ST-only producers", and demonstrated that a specialized nucleotide sequence called the "Ent gene" of a conjugative plasmid controls the transmission of two distinct plasmids, LT and ST.

E. coli LT which has a common molecular weight of about 20,000–24,000, is inhibited by antibodies to the toxin of *V. cholerae* (CT). When LT and CT are assayed with the agar-gel diffusion test of Ouchterlony using antiserum to either toxin, precipitin lines do not cross, but form a spur indicating the presence of homologous amino acid sequences in these enterotoxin molecules.

Because classical animal models, e.g., the ligated ileal loop model and the infant rabbit model, are unable to discriminate LT activity from ST, the following assay systems that are sensitive to LT, but not to ST, were employed for further studies.

1. Rabbit-skin capillary permeability test: The test samples are injected s.c. into multiple sites of shaved rabbit skin. When an appropriate dye, e.g., Evans blue or protamine blue is injected i.v. thereafter, quantitatively measurable blue-colored skin areas appear, because LT or CT enhance the permeability of blood capillary walls.

2. Tests for morphological changes of tissue-culture cells: Mouse adrenal Y1 cells, an established tissue-culture cell line originated from mouse adrenal tis-

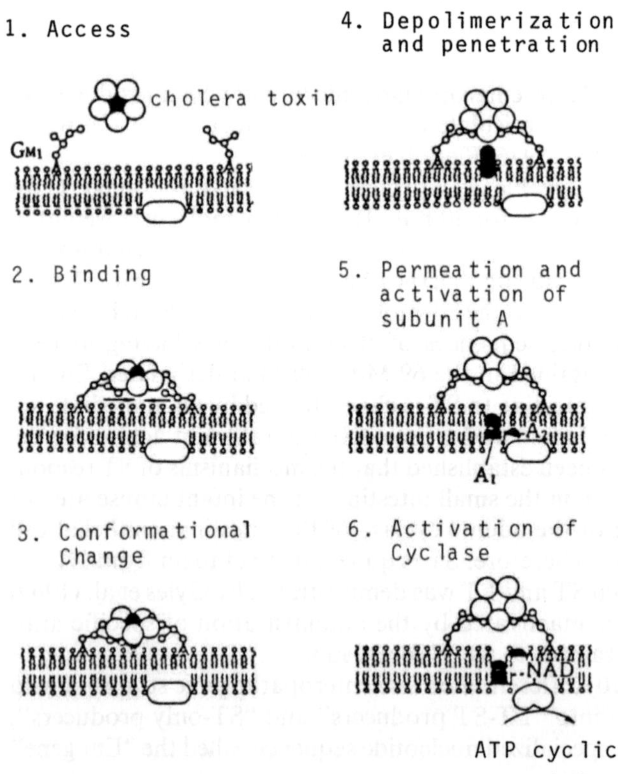

Fig. I.8. The activation process of adenylcyclase by the binding of cholera enterotoxin (CT). [Adapted from T. Yamakawa (1981) Biochemistry of glycolipid, "Science (Japanese edn)" No. 1]

sue, are cultured (Donta et al. 1974; D. A. Sack and R. B. Sack 1975) and when samples are added to the culture plate adrenal Y1 cell monolayers change their morphology into rounded, separated cells.

3. Tests for stimulation of an adenyl cyclase system: The increase of adenyl cyclase activity of various kinds of animal cell membranes, e.g., intestinal, leukocytic, liver, pigeon erythrocyte ghost, cardiac muscle or fat cells, is measured after incubation with the test sample. For example, the adenyl cyclase system of fat cells is activated to be measured by conversion of [^{32}P]-labeled ATP to [^{32}P]-labeled cyclic AMP.

The results obtained by physicochemical measurements and the above newly refined assay methods to compare LT and CT, are summarized in Table I.2. Although these two enterotoxins have essentially similar characteristics, it is remarkable that CT is 10 to 2,000 times more active than LT in its biological properties. The toxic mechanisms of CT have been elucidated by Gill and Meren (1978) and are illustrated diagrammatically in Fig. I.8.

As aforementioned (p. 22), a single CT molecule excreted from the vibrio consists of one subunit A surrounded by five to six subunits B (Fig. I.8). According to this system, the latter designated as SB binds to the specific epithelial cell receptor GM_1, which is a ganglioside glycolipid exposed on the gut surface having the following chemical formula:

$$\begin{array}{l} \text{Gal}\beta 1 \rightarrow 3\,\text{GalNAc}\beta 1 \rightarrow 4\,\text{Gal}\beta 1 \rightarrow 4\,\text{Glc}\beta 1 \rightarrow 1\,\text{cer} \\ \qquad\qquad\qquad\qquad\qquad\quad 3 \\ \qquad\qquad\qquad\qquad\qquad\quad \uparrow \\ \qquad\qquad\qquad\qquad\quad 2\,\text{NeuAc} \end{array}$$

This binding results in the conformational change of the CT molecule and subunit A is inserted into the membranous layer of epithelial cells. Then A_1 is activated by dissociating from A_2 due to a protease action and functions as an ADP-ribosyltransferase according to the formula:

$$\text{NAD}^+ + \text{Protein} \rightarrow \text{ADP-ribosyl-protein} + \text{nicotineamide} + \text{H}^+$$

(Gill and Meren 1978). The protein in this formula is a GTP-binding protein with a molecular weight of 42,000. This reaction is followed by a rapid increase in the intracellular concentration of cyclic AMP, acting to enhance the outflow of Cl^- ion accompanied by the inhibition of Na^+ ion influx. The final outcome is loss of fluid and the excessive dehydration of the body via the intestinal epithelium. The typical "rice-water stool" is, in fact, a water suspension of vibrios and epithelial cells detached from the intestinal wall (Wake, A. personal observation).

Because the heat-labile protein enterotoxins of *V. cholerae* and of *E. coli* are immunologically cross-reactive and have similar modes of action, it is of interest to determine whether or not the production of enterotoxin by *V. cholerae* is regulated by plasmids or by a bacterial chromosomal genes. To determine if a chromosomal gene was the determinant, Vasil et al. (1975) performed transfer experiments with the *Vibrio cholerae* chromosome, based on the assumption that the virulence determinant *tox*$^+$ that is responsible for CT production must be localized on the chromosome if the *tox*$^+$ gene was mobilized by the vibrio sex factor P established by Bhaskaran (1974). For this purpose, a streptomycin (SM)-resistant *V. cholerae* strain RV-102 (P$^+$) *str*^{r} *tox*$^-$ was conjugated with the SM-sensitive strain 569B (P$^-$) *str*^{s} *tox*$^+$ to obtain 569B (P$^+$) *arg*$^+$ *his*$^+$ *ilv*$^+$ *str*^{s} *tox*$^+$. The latter served as the donor strain, and another auxotrophic *V. cholerae* strain RV-31 (P$^-$) *arg*$^-$ *his*$^-$ *ilv*$^-$ *str*^{r} *tox*$^-$, was the recipient. These experiments demonstrated that a gene-controlling toxinogenesis in *V. cholerae* can be transferred by conjugation. Bacterial genetic segregation analysis during recombination in this donor-recipient system produced data demonstrating the linkage of the *tox* gene with a gene required for histidine biosynthesis, *his*. This result is not consistent with control of toxinogenesis by a plasmid and provides strong evidence for the chromosomal location of the *tox* gene.

In contrast to the chromosomal localization of the CT-producing virulence determinant in *V. cholerae,* the plasmid localization of LT- and/or ST-producing virulence determinants was demonstrated in enterotoxinogenic *E. coli.*

However, CT and LT molecules have similar amino acid sequences, because they are encoded for by almost homologous nucleotide base sequences, although the former encoding CT is located on the chromosome of *V. cholerae,* and the latter encoding LT, is on a plasmid of *E. coli.* For example, Dallas and Falkow (1980) showed that LT synthesized by using minicells (Chap. 3) containing the Ent plasmid was composed of two distinct proteins with molecular weights of 25,000 and 11,500, very similar to those of the A and B subunits, of CT, respectively. Figure I.9 illustrates these similarities with regard to LT-B and CT-B; and Fig. I.10 presents the nucleotide base sequence coding for LT-B.

Furthermore, Takeda et al. (1981) succeeded in reconstituting hybrid toxins between subunits of LT and CT in vivo after separation of the A and B subunits by treatment with 6 M urea solution in 0.1 M propionic acid (pH 4.0). The hybrid toxins showed a toxicity similar to that of the parent toxins from which the A subunits were derived.

On the other hand, diversities at the molecular level between CT and LT; and differences among three such enterotoxins as CT, LT derived from the porcine toxinogenic *E. coli* (Clements and Finkelstein 1979) and the one derived from the human toxinogenic *E. coli* (Honda et al. 1981 a; Honda et al. 1981 b), have been shown by serological techniques, i.e., immunoaffinity chromatography or agar-gel immunodiffusion assay.

The nucleotide base sequences encoding the A and B subunits of each enterotoxin have been determined either within chromosomal or plasmid DNA by the use of techniques described in Chap. 3 (Shipley et al. 1979). Thus, this approach to identify the homologous nucleotide base sequences for the production of heat-labile enterotoxins in various diarrhea-causing bacterial species is progressing. If these results can be extended, the distribution and microevolution of genetic traits that are responsible for the pathogenicity of

	LT	CT		LT	CT		LT	CT		LT	CT		LT	CT
	Met			Ile	Ile	30	Ser	Ser		Ser	Ser	80	Thr	Ala ◆
	Asn			Thr	Thr		Met	Leu ◆		Gln	Gln		Lys	Lys
	Lys			Glu	Asp ◆		Ala	Ala		His	His		Ile	Val ◆
	Val			Leu	Leu		Gly	Gly		Ile	Ile		Asp	Glu ◆
	Lys			Cys	Cys		Lys	Lys		Asp	Asp		Lys	Lys
	Cys		10	Ser	Ala ◆		Arg	Arg	60	Ser	Ser		Leu	Leu
	Tyr			Glu	Glu		Glu	Glu		Gln	Gln		Cys	Cys
	Val			Tyr	Tyr		Met	Met		Lys	Lys		Val	Val
	Leu			Arg	His ◆		Val	Ala ◆		Lys	Lys		Trp	Trp
	Phe			Asn	Asn		Ile	Ile		Ala	Ala		Asn	Asn
	Thr			Thr	Thr	40	Ile	Ile		Ile	Ile	90	Asn	Asn
	Ala			Gln	Gln		Thr	Thr		Glu	Glu		Lys	Lys
	Leu			Ile	Ile		Phe	Phe		Arg	Arg		Thr	Thr
	Leu			Tyr	His ◆		Met	Lys ◆		Met	Met		Pro	Pro
	Ser			Thr	Thr		Ser	Asn ◆		Lys	Lys		Asn	His ◆
	Ser		20	Ile	Leu ◆		Gly	Gly	70	Asp	Asn ◆		Ser	Ala ◆
	Leu			Asn	Asn		Glu	Ala ◆		Thr	Thr		Ile	Ile
	Tyr			Asp	Asn ◆		Thr	Thr		Leu	Leu		Ala	Ala
	Ala			Lys	Lys		Phe	Phe		Arg	Arg		Ala	Ala
	His			Ile	Ile		Gln	Gln		Ile	Ile		Ile	Ile
	Gly			Leu	Phe ◆	50	Val	Val		Thr	Ala ◆	100	Ser	Ser
1	Ala	Thr ◆		Ser	Ser		Glu	Glu		Tyr	Tyr		Met	Met
	Pro	Pro		Tyr	Tyr		Val	Val		Leu	Leu		Lys	Ala ◆
	Gln	Gln		Thr	Thr		Pro	Pro		Thr	Thr		Asn	Asn
	Thr	Asn ◆		Glu	Glu		Gly	Gly		Glu	Glu			

Fig. I.9. Comparison of the amino acid sequence of LT-B and CT-B. The *arrows* indicate non-homologous amino acids. Amino acids are numbered starting with the proposed N-terminal residue of mature LT-B (Dallas and Falkow 1980)

HinfI *EcoRI*
5′ GTTGACATATATAACAGAATTCGGGATGAATT ATG AAT AAA GTA AAA TGT TAT GTT TTA
Met Asn Lys Val Lys Cys Tyr Val Leu

5′ TTT ACG GCG TTA CTA TCC TCT CTA TAT GCA CAC GGA GCT CCC CAG ACT ATT
Phe Thr Ala Leu Leu Ser Ser Leu Tyr Ala His Gly Ala Pro Gln Thr Ile

5′ ACA CAA CTA TGT TCG GAA TAT CGC AAC ACA CAA ATA TAT ACG ATA AAT GAC
Thr Glu Leu Cys Ser Glu Tyr Arg Asn Thr Gln Ile Tyr Thr Ile Asn Asp
10 20

HinfI
5 AAG ATA CTA TCA TAT ACG GAA TCG ATG GCA GGC AAA AGA GAA ATG GTT ATC
Lys Ile Leu Ser Tyr Thr Glu Ser Met Ala Gly Lys Arg Glu Met Val Ile
30

MspI *HincII*
5 ATT ACA TTT ATG AGC GGC GAA ACA TTT CAG GTC GAA GTC CCG GGC AGT CAA
Ile Thr Phe Met Ser Gly Glu Thr Phe Gln Val Glu Val Pro Gly Ser Gln
40 50

HinfI
5′ CAT ATA GAC TCC CAG AAA AAA GCC ATT GAA AGG ATG AAG GAC ACA TTA AGA
His Ile Asp Ser Gln Lys Lys Ala Ile Glu Arg Met Lys Asp Thr Leu Arg
60 70

HinfI
5 ATC ACA TAT CTG ACC GAG ACC AAA ATT GAT AAA TTA TGT GTA TGG AAC AAT
Ile Thr Tyr Leu Thr Glu Thr Lys Ile Asp Lys Leu Cys Val Trp Asn Asn
80 90

5 AAA ACC CCC AAT TCA ATT GCG GCA ATC AGT ATG AAA AAC TAG
Lys Thr Pro Asn Ser Ile Ala Ala Ile Ser Met Lys Asn STOP
100

Fig. I.10. Nucleotide sequence of *elt*B. The nucleotide sequence was determined using the chemical method of Maxam and Gilbert. DNA strands were radiolabeled by a kinase reaction or by end repair using DNA polymerase I Klenow fragment. The five *underlined* nucleotides are complementary with 5 of the first 12 nucleotides of the 16 S ribosomal RNA. The amino acid labeled *1* is the proposed N-terminal amino acid residue of mature LT-B. Several restriction enzyme recognition sites are shown (Dallas and Falkow 1980)

Table I.3. Comparison of the virulence of motile and nonmotile strains of classifical *V. cholerae* in mice

Challenge organisms	Serovar	Motility	No. of suckling mice[a]	% Mortality[b]	
				36 h	2 weeks
CA 411	Ogawa				
WT		+	37	100	100
M-5		−	20	10	20
M-6		−	20	0	15
CA 401	Inaba				
WT		+	32	97	100
M-1		−	28	32	43
M-4		−	30	0	17
M-5		−	30	63	73
M-6		−	20	30	45
569 B	Inaba	Slight	25	40	40

[a] 8-day-old ICR mice.
[b] Oral challenge dose per mouse 5×10^7 to 1×10^8 CFU.

microorganisms causing diarrhea may be elucidated throughout the microbial world.

However, enterotoxins are not the sole virulence factors in the causation of diarrheal disease because overall virulence is composite of several virulence factors. Virulence factors that are responsible for successful colonization of the intestine by enteropathogenic *E. coli* have been reported. These play roles in addition to ST and LT which are the distinct virulence factors responsible for causing the diarrhea. Plasmids mediating ST production and those mediating the colonizing ability of enteropathogenic *E. coli* – e.g., K88 antigen for the porcine-enteropathogenic strain (Wittig 1965; Ørskov and Ørskov 1966); K99 antigen for the bovine-enteropathogenic strain (Ørskov et al. 1975; Wilson and Hohmann 1974; Nazy 1980) and "colonizing factor (CF)" for the human-enteropathogenic *E. coli* (Evans et al. 1975) – have been cloned and analyzed employing the techniques described in Chap. 3.

The results of these studies, also, enable us to conclude that similar types of virulence factors expressed by a variety of microorganisms are encoded, respectively, by similar nucleotide base sequences, although there are minute diversities among them which characterize their specific host-parasite relationships.

Even with the elucidation of plasmid-mediated virulence factors that are responsible for the adhesion of *E. coli* (e.g., K88 and CF antigens), it has become increasingly difficult to definitively correlate specific adhesive properties of the enteropathogenic *E. coli* organisms with their pathogenicity. This may be due at least in part to the role of the mucus gel excreted from goblet cells of the host intestine in the establishment of bacterial colonization which may be of considerable importance in pathogenesis (Freter 1982).

Because the mucus gel is not a bacterial factor, but one of the host-defensive barriers, this problem will be considered in Part II.

Consistent with this view, Schrank and Verwey (1976) showed that *V. cholerae* organisms were present within the small intestinal mucus gel of rabbits within 2 h after experimental infection. In contrast, they found carbon particles to be excluded by the mucus gel in their studies. They considered bacterial motility to be the mechanism by which cholera vibrios penetrate this mucus layer. The cellular organelle for the *V. cholerae* motility is a single long flagellum originating from within the cytoplasm and extruding through the cell wall at one pole of the bacillus.

Through the entire length of rabbit ileum and jejunum, the thickness of this mucin-rich, mucus gel was estimated to be about 0.1 mm, irrespective of whether animals are fasted or not. Driven by quick spiral movements of the flagellum, the vibrio moves along lines of least resistance created by the alignment of the glycoprotein components of the mucus before reaching the brush border. The histopathological studies of Schrank and Verwey (1976) estimated that the time needed for the successful penetration of the mucus by the vibrios to arrive at the intervillus spaces is 2–3 h and to reach the crypts, 4 h. The effect of CT activity became apparent after this lag time to allow penetration of bacteria which was followed by the production of intraintestinal fluid.

The fact that CT is not the sole virulence factor for *V. cholerae* and that its motility had a role was demonstrated by the following experiments of Guentzel and Berry (1975) who emphasized the importance of the motility of monoflagellated wild *V. cholerae* for successful colonization. Wild-type *V. cholerae* of classical and El Tor biotypes and their respective nonmotile derivative mutants were compared for lethal virulence after oral challenge with 5×10^7 to 1×10^8 organisms in fasted 8-day-old infant mice. The highly enteropathogenic, slightly motile classical Inaba strain 569B was also included for comparison. The results shown in Table I.3 indicate that loss of motility by the vibrios is accompanied by a marked reduction in virulence. The mortality (at 36 h) caused by the nonmotile strains except one was 32% or less, whereas the motile wild strains produced nearly 100% deaths. In Table I.3, however, successful colonization by some nonmotile vibrios was suggested, because their mortality rates increased 2 weeks after challenge. The mortality rates caused by motile vibrios were high 36 h after challenge, while the rate caused by highly enterotoxinogenic, but slightly motile, *V. cholerae* 569B was lower. This evidence indicates that an initial and critical phase of cholera pathogenesis is a superficial infection with close association of vibrios with the intestinal tissue of the host, related to motility.

For confirmation of these results, intestinal segments were prepared from 10-day-old mice and exposed to the separately washed suspensions of motile or nonmotile vibrios labeled with [U-^{14}C] glucose at 37 °C for 15 min. The respective segment was washed in physiological saline for 5 min to remove loosely bound vibrios and then the radioactivity was measured. The ratio of the bound organisms shown by segments infected with wild-motile vibrios ranged from $^1/_{12}$ to $^1/_{20}$ of the number of vibrios contained in the suspension. The adsorption capacity of the slightly motile strain Inaba 569B, a mutant characterized by extraordinary high production of CT (p. 22), was about $^1/_{20}$ of that of a wild-Inaba strain CA401, with ordinary CT production and normal motility. The lethal virulence of these two strains in the suckling mouse in vivo infection model was calculated as approximately 100-fold higher in CA 401 than 569B, in spite of the higher CT production in 569B (Table I.3), and appeared to be related to the adherence capacity of the organisms. Because 569B is a high CT-producing mutant strain, its mechanisms for motility might be adversely modulated by the competitive or allosteric effect of CT-production mechanisms; thus, the in vivo lethal virulence might be the integrated effect of CT production and motility which produced the different results obtained with CA 401 and 569B.

However, Freter (1982) regarded this straightforward interpretation of the importance of *V. cholerae* motility as a virulence factor as rather problematic, when Jones et al. (1976) described another vibrio virulence factor "adhesin" as playing a role. Jones and Freter prepared brush-border preparations and integrated glycocalyx consisting of glycoproteins and glycolipids from rabbit small intestine epithelial cells to mix with *V. cholerae* suspensions. Their conclusions were based upon the calculations of the number of vibrios attached to the brush-border cells at an equilibrium state. However, these conclusions were modified when recent data from their laboratory implicated the chemotactic attraction of motile bacteria into the mucus gel as a major force aiding *V. cholerae* organisms in the penetration of the intestinal mucous layer. Thus, the unique mutual interaction between *V. cholerae* organisms equipped with suitable virulence factors and host mucus gel equipped with peculiar rheological properties, is a prequisite for the successful adhesion of vibrios to the brush border where the role of adhesin becomes apparent. On the other hand, Finkelstein and Hanne (1982) identified adhesin as the hemagglutinin produced by *V. cholerae* which is sensitive to the action of protease produced by the same microorganisms. Thus, vibrios are eluted from the brush border after their adhesion by the dissociating action of their protease to result in a constant turnover maintaining the equilibrium state between adhesion of new vibrios and elution of dissociating ones (Jones et al. 1976).

After adhesion, the vibrios require metallic iron for multiplication and colonization. In common with most biological systems, iron is essential for *V. cholerae* organisms because of its remarkable capacity to engage in electron transport reactions, one of the fundamental life-supporting mechanisms (Neilands 1957). Vibrios that are deficient in the ability to acquire iron in a host were found to be 100- to 5,000-fold less virulent for chicken embryos.

Payne and Finkelstein (1977) found that many Gram-negative bacteria including *V. cholerae* that show virulence in the chicken embryo model, absorb Congo red dye from the medium onto their cell surfaces during their growth which was correlated with their ability to acquire iron. Therefore, when such bacteria were cultivated on an appropriate complex agar medium containing

Table I.4. Enhancement of virulence for chick embryos of nonpigmented mutants by iron

Strain	Route of inoculation	LD_{50}[a] at 24 h	
		Control	With added iron[b]
Sh. flexneri 2457 T	Allantoic[c]	8.7×10^1	8.2×10^1
Sh. flexneri 2457 O (P−)	Allantoic	4.5×10^5	4.1×10^2
V. cholerae 569 B	Allantoic	6.3×10^1	6.5×10^1
V. cholerae 569 B (P−)	Allantoic	8.0×10^3	3.8×10^1
E. coli U 169	CAM[d]	2.0×10^1	2.0×10^1
E. coli U 169 P−	CAM	3.1×10^4	1.2×10^2
N. meningitidis B-11	CAM	2.8×10^3	8.7×10^2
N. meningitidis B-11 P−	CAM	3.0×10^5	2.3×10^3

[a] LD_{50}, mean lethal dose.
[b] Imferon, 500 μg iron.
[c] Challenge organisms were inoculated into the allantoic cavity.
[d] Challenge organisms were inoculated on the chorioallantoic membrane.
Cited from S. M. Payne and R. A. Finkelstein (1977) with modifications.

0.1 to 1% Congo red dye, pigmented colonies were formed, to be designated P+. In contrast, those bacteria deficient in the ability to acquire iron do not absorb the dye; therefore, nonpigmented colonies are formed, to be designated P−.

This finding was not unique to *V. cholerae,* and Payne and Finkelstein (1977) isolated P− mutants of *Shigella flexneri, V. cholerae, E. coli,* and *Neisseria meningitis* from their P+ parents by natural segregation or treatment with a mutagenic agent. A bacterial suspension of each P+ parent or P− mutant organisms was inoculated into the chicken embryo model with addition of iron. Table I.4 summarizes the results of these experiments. Thus, it was concluded that the virulence of P− mutants of these bacteria could be enhanced by the addition of iron into the inoculum, but that of the P+ parent could not. This indicates that the ability of *V. cholerae* to acquire iron in a host is important for its virulence.

The virulence factor which is responsible for the acquisition of iron by *V. cholerae* was demonstrated by Payne and Finkelstein (1978 a) to be siderophore production. Siderophore, formerly called siderochrome, is the collective name for numerous iron-transporting substances produced by a wide range of microorganisms. Because siderophore molecules of most microorganisms are known to be synthesized under iron-poor cultural conditions, the *V. cholerae* 569B strain was grown in a deferrated synthetic medium and the harvest was extracted with acetyl acetate in which siderophore molecules are generally soluble. This extract was as effective as iron in enhancing growth of *V. cholerae* in the deferrated medium, indicating that the siderophore molecules bound the residual iron contained in the medium and transported it into the bacterium.

Because the siderophore produced by *E. coli* or *Salmonella typhimurium,* i.e., enterochelin, has been known to have the molecular structure of a cyclic trimer of 2,3-dihydroxy-benzoyl-serine (O'Brien and Gibson 1970), it is of interest to know whether *V. cholerae* belonging to another distinct genus produces the same molecular species as its siderophore. For this purpose, both the *V. cholerae* extract and a reference material of the enterochelin were analyzed by paper chromatography.

Although both samples were separated into two fractions, which were detectable by ultraviolet absorption or by spraying the paper with ferric chloride, the spot on the chromatographic paper produced by *V. cholerae* extract did not comigrate with enterochelin. Generally speaking, the bacterial siderophores can be classified into either a hydroxamate or phenolic type. *V. cholerae* was found to produce the latter type similar to that of *E. coli* and *Sal. typhimurium,* but these experiments indicate that its molecular structure is slightly different from enterochelin.

To summarize the above discussion, the pathogenesis of *V. cholerae* infection, which is selected as the prototype of superficial intestinal infections (Finkelstein 1982), is a consequence of the virulence factors: CT, motility, adhesion, and siderophore production. A change in one of these factors as well as certain deviations from the normal functions of host-defensive barriers may result in variations in the severity of the infection as indicated by the degree of the dehydration of the host. In fact, cholera is characterized by a great variation in the severity of symptoms after infection, i.e., many symptomless carriers and patients with only mild diarrhea can be seen among the people living in the area during an epidemic (Dizon et al. 1967).

Among these virulence factors of *V. cholerae,* CT belongs to those that are expressed by a variety of microorganisms and are encoded by similar homologous nucleotide base sequences (p. 26), as was concluded from the studies on the localization of the virulence determinants in *E. coli* and *V. cholerae* (0:1 serovar) at the nucleotide base sequence level.

The following laboratory experiments suggest that similar homologous nucleotide base sequences are present in the *V. cholerae* which are nonagglutinable by anti-0:1 lipopolysaccharide serum (NAG vibrios). Wake et al. (1954) induced lethal diarrhea and hemorrhagic fluid accumulation in infant rabbits by intraluminal injection of three strains of these "NAG" vibrios, NCTC 3661, NCTC 4714, and NCTC 4715. In these experiments, the vibrio did not express virulence at the time of the first inoculation with broth cultures of organism, but fluid accumulation and death resulted after 2–3 intestinal passages through nondiarrheal animals. This suggests that the genes responsible for enterotoxin production were in a repressed state in NAG vibrios. The symptoms obtained finally were indistinguishable from those of classical and El Tor biotypes, except for the rather hemorrhagic nature of the accumulated fluid.

In relation to this hemorrhagic phenomenon, Robins-Browne et al. (1977) found that a NAG *Vibrio cholerae* strain, isolated from the blood of a child with Kwashiokor and fever, showed an invasive potentiality for the mucosal epithelium and the lamina propria of rabbit ileum. This slight invasiveness shown by NAG vibrio strains has never been observed in infection with 0:1 serovar *V. cholerae* of either classical or El Tor biotype.

Aside from this divergence, the enterotoxin isolated from NAG vibrios caused elongation of Chinese hamster ovary tissue-culture cells and enhanced the permeability of the rabbit-skin blood capillaries in the assay methods for heat-labile enterotoxin (Robins-Browne et al. 1977). Furthermore, Zinnaka and Carpenter (1972) and Ohashi et al. (1972) performed cross-neutralization tests between NAG enterotoxin and CT with antiserums. These enterotoxins shared common antigenic determinants, but partial serological differences were also detected. This evidence suggests that the nucleotide base sequences encoding heat-labile enterotoxin production are distributed among vibrios of variable serovars, irrespective of the presence or absence of 0:1 lipopolysaccharides on their cell surfaces.

This wide distribution of similar homologous nucleotide base sequences encoding heat-labile enterotoxins among the vibrio varieties, provides an explanation for the fact that various vibrios belonging to divergent biotypes have caused epidemics with historical time courses as will be described as follows.

The wavelike occurrences of cholera pandemics that have killed some ten millions of people until the twentieth century are well known. It had been believed that the causative organisms of these pandemics were classical types divisible into three 0:1 serovars, which had originated from the reservoir of the basin of the river Ganges in India, and repeatedly spread throughout the world. Nowadays, it has been disclosed that these serovars: Ogawa, Hikozima, and Inaba are determined by the quantitative proportions of two distinct lipopolysaccharide-0:1 antigens A and B.

However, the causative agent of the seventh pandemic which originated in Indonesia in 1961 was different from those of the past pandemics because it was hemolytic for sheep erythrocytes, although it shared the common lipopolysaccharide 0:1. Choleralike diseases due to hemolytic vibrios with 0:1 serovars had been recorded previously in the Arabian Peninsula, Thailand, and Celebes Island of Indonesia. But, these outbreaks were endemic in nature before 1961. These types of vibrios were designated "El Tor" (the name of a port in the Arabian Peninsula) and were included in *Vibrio cholerae* as one of the biotypes as a result of this widespread epidemic by the decision of the WHO (World Health Organization). The seventh pandemic is still continuing, and the El Tor biotype has frequently been imported into Japan, America, and Europe. Even in India, where the "classical biotype" has prevailed, the El Tor biotype is substituting for it in current outbreaks. In addition, the important biological characteristic of hemolysin production has been disappearing from the El Tor biotype as time has passed, but another important characteristic for this biotype, the hemagglutination of sheep erythrocytes has been preserved to serve for the identification of this biotype.

This "transition phenomenon" as used in the terminology of botanical ecology to mean sequential changes of living species in a certain area with the historical time course, has proceeded further with reports of scattered epidemics of choleralike diseases in many countries caused by vibrios nonagglutinable by anti-0:1 lipopolysaccharide serum. For example, a large-scale epidemic caused by a nonagglutinable (NAG) vibrio involving more than 600 patients with more than 100 deaths occurred in Sudan, Africa, in 1968 (Kamal 1971; World Health Organization 1969; Zinnaka 1971).

Thus, this "Example 2" dealing with two diarrhea-causing bacterial species: *V. cholerae* and *E. coli* that belong to divergent genera, can be summarized as follows: first, the bacterial virulence, responsible for the diarrhea due to superficial infection with these pathogens is an integrated expression of multiple virulence factors; second, the main virulence factor CT or LT, directly responsible for the diarrheal mechanisms, is genetically coded by similar homologous nucleotide base sequences, although the location is on the chromosome in *V. cholerae* and on a plasmid in *E. coli;* third, these CT- or LT-encoding homologous nucleotide base sequences have been identified in varieties of *V. cholerae* that have been a "transitionlike phenomenon" in the history of epidemics which have appeared sequentially in human society; and fourth, the LT-encoding homologous sequences of *E. coli,* on the other hand, show some diversity depending on their host-animal species of origin.

Therefore, "Example 2" may support the hypothesis that there is a particular group of homologous nucleotide base sequences that are virulence determinants which have been translocated in some way among microorganisms during a long evolutionary history of specific host-parasite relationships. This hypothesis indicates the importance of an understanding of microbial pathogenicity at the molecular as well as at the epidemiological level.

Chapter 5 Exudative (Partially Invasive) Infections

Pathogenic organisms, causing exudative infections penetrate into mucous epithelial cells or lodge within minute intercellular spaces at natural openings of skin tissue, such as sebaceous glands or those produced by traumatic injury, and colonize the tissues. However, internal sites of the body can also be infected with these organisms and become the foci of exudative infection once their colonization is established.

Irrespective of the site, the host organisms respond by inducing inflammatory reactions as one of the defensive mechanisms as will be described in Part II. As a result, exudates, containing humoral and cellular defensive factors, develop which are more or less divergent in their composition depending on the pathogen. There is usually no significant proliferation of host cells. The lesion is characterized as an exudative-inflammatory versus a proliferative-inflammatory infection (Kobayashi 1944). The virulence factor(s) of these pathogens confer(s) on them limited invasiveness, e.g., penetration into the epithelial cells without further expansion of the inflammatory area or dissemination throughout the body. The balance between the grade of invasion and the resistance of the host determines the volume of the lesion, within which the host-tissue cells and the phagocytes recruited by the chemotactic factors produced by inflammation are destroyed as well as are the invading organisms. Thus, the involved tissue becomes necrotic, and develops into an ulcer.

The necrotic cell mass together with living and dead pathogens is surrounded by replicating connective tissue and is gradually demarcated from the healthy tissues. The isolated content forms pus, a mixture of dead host cells and pathogens, which is removed through drainage or disposed of by the scavenger cells, macrophages.

The critical reason why the body can localize and wall off the exudative infection seems to reside in the fact that pathogens are sensitive to the destructive and/or neutralizing actions of humoral or cellular host-defensive mechanisms that will be described in more detail in Part II.

Example 1: Staphylococcus aureus

The most common causative agent of exudative infections, such as pustules, cutaneous abscesses, and some cases of impetigo, *Staphylococcus aureus,* is chosen as an example because of its common occurrence in disease, and because of the complex mechanisms of its pathogenesis. Staphylococci may also invade and induce severe infections, such as abscesses within kidney, liver, and other internal organs, and even produce septicemia.

In spite of limiting this discussion only to exudative infections, the situation is complicated because many different types of extracellular products (Wadstrom 1974) are produced by *Staph. aureus* and many of these products are toxins and enzymes. Wiseman (1975) concluded that perhaps it is best to consider that virulence in staphylococci is associated not with a single factor, but with a broad spectrum of aggressin-virulence factors supporting the invasiveness of the cocci. For the purpose of this concise review, the discussion will be confined to coagulase, protein A, and some toxins that play a role in the pathogenesis of staphylococcal exudative infection. The enterotoxin of staphylococci will not be considered because it has no role in exudative infections.

Coagulase production by pathogenic *Staph. aureus* has been used as the most acceptable marker of virulence, because almost all isolates from pathogenic lesions show this characteristic. Coagulase may play a role in colonization of the organisms by confining the cocci within the coagulated gel mass of the exudate, thus, protecting them from the action of infiltrating polymorphonuclear phagocytes, PMN (refer to Part II).

Spika et al. (1981) suggested that the cell wall-bound protein A of *Staph. aureus* may cover up alternative-pathway complement-activating sites within the peptidoglycan matrix of the staphylococcal wall which would hinder the alternative-pathway activation of complement by *Staph. aureus* and result in decreased opsonization of these bacteria in the absence of an intact classical pathway.

Among the nonenteric toxins designated alpha, beta, gamma, and delta (Rogolsky 1979), alpha-toxin has always been considered to play a significant role in the pathogenesis of staphylococci, but its exact role has never been defined. Its main significance in pathogenicity is that of producing tissue damage after the establishment of a focus of infection. Although it is frequently designated alpha-hemolysin because of its hemolytic effect on rabbit erythrocytes, alpha-toxin is cytotoxic and cytolytic to a wide variety of cell types – platelets, hepatocytes, white blood cells, human diploid fibroblasts, HeLa cells, and Ehrlich ascites carcinoma cells. Furthermore, it is dermonecrotic, neurotoxic, and lethal to experimental animals. These actions on mammalian cells suggest that this toxin damages the plasma membrane of sensitive cells and that it should better be called a cytolysin.

Alpha-toxin is a protein whose amino acid composition has been defined (Wiseman 1975). However, the estimates for the molecular weight of the protein vary from 26,000 to 39,000, probably due to presence or absence of a smaller isomer with identical antigenicity. The reported localization of the virulence determinant encoding alpha-toxin production varies. Depending on the *Staph. aureus* strains used for genetic analyses, chromosomal as well as extrachromosomal controls of alpha-toxin have been described by different authors. The data from the studies of Witte (1976) indicate that there is either more than one genetic determinant for alpha-toxin synthesis or that the gene for toxin production is associated with a transposon.

Beta-toxin is a protein having an approximate molecular weight of 30,000 of known amino acid composition (Bernheimer et al. 1974). It is designated beta-hemolysin and is hemolytic and most active on ovine and bovine erythrocytes, its mode of action clearly indicates that it is a sphingomyelinase. Therefore, its damaging effects to the plasma membranes of erythrocytes, PMNs, and macrophages of various animal species depend on their content of the substrate-sphingomyelin. The contribution of beta-toxin to the pathogenesis of staphylococci is also attributable to its cytolytic action.

Delta-toxin is characterized by its wide spectrum of cytolytic activity, chemical architecture, high-surface activity, and reaction kinetics. For example, molecular weights ranging from 68,000 to 200,000 have been reported and the antigenicity of the toxin still remains in doubt in spite of some studies reporting the production of anti-delta-lysin antibody. Delta-toxin disrupts biological membranes by means of a detergentlike action, the membrane-receptor site for which is believed to be a straight-chain fatty acid with 13 to 19 carbons. Without question this activity contributes to the invasiveness of *Staph. aureus*. However, delta-toxin also causes histological damage to the guinea pig ileum at high doses and elevates cyclic adenosine 3′, 5′-monophosphate levels of the ileum at low doses (O'Brien et al. 1976). The mechanism of this cholera-toxinlike action of delta-toxin is different from that of cholera toxin (CT, p. 22) because of its delayed-lag period, but these findings indicate that delta-toxin might play a role in the pathogenesis of intestinal diseases ranging from mild diarrhea to severe enteritis.

The production of another cytolysin, gamma-toxin by *Staph. aureus* has been established by its unique spectrum of activity, i.e., hemolytic to rabbit, human, sheep, goat, dog, and fowl erythrocytes, but not those of the horse. Little is known about the mode of action of gamma-toxin. Gamma-toxin does not share common antigenicity with the other staphylococcal cytolysins. It can be separated into an active component I with a molecular weight of 29,000 and component II with that of 26,000. Amino acid analysis reveals that gamma-toxin resembles alpha-toxin because it has high levels of aspartic acid and lysine and no detectable cystein.

The determinants encoding the production of these latter toxins have not been localized.

Leukocidin described first by Panton and Valentine (1932) and designated Panton-Valentine leukocidin in honor of its discoverers by Wright (1936) is another extracellular staphylococcal product of unique characteristics that exerts an effect on leukocytes, but not erythrocytes. Except for PMNs and macrophages, no other cell types of rabbits or humans appear to be susceptible. Almost all knowledge about the molecular properties and the extremely complex mode of action of this toxin are derived from the serial studies of Woodin and co-workers (Rogolsky 1979). This leukocidin consists of two protein components designated F (fast) and S (slow) on the basis of their migration on carboxymethyl-cellulose columns. The molecular weight of F was determined to be 32,000 and that of S, 38,000. Each component itself is nontoxic, but they act synergistically to induce leukocytosis in the host. Noda et al. (1980) suggested that phosphatidylcholine in the rabbit leukocyte membranes might be the receptor for the component F, and ganglioside GM_1 (p. 25), for S. The sequential biochemical interactions between rabbit-leukocyte membranes and the two leukocidin components have been studied in detail by Hirata et al. (1979) and Noda et al. (1981a, b) following Woodin and Wieneke (1966a, b, 1967, 1969–1971). Probably due to a cell-membrane-perturbing action of the *Staph. aureus* leukocidin, rabbit PMNs containing the organisms coated by protein A which inhibits the activation of the alternative complement pathway, do not kill the bacteria readily after phagocytosis and, thus, large numbers of bacteria are maintained in the circulating blood after the i.v. injection of organisms into rabbits (Rogers 1956). These bacteremic, intracellular organisms may deposit in kidney glomeruli and other organs to induce local exudative inflammation and form abscesses.

In addition to these staphylococcal products enumerated above, hyaluronidase and staphylokinase are considered to play roles in the invasiveness and promotion of exudative infection with *Staph. aureus*. However, it has not been possible to clearly establish a definite role for any of these factors in clinical infections, in spite of the accumulated knowledge of their purification, pharmacological properties, and mechanisms of genetic control.

Among the particular virulence factors of *Staph. aureus* associated with pathogenesis in man, the one termed "exfoliatin" (Kapral and Miller 1971) or "exfoliative toxin" (Melish et al. 1972) is worth discussing. Its definitive pathological effects, its discrimination from the other substances described, as well as the genetic control of its production provides an example of a well-established virulence factor in contrast to the present state of knowledge concerning the several staphylococcal toxins mentioned previously.

The first description of an outbreak of exfoliative dermatitis can be traced back to 1878 (Ritter von Rittershain 1878). This disease is generally called toxic epidermal necrolysis (T.E.N.) or Ritter's disease of babies or Lyell's disease of young children.

The primary lesion consists of skin which loosens and may be pushed away with the finger – "scalded skin", called a positive Nikolsky sign. Pathohistological pictures of biopsy specimens show an interepidermal split at the level of the stratum granulosum with the deeper layers of the epidermis intact. A few cells from the stratum granulosum adhere to the stratum corneum, which loosens. PMNs are virtually absent from the "area", but there is a sparse lymphocytic reaction. Although no organisms are seen in the section of the skin, coagulase-positive *Staph. aureus* can be isolated from the skin, nose, throat, or eye, – usually from more than one of these sites. The same organism can be isolated from household contacts of the patients and can be confirmed to be the same organism by phage typing (Lyell et al. 1969). After observing 17 infants and children with a generalized exfoliative dermatitis, Melish and Glasgow (1970) included a generalized erythematous rash resembling scarlet fever as one of the symptoms of T.E.N. and named this disease staphylococcal scalded-skin syndrome (SSSS).

Lyell et al. (1969), Melish and Glasgow (1970), Melish et al. (1972), Dajani and Wannamaker (1969), Dajani (1972), Kapral and Miller (1971), Rogolsky et al. (1974), and Warrent et al. (1975) have all reported that all *Staph. aureus* causing these symptoms belong to phage-typing group 2. However, Keyhani et al. (1975) and Kondo et al. (1973) reported exfoliatin production by phage group 1 *Staph. aureus* and the isolation of several nonphage group 2 *Staph. aureus* from patients with SSSS. Only phage group 2 *Staph. aureus* has been studied in relation to the genetic control of exfoliatin production, as will be discussed later.

Newborn mice less than 7 days of age are the only available experimental model for study of this disease. Characteristic reactions to the staphylococci occur in mice up to about 5 days of age, when hair begins to develop. In the test, *Staph. aureus* is cultured in broth and inoculated into groups of newborn suckling mice either by the subcutaneous (s.c.) or the intraperitoneal (i.p.) route. The response of the animals to injection of staphylococci is dose-related. The "effective dose" as stated by Melish et al. ranges from 10^6 to 10^9 organisms.

Cultures of the denuded skin surface are positive for staphylococci, but fluid aspirated from intact bullae is consistently found to be sterile. Animals generally die shortly after complete exfoliation, usually 20 to 24 h after infection.

Usually, the effective i.p. dose is 10 times higher than the s.c. one. In the i.p. infection experiment, the epidermis shows a positive Nikolsky sign 12 to 16 h after injection; first on the back and then on the abdomen. Four hours later bullae and spontaneous epidermal wrinking appear, followed by extensive exfoliation. Denuded skin surfaces are positive for staphylococci.

In the s.c. infection experiment, clusters of cocci are seen deep in the dermis beneath the superficial muscle layer at the site of inoculation 20 h after injection. As a particularly striking histopathological characteristic, no evidence of local inflammatory response to the organism is recognized. The histological pattern in the experimental syndrome is identical to that described for the human disease. Some evidence of degenerative changes appears in the basal epithelium, and marked degeneration may be seen in cells of the dermis. The failure of the newborn mouse to have an adequate inflammatory response to an effective dose of staphylococci appears to provide optimum conditions for the growth of bacteria and the production of the toxin (Melish and Glasgow 1970).

Kapral and Miller (1971) purified and characterized exfoliatin by repeated ultrafiltration and DEAE Sephadex colum chromatography and assayed its biological activities using this animal model. On the other hand, Melish et al. (1972) grew the phage group 2 *Staph. aureus* within sacs implanted in the peritoneal cavities of rats and rabbits and separated exfoliatin from the contaminating alpha-toxin by electrofocusing. Approximately 0.5 μg of the material was sufficient to produce exfoliation in a newborn mouse.

Such efforts to purify exfoliatin using a variety of isolation procedures performed by a number of investigators resulted in controversy about its physicochemical properties. In fact, the phage group 2 *Staph. aureus* produced at least two distinct types of toxin. For instance, molecular weights were divergently calculated from 23,500 to 33,000 and isoelectric points were from 4.0 to 7.0. Some investigators reported the toxin as thermoresistant, others, thermolabile. Kondo et al. (1975) concluded that his strain ZM synthesized only one distinct type of toxin that could have multiple polymeric forms with different electrophoretic migration patterns.

However, later studies from various laboratories came to the conclusion that exfoliatin-producing *Staph. aureus* synthesize at least two antigenically distinct types of exfoliatin, to be designated ET A and ET B, which differ in physicochemical properties. There appears to be no clear relationship between a specific phage group and the type of exfoliatin produced. Sakurai and Kondo (1978) showed that ET A is a metallotoxin containing one molecule of copper that was essential for toxin activity. ET A is thermostable, while ET B is thermolabile. Furthermore, the molecular diversity of exfoliatin, shown by Ouchterlony immunodiffusion tests, demonstrated that at least two distinct species of antibody are induced against the exfoliatin produced by staphylococcal phage group 1, 2, and 3 strains.

Concerning the precise mechanism of exfoliation action, little is known except that exfoliation perturbs the forces of adhesion between the susceptible cells of the stratum granulosum. Neither surface receptor nor target site for exfoliation have been identified on the host tissues. The involvement of some enzymatic mechanisms to degrade an intercellular cementing component or the enhancement of tissue-fluid secretion into intercellular spaces of the stratum granulosum have been speculated at the possible mechanisms of action (Rogolsky 1979).

The genetic regulation of exfoliatin production by *Staph. aureus* has been extensively studied. First, the possibility that the synthesis might be under the control of a phage in a manner similar to *C. diphtheriae* was ruled out because mitomycin C and ethidium bromide that can cure infection with lysogenic phages did not cause exfoliatin-producing *Staph. aureus* to lose this property. Furthermore, exfoliatin-nonproducing strains were found to harbor phage, in spite of the loss of toxin production. Finally, all trials of lysogenic conversion of a nontoxinogenic strain to toxinogenic, as well as attempts to enhance the exfoliation activity of toxinogenic strains after phage induction failed.

Definitive evidence for the plasmid-mediated nature of the genetic control of exfoliatin production was obtained by Warren et al. (1975) who isolated the responsible plasmid with a molecular weight of 33 Mdal. However, at least two phage group 2 *Staph. aureus* strains continued to produce smaller amounts of exfoliatin even after being cured of the 33 Mdal plasmid infection by cultivation at 44 °C. One strain reduced its exfoliatin production twofold, another strain, ninefold.

Furthermore, Keyhani et al. (1975) found that two plasmidless phage group 2 *Staph. aureus* strains synthesized a smaller quantity of exfoliatin under the control of a *tox*$^{+}$ gene located on the bacterial chromosome. These two plasmidless strains produced neither detectable exfoliatin by the standard assay procedure nor SSSS in the animal model, but were found to make minute quantities of toxin after their culture supernatant fluids were concentrated and inoculated into newborn mice. In addition, Rosenblum and Tyrone (1976) isolated from patients with SSSS a *tox*$^+$ phage group 2 strain which has only chromosomal determinants to regulate exfoliatin synthesis, and another strain having both a 33 Mdal plasmid and chromosomal determinants.

Thus, it has been established that phage group 2 strains of *Staph. aureus* make two antigenically distinct types of exfoliatin, i.e., ET A and ET B, and that control of exfoliatin synthesis in these strains can be either chromosomal, extrachromosomal (i.e., plasmid), or a combination of both.

Consequently, an attempt was made to determine if the nucleotide sequences of the chromosomal and the plasmid genes for exfoliatin shared homology. The results of these experiments using the DNA-DNA hybridization technique (Chap. 3), indicated that DNA from the plasmid encoding the exfoliatin biosynthesis did not share any detectable homology with the chromosome of the strain producing exfoliatin without the plasmid (Wiley et al. 1976). The same group of investigators (Wiley and Rogolsky 1977) purified exfoliatin from *Staph. aureus* strains known for its genetic control, i.e., plasmid, chromosomal, or both, after isoelectric focusing on glycerol density gradients and compared the bandformations after SDS-polyacrylamide slab-gel electrophoresis together with crude preparations before purification. A strain having both chromosomal and plasmid determinants for exfoliatin synthesis showed two distinct bands. The crude preparation from the strain known for the chromosomal control of production showed only one band that directly aligned with the electrofocused exfoliatin that can be identified as ET A. Similarly, the band formed by the crude preparation from the strain known for the plasmid control was identified as ET B. As mentioned before, ET A is heat-stable and ET B, heat-labile. The assay for heat stability and Ouchterlony immunodiffusion tests carried out with antisera against electrofocused preparation from all strains and purified ET A and ET B, confirmed the validity of these findings.

In summary, the chromosomal and plasmid genetic determinants for exfoliatin synthesis in phage group 2 *Staph. aureus* represent different genes. Each genetic locus produces a distinct molecular species of exfoliatin. ET A is coded by a chromosomal virulence determinant, ET B by a plasmid virulence determinant.

In addition to these nonenteric toxins produced by *Staph. aureus,* the possibility of involvement of iron uptake in invasiveness has been studied, because the presence of an iron-binding protein, lactoferrin within the phagosome of PMNs may be essential for the bactericidal effect of the host cells on *Staph. aureus* by removing any free ionic Fe from the phagosome contents (Bullen et al. 1978). Gladstone and Walton (1971) found that the hematin equivalent to 0.46 mM Fe would reverse the bactericidal effect of 10^7 PMNs per ml against *Staph. aureus,* while $FeSO_4$ equivalent to 1.8 mM Fe was far less effective. However, the reducing effects of iron-containing compounds on the bactericidal activity of PMNs may be due to the direct neutralization of antibacterial activity of cationic proteins present in the phagosomes (Gladstone and Walton 1971). At present, no evidence for the siderophore production by *Staph. aureus* is available.

Example 2: Shigella dysenteriae

Shigella dysenteriae is characterized by the dominant role that mucosal invasion plays in its pathogenesis in contrast to a usual lack of toxin production. There are four groups (subspecies) of *Sh. dysenteriae: Sh. dysenteriae* (A), *Sh. flexneri* (B), *Sh. boydii* (C), and *Sh. sonnei* (D).

It is of interest epidemiologically that the etiological role of the various respective groups of *Shigella* is related to the country in which the disease is studied ranging from the least developed with *Sh. dysenteriae* dominant to *Sh. sonnei* which is the most common in advanced societies. This presently unequal distribution of each *Shigella* group with reference to world geography has resulted from changes in the dominant *Shigella* group in a given country following the development of public health measures to control gastrointestinal diseases. In other words, the organisms causing infection have changed in parallel with the advances in public health measures in each country. For example, the first dysentery bacilli found by Kiyoshi Shiga in Japan (Shiga 1898) was *Sh. dysenteriae,* which was known as the producer of a neurotoxin. This was followed by a period when *Sh. flexneri* prevailed just after the Second World War. However, since 1970, over 90% of the dysentery bacilli isolated in Japan have been shown to belong to *Sh. sonnei* group (Nakamura 1973), although occasional isolations of the *Sh. flexneri* group have been reported which appear to be imported from developing countries (Fig. I.11).

Another remarkable fact is that this change in the dominant *Shigella* group has been paralleled with a decrease in the virulence of *Shigella*. The clinical symptoms of classical bacillary dysentery as observed earlier characterized by acute colitis accompanied with fever, frequent diarrhea with tenesmus, and excretion of bloody, purulent, mucous stools as described in textbooks, have changed to milder forms of enteritis as the less virulent strains of *Shigella,* e.g., *Sh. sonnei* became dominant. In parallel, the older typical histopathological pictures of pseudomembranous or ulcerative colitis have changed into catarrheal enteritis with rare cases of ulcer formation in more than 95% of recent dysentery patients (Nakamura 1973). Although the present common use of antibiotics cannot be disregarded as having had an influence on the severity of the disease, this phenomenon might connote some change in the molecular biology of these pathogens as well as the change in the etiologically dominant species of *Shigella*.

Alternatively, this long-term changing pattern of symptomatology as well as of the incidence of the pathogenic subspecies of *Sh. dysenteriae* as etiological agents might reflect a general ecological transition in host-parasite relationships. It is noteworthy that similar phenomena are being recognized in *V. cholerae* which has already been discussed.

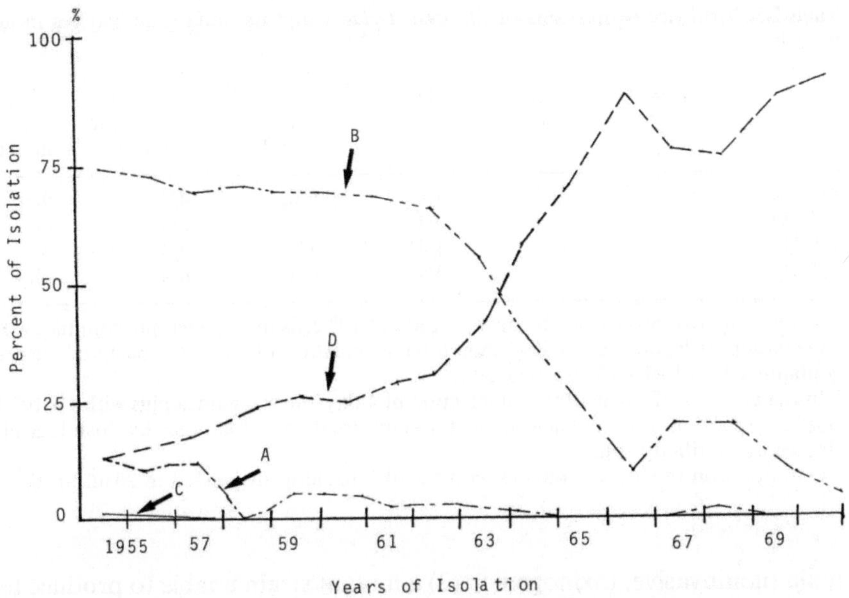

Fig. I.11. Periodical transitions of isolation rate (%) of respective groups belonging to *Sh. dysenteriae* in Japan. **A** Shigella dysenteriae; **B** Shigella flexneri; **C** Shigella boydii; **D** Shigella sonnei. (Nakamura 1973)

The recent spread of multiple-drug-resistance plasmids among *Enterobacteriaceae,* also, is worth consideration, because most organisms belonging to the *Shigella,* which are common, are thought to harbor these plasmids as a result of the selective pressure of antimicrobial drugs. These plasmids may interact with other genetic materials of the bacilli, be they plasmid of the other species or the chromosomes.

An approach to elucidate the possible role of the molecular biology of *Shigella* in these changing patterns may be to define the virulence factors of these pathogens by using suitable experimental model systems with modern techniques of genetic analysis.

For the demonstration of the relative importance of mucosal invasion versus toxin production in the pathogenesis of bacillary dysentery, Gemski et al. (1972) and Levine et al. (1973) employed *Sh. dysenteriae* 1 (synonym of subspecies A or Shiga bacillus) as the typical wild strain (prototype) with full virulence factors, i.e., capable of penetrating epithelial cells and producing toxin (neurotoxin). This neurotoxin causes paralysis and death when injected into mice or rabbits (Engley 1952), and is cytotoxic to monolayer-tissue cultures of some mammalian cell lines (Viscari et al. 1960). Both Gemski and Levine have compared the virulence factors of the following strains in various experimental models in animals: (1) a wild-type strain capable both of penetrating the intestinal mucosa and producing toxin (invasive, toxinogenic); (2) a naturally occurring mutant unable to invade, but capable of elaborating a high yield of

Table I.5. Virulence expressions of *Sh. dysenteriae* 1 and its mutant derivatives in animal models

Strain	Invasion	Toxin production	Sereny test [a]	Rabbit ileal loop [b]	Invasion of guinea-pig intestine [c]	Clinical disease in monkeys [d]
3818-T	+	+	4/4	6/6	4/4	9/15
3818-O	−	+	0/4	4/4	0/4	0/15
725-78	+	−	4/4	7/8	4/4	3/ 6
735-19	−	−	0/4	0/4	0/4	Not done

[a] Keratoconjunctivitis caused by dropping about 10^8 cells into guinea-pig conjunctival sacs.
[b] Dilatation of ligated rabbit ileal loop after inoculation of 1 ml of a bacterial suspension containing 1×10^8–1×10^9 washed cells.
[c] Invasion was confirmed after oral infection of 4-day-starved guinea pigs with 1×10^8/10 ml broth suspension of the bacteria and opium treatment followed by histological and fluorescent-antibody study.
[d] Oral infection of rhesus monkeys with 5×10^{10} bacteria suspended in 20 ml broth.

toxin (noninvasive, toxinogenic); (3) a mutant strain unable to produce toxin, but capable of cell penetration (invasive, nontoxinogenic); and (4) a mutant strain unable either to penetrate or elaborate toxin (noninvasive, nontoxinogenic). The results obtained by both investigators showed that the ability to penetrate and multiply in the colonic mucosa was most important in causing experimental disease (Table I.5), although the role of toxin in pathogenesis could not be excluded.

These studies on the virulence mechanism of *Sh. dysenteriae* 1 have revealed that a small plasmid of 6 Mdal size specifies one or more functions essential for O antigen production and bacterial invasion, after application of the techniques described in Chap. 3 (Watanabe and Timmis 1984, Watanabe et al. 1984).

Intestinal epithelial cell penetration was recognized by LaBrec et al. (1964), Voino-Yasenetsky and Khavkin (1964), and Ogawa et al. (1966) as the most essential step in the pathogenesis of bacillary dysentery. In a comparison of the susceptibilities of cell cultures from various mammalian tissues (established lines or primary to 20 subcultures of different tissue cells derived from man, monkey, guinea pigs, or mouse organs) to infection by *Sh. flexneri* 2 a, 3 a, and *Sh. dysenteriae,* Ogawa et al. (1967) confirmed that animal species- or organ-specificity as found in the natural *Shigella* infection does not show any consistent correlations with the infection rate in cells from the corresponding animals. Thus, *Shigella* infection at the tissue-culture level does not correspond to either natural or experimental model infections.

By the use of these tissue-culture systems, however, an unexpected intracellular behavior of *Shigella flexneri* was observed under phase-contrast microscopy and recorded by time-lapse cinematography (Ogawa et al. 1968). After the invasion of tissue-cultured HeLa S3 or Henle cell-line monolayers, virulent intracellular *Shigella* organisms exhibited vigorous movement accompanied by a polarity. This movement was independent of visible cellular organelles. Neither the protoplasmic liquid flow nor Brownian movement correlated with this behavior. Although the physiological or physicochemical mechanisms of this movement are still unknown, the transmission of bacteria from the parasitized cells to the neighboring healthy cells (intercellular transmission) seems to be mediated by the expulsionlike action of the epitherial cell microfibrillar protrusions containing bacilli which are accepted by the receptor cells. By this movement, dysentery bacteria may overcome the defense mechanisms of the intestinal epithelial mucous membrane and penetrate the tissues.

Serény (1957) devised a simple experimental system for the study of bacterial invasion with intracellular penetration which is the most important indicator of this type of virulence. This method uses the guinea pig in which drops of bacterial suspensions (0.05 ml containing approximately 10^6–10^9 cells) are deposited into the conjunctival sacs of the animals. They are examined over a period of 72 h for evidence of keratoconjunctivitis. Edematous swelling of the palpebral conjunctiva and exudative inflammation become recognizable after 24 h and spread into the corneal tissues. The presence of bacteria can be demonstrated within epithelial cells of these tissues by the fluorescent-antibody technique. At the peak of this reaction, a remarkable intracellular bacterial multiplication is recognized. However, the distribution of bacteria is strictly limited to the epithelial cells and penetration into the deeper tissues beyond the basal membrane is not seen. This assay method is called the Serény test (Table I.5).

The ligated rabbit ileal loop method (De's test), also, is applicable for the study of shigellosis (Gohda et al. 1973). Ogawa and Nakamura (1969) were successful in infecting rabbits, an animal not sensitive to infection with dysentery bacilli under natural conditions. They administered 2 ml of a bacterial suspension containing 10^8–10^9 ml organisms of virulent *Sh.flexneri* into a temporarily ligated colon segment after washing out the intestinal contents of the rabbit which had been starved for 48 h. The clips for the distal and proximal parts of the colon employed to prepare the segment were removed later after administration of bacteria. The fact that *Shigella* could infect even the naturally insusceptible rabbit by the retention of a bacterial suspension for 5 h within the colon segment between the clips, indicates that the mechanism of resistance of animals like rabbits is not the lack of the acceptability of the organism by the intestinal epithelial cells. The difficulty is the retention of the bacteria at a local site in the colon for sufficient time to permit adequate multiplication to attain the number of *Shigella* sufficient to cause infection. In other words, the active peristalsis of the rabbit colon which accelerates the removal of the organisms from the colon rather than the susceptibility of the animal, constitutes itself as a natural resistance mechanism against shigellosis.

In another animal, Honjo et al. (1964) succeeded in inducing typical shigellosis in 50–60% of healthy cynomolgus monkeys by intragastric administration of 10^9–10^{10} *Sh.flexneri* bacilli, although lower challenging doses failed to cause symptoms. The inflammatory lesions were found to be limited to the large intestine beyond the ileocecal junction. No bacterial penetration into organs other than the intestine was detected. Takasaka et al. (1970) succeeded in reducing the infecting dose of viable *Shigella* bacteria for these animals to as low a dose as 10^4–10^5 bacteria by using fresh stool from patients as the inoculum. These stool organisms possess maximal virulence as Nakamura (1967) using tissue-culture assays and Serény tests showed that the virulence of dysentery bacteria is rapidly lost after repeated subcultivations.

All these observations are consistent with the possibility that virulent wild strains of *Sh. dysenteriae* harbor plasmid(s) that control some steps of their pathogenesis. The validity of this view was demonstrated by genetic analysis using sexual mating (Chap. 3).

Formal et al. (1971) mated a streptomycin-resistant, gal$^-$ mutant of *Sh. flexneri* 2a that causes keratoconjunctivitis in guinea pigs with the various *E. coli* Hfr strains that are able to transfer their chromosome into the recipient *Sh. flexneri* 2a with high frequency, and selections were made for the following donor carbohydrate utilization markers: lac$^+$, mal$^+$, fuc$^+$, and gal$^+$. The resulting *Sh. flexneri-E. coli* hybrids which had inherited either lac or gal chromosomal markers of *E. coli* lost the capability of producing a positive Serény test. These results indicated that a locus responsible for virulence as operationally defined by the Serény test, is genetically linked to the lac-gal chromosomal segment and its loss occurred by its replacement with the *E. coli* locus. Another genetic locus marked by the requirement for adenine being designated *pur* E, has been found to locate on the same lac-gal chromosomal segment. *Sh. flexneri-E. coli* hybrids of different genotypic classes derived from mating of *E. coli* Hfr with *Sh. flexneri* showed uniformly negative Serény tests whenever the *pur* E$^-$ locus of *E. coli* is incorporated into the *Sh. flexneri* genome. Moreover, hybrid classes that have retained their native *Sh. flexneri pur* E$^+$ allele, for the most part showed positive Serény tests. On the basis of these findings, Formal et al. (1971) concluded that the virulence determinant responsible for the positive Serény test designated *kcp* A locus is closely linked to the *pur* E allele. As the next step, phage P1 vir was grown on the wild-type *pur* E$^+$ *kcp* A$^+$ virulent *Sh. flexneri* 2a strain and the lysate was employed in transductions using the *Sh. flexneri pur* E$^-$ avirulent hybrid strain, selection being made for the *pur* E$^+$ transductant. The resulting transductants were then tested for their ability to evoke keratoconjunctivitis. Sixty-two percent of purine-independent transductant clones caused a positive Serény test, indicating that the *kcp* A locus is cotransduced with the *pur* E locus. Prior to this study, a similar genetic study using a conjugal transfer of *E. coli* Hfr chromosome into *Sh. flexneri* 2a had revealed that the chromosomal segment encoding xylose and rhamnose fermentation (*xyl-rha* region) is involved in the conservation of the ability to cause a positive Serény test and the penetration into tissue-cultured mammalian cell lines. The importance of the ability to multiply in the tissue after invasion in the determination of the virulence of dysentery bacilli was also stressed by Formal et al. (1965).

Thus, the chromosomal location of the virulence determinant responsible for the invasiveness into the epithelial mucosa was established using the chromosomal transfer technique mediated by bacterial conjugation and phage transduction during the period when the technique of isolation and characterization of extrachromosomal plasmids (Chap. 3) was not well defined.

However, the use of various experimental animal and tissue-culture models disclosed that virulence factors of dysentery bacteria are rapidly lost on repeated subcultivations which is paradoxical considering their proposed chromosomal localization.

This controversy was resolved recently by the following new approach. With the direct observation of *Sh. dysenteriae* colony morphology and the serological analysis of bacterial surface antigenic structures, the spontaneous segregation of avirulent form II serotypes from the virulent serotype form I in *Sh. sonnei* as well as the opaque (O-type) avirulent colonial forms from the virulent parental translucent (T-type) colonies in *Sh. flexneri* have been recognized at a frequency of 1 per 10^4 to 10^5 cell divisions (La Brec et al. 1964).

By the application of modern techniques for the isolation and characterization of extrachromosomal plasmids (Chap. 3), Kopecko et al. (1980) detected a large plasmid in *Sh. sonnei* isolates obtained from different areas of the world. *Sh. sonnei* organisms harboring this ~120 Mdal nontransmissible plasmid expressed the surface form I antigen; but organisms expressing form II antigen were deficient in this plasmid. However, recombinants of this plasmid with the self-transmissible plasmid F lac or F′ could transfer the form I trait to other *Sh. sonnei, Sh. flexneri,* and *Salmonella typhi.* Furthermore, a large cointegrate molecule constructed of the form I plasmid and F′ ts lac:Tn3 plasmid was successfully used to reintroduce the form I trait into form II *Sh. sonnei* cells as well as into *Sh. flexneri, E. coli, Salmonella typhimurium,* and *Serratia marcescens.* These data suggest that the form I plasmid carries structural genes, necessary for form I antigen expression (Kopecko et al. 1980; Sansonetti et al. 1981; Kopecko et al. 1981), but is deficient in regulatory genes. The form I plasmids of several *Sh. sonnei* strains were tagged with the antibiotic resistance transposons Tn3, Tn5, or Tn10. Transposon-tagged form I plasmids were not self-transmissible, but could be mobilized by the plasmid R368. Form II *Sh. sonnei* transconjugants for the form I plasmid acquired both virulence and the ability to synthesize form I antigen, establishing that this 120 Mdal form I plasmid is physically unstable in any of several host bacteria and suggest that it is a member of the F1 incompatibility group, according to the classification of plasmid species. On some occasions, however, this tagging of form I plasmid with these transposons resulted in the loss of the invasive virulence responsible for a positive Serény test and penetration into tissue-culture monolayers without affecting form I antigen expression, probably due to the insertion of a transposon at the locus of the virulence determinants. As a result of these experiments, it was concluded that one or more *Sh. sonnei* virulence properties may be plasmid-borne.

While the presence of the large (~120 Mdal) virulence plasmid correlates to the form I antigen strictly in *Sh. sonnei,* spontaneous transition from a translucent (T) to an opaque (O) colonial morphology, which correlates with the loss of virulence in *Sh. flexneri*, is accompanied by the loss of a larger plasmid of ~140 Mdal size (Kopecko et al. 1979). Sansonetti et al. (1982) examined the agarose-gel electrophoretic profiles of plasmids isolated from six serovars of *Sh. flexneri* (1 b, 2 a, 3 a, 4 b, 5, and 6, respectively) after verifying their invasive virulence by the Serény test and the penetration assay into HeLa-cell monolayer cultures. Irrespective of serovar, all these strains were found to harbor a large plasmid 140 Mdal in size, although some strains carried additional plasmid species. Spontaneous variants of strains of serotypes 1, 2, and 5 had lost this 140 Mdal plasmid and had concomitantly become avirulent, i.e., could neither invade a HeLa-cell monolayer nor produce keratoconjunctivitis in guinea pigs.

The trial to monitor the transfer of the 140 Mdal plasmid was carried out as follows:

A kanamycin-resistant strain M90T (serovar 5) carried the 140 Mdal plasmid which was not self-transmissible. In order to mobilize this plasmid, each one of several different conjugative plasmids belonging to three incompatibility groups was introduced into M90T by conjugation with avirulent derivatives of the heterologous serovars 1 and 2 which had lost the comparable large plasmid. The transconjugants of both serovars which had received the large plasmid (pWR 110) were selected by the kanamycin resistance. The resulting transconjugants were shown to have regained virulence. These facts demonstrate that the 140 Mdal plasmids are functionally interchangeable and strongly suggest that this *Sh. flexneri* plasmid encodes or regulates functions necessary for epithelial penetration. Unlike *Sh. sonnei* form I plasmids, which are lost from some strains at a frequency of approximately 50%, the *Sh. flexneri* virulence plasmids appear rather stable (i.e., 2.4% rate of loss per generation in the M90T strain). Smooth *Sh. flexneri* colonies which had lost both virulence and the 140 Mdal plasmid were not altered in their group and type of antigenic specificities. Therefore, this *Sh. flexneri* plasmid does not appear to encode any lipopolysaccharide component, unlike the *Sh. sonnei* form I plasmid. In spite of this divergence, both of these virulence plasmids may encode some common virulence function, i.e., the capability to cause a positive Serény test and to penetrate into the mammalian tissue-culture monolayers.

A comparative analysis of the nucleotide base sequence homology (Chap. 3) might reveal some homologous sequences within the DNA of ancestor *Shigella* organisms that have been preserved during the long-term microevolution of host-parasite relationships in *Sh. flexneri* in the same way that certain homologous sequences were found in *V. cholerae* and enterotoxinogenic *E. coli* to encode heat-labile enterotoxin production (Chap. 4, Example 2). These presumptive homologous sequences might encode an easily scorable *virulence factor* that is responsible for the positive Serény test and penetration into tissue-culture cells. However, as stated before, the ability to cause a positive Serény test and penetration into the tissue-culture cells is not the sole determinant of the invasiveness of virulent *Shigella* organisms and does not reflect the organism's ability to multiply in tissue (Formal et al. 1965).

Sansonetti et al. (1982) looked for the essential growth factors that selectively stimulate the multiplication of virulent *Shigella* organisms in tissue. Irrespective of the presence of the 140 Mdal plasmid, both the parental and isogenic plasmidless derivative strains grew equally well in glucose minimal medium containing $FeCl_3$, or in minimal medium containing the iron chelator α,α'-dipyridyl. Furthermore, Miles and Khimiji (1975) detected growth-stimulating siderophore production by both smooth and rough *Shigella* organisms. Because the latter organisms are avirulent, siderophore production does not confer virulence in *Shigella* organisms.

The molecular nature of the product(s) of the 140 Mdal plasmid that are effective in conferring on *Sh. flexneri* the ability to invade host mucosal cells remains to be determined.

To conclude this discussion, Chap. 5, Exudative (Partially Penetrating) Infections, it is important to note that this type of infection results in several complex steps of tissue damage, in a limited localized area, and that the main virulence determinants related to at least one of the essential steps in inducing the typical symptoms are found to localize on both the chromosome and extrachromosomal plasmids. The nature of the resulting effects on the host depend on the state of the defense mechanisms of the host, which is considered in Part II.

Chapter 6 Invasive (Penetrating) Infections (Except Yersiniosis)

There is a wide range of microorganisms: viruses, rickettsia, chlamydia, mycoplasma, bacteria, fungi, and protozoa that produce damage to internal tissues and organs by their ability to break through or circumvent the multifaceted defensive barriers at the various external and internal body surfaces of the host. Some of these microorganisms can be transmitted via fertilized eggs of the host or are acquired during intrauterine differentiation of individuals when the host-defensive barriers are immature. In fact, there exists no definitive borderline of exogenous versus endogenous disease in the case of certain viral genetic materials which have been incorporated into the chromosomes of zygotic cells and later initiate active pathogenic processes at a certain stage of postnatal life.

From the general point of view, however, the category named "invasive (penetrating) infections" versus the preceding "superficial" and "exudative (partially penetrating) infections" deals with pathogenic organisms that are equipped with characteristic virulence determinants that play essential roles in circumventing the fully-developed defensive mechanisms of the *mature* host.

These organisms enter the host by invading and passing through the mucous membrane cell layers or infecting through lesions of the injured skin or mucous membranes. In some cases insect or tick bites serve as the point of entrance.

After entry, the facultative intracellular microorganisms multiply intracellularly resisting the inhibiting factors of extracellular body fluids. In addition, these organisms can survive intracellular-destructive mechanisms also after phagocytosis; therefore, they can be transported to the deeper tissues or organs by the migration of the phagocytes. In the cases of obligatory intracellular microorganisms, e.g., viruses, rickettsia, chlamydia, some bacteria, and protozoa, they directly enter the target cell via absorption to specific receptors followed by active penetration or passive phagocytosis and there reach their niche for multiplication.

A characteristic feature is that this in vivo multiplication of these microorganisms does not stimulate, but rather inhibits, the induction of exudative inflammation with the exception of facultatively intracellular bacteria and some rickettsia, e.g., *Rickettsia tsutsugamushi*. If the pathogenic microorganisms are deficient in the virulence determinant(s) which are responsible for further invasiveness, this characteristic constitutes an effective defensive barrier, as will be described later. Therefore, the degree of the inflammatory reaction at the local site of invasion of these pathogenic bacteria has a negative correlation with the degree of virulence of a given bacterial strain. In addition, the reaction of the host-immune system to these microorganisms frequently produces local adverse effects through the induction of proliferative inflammatory lesions, like granulomas, necrosis, and/or degeneration by the deposition of antigen-antibody immune complexes that may involve the activation of complement systems to produce local damage as was already discussed (p. 35).

On the other hand, the release of toxic substances, e.g., endotoxins from the pathogenic bacteria that are degraded during the interaction with the host, and the progressive tissue damage due to bacteria disseminated from the local inflammatory foci finally cause the collapse of all facets of host-defense mechanisms. In other words, these pathogenic microorganisms are not confined within the local foci, and the blood-stream of the host may become flooded with large numbers of pathogenic organisms. Death of the host results either by intoxication with endotoxin (endotoxemia) or dysfunction of vitally important organs, e.g., heart, kidney.

These serious outcomes can be avoided, if the pathogenic organisms fail to express their potential virulence factors fully before the specific immune defense mechanisms of the host reach their effective levels (see disease formula, p. 11); or if the pathogenic organisms have lost at least part of their genetic virulence determinants beforehand, i.e., they have mutated into nonvirulent or weakly virulent strains. In the latter case, only local infection may occur with recovery as stated before, but this will result in specific protective immunity against a second infection with fully pathogenic wild strains. This is the basis for the development of attenuated strains for use as living vaccines which has been highly successful in poliomyelitis, etc.

Each pathogenic organism causing invasive infection possesses unique ways to achieve penetration into internal tissues of the host either by circumventing the local limiting action of exudative inflammation as the defensive mechanism of the host or by *not*-stimulating host defense mechanisms (Smith 1972). Sometimes this occurs by the parasite masquerading as host-self.

In both cases, the microorganisms initiate multiplication at the site(s) to which they were disseminated and damage the host by unique mechanisms. Therefore, the difference between exudative (partially invasive) infections versus invasive (penetrating) infections resides within the distance the parasite invades from the body surface into the circulatory system and internal organs.

It is to be noted that invasive (penetrating) infections can be induced by the causative pathogens of exudative infections when the host is infected through artificial routes or is deficient in some facet of the defensive mechanisms either hereditarily or through acquired deficiencies. In contrast, the characteristic causative pathogens of invasive infections do not require such host deficiencies, because they are equipped with virulence determinants, whose genetic expression permits penetration, i.e., spreading becomes possible.

We now turn to a discussion of intracellular parasitism as representative of the ultimate sophistication in parasitism. Some data on viruses and rickettsia will be summarized here as examples of obligate intracellular parasites. Among the three bacterial examples of intracellular parasitism which are to be discussed, the virulence determinant(s) responsible for this capability have not been elucidated yet. Spirochetes will also be introduced because of the unique manner in which they penetrate the host as well as their property of not-stimulating the host-immune system.

Viruses

The general concept of viruses as microorganisms was already mentioned briefly (p. 3) and the existence of endogenous viruses was also described previously (p. 3). There are groups of RNA and DNA viruses that have caused epidemics and pandemics as well as viruses that are very common, but exist without such dramatic outbreaks of disease.

Pathogenetically, viral infections can be divided into several groups, irrespective of their biological classification, depending on their infection routes and vectors, e.g., respiratory viruses, enteroviruses, and arthroviruses. Furthermore, the characteristic pathogenesis unique to each viral group is determined by each agent's preference for attachment sites exposed on certain cell surfaces of the host.

In contrast with such specific interrelationships between the viral species and the susceptible host cells composing the living animal body, the range of cells which the virus can infect shows a broader spectrum if the tissues are cultured in vitro, i.e., viruses can infect cultured cells that have been derived from a nonsusceptible host animal. With regard to the studies of viral pathogenesis, however, reliance on the tissue-culture systems is inadvisable because of the dedifferentiation and other changes that occur in the cultured cells. Also the virus cultivated in vitro may be altered in its biological characteristics to result in the attenuation of its virulence for the natural host (Smith 1972).

Viruses absorbed onto the receptors penetrate their specific cells by the mechanism called receptor-mediated endocytosis. The release of viral nucleic acids, DNA or RNA, from the virion, i.e., uncoating, occurs after penetration into the cells. Another point of difference of viruses from other intracellular parasitic microorganisms resides in the fact that replication or production of viral nucleic acids and of the coat proteins proceeds separately, being coded by the genetic information in the viral nucleic acids, but utilizing energy sources and amino acids, etc. of the host cells. The intracellular site where these viral constituents assemble to form complete viruses is sometimes within cytoplasm or nucleus or it may occur at the nuclear or cytoplasmic membranes depending on the viral species. Thus, some viruses form intranucleic or intracytoplasmic inclusion bodies, crystallinelike structures of clumped viruses that are visible under light microscopes after staining with fluorescent antibodies or suitable dyes. The complete viruses are released from the cell one by one through budding, from the cytoplasmic membranes or through channels to the cell membrane or by disintegration of the host cell. In the former case, the host cell may live producing a small number of viruses for a long time to result in a persisting infection. Sometimes viruses remain intracellularly without being

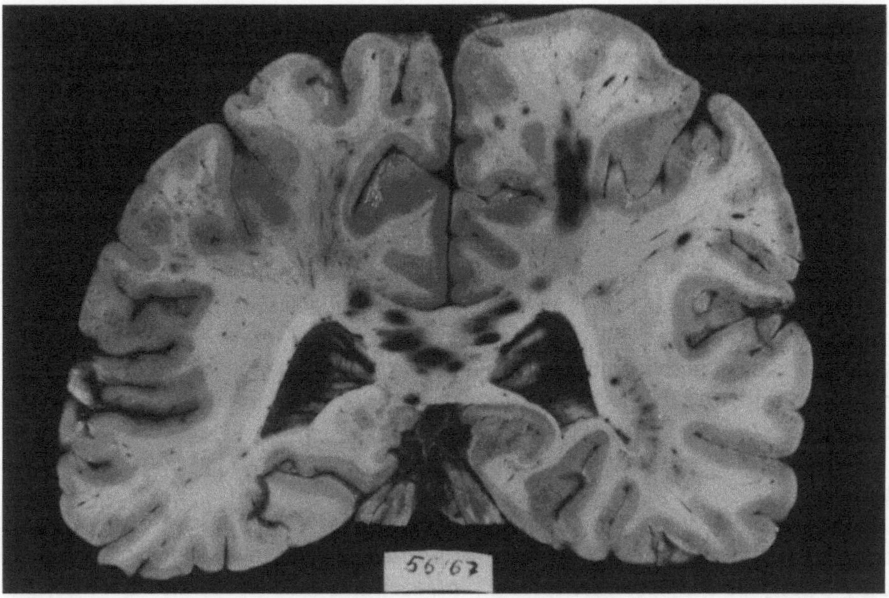

Fig. I.12. Multiple hemorrhage in the splenium corporis callosi and in the white substance of parieto-occipital lobe. [From Fig. 9 in Marburg Virus Disease, Martini, Siegert (eds) Springer, Berlin Heidelberg New York, p. 58]

released for certain lengths of time and are thus cryptic and the host is symptomless.

Viruses may spread throughout the body by viremia, with the viral particles reaching the receptors of new host cells or they may spread along the axons of neurons, e.g., rabies virus.

DNA and RNA viruses replicate themselves, i.e., DNA to DNA directly or via RNA, or RNA to RNA, by viral-encoded polymerases. Some oncornavirus (i.e. oncogenic RNA virus) nucleic acids are incorporated into the chromosomal DNA of the host after transcription from virus RNA to provirus DNA. A specific enzyme, reverse transcriptase, translates the information of RNA nucleotide sequences into those of DNA nucleotide sequences prior to this integration into the host-cell DNA.

Viruses produce damage to the host tissue and organs via various ways; first, by their cytopathogenicity, i.e., by destroying cells to result in tissue injury. This produces damage to host organs through various pathogenic processes, e.g., degeneration of vitally important cells, such as nerve cells (poliomyelitis); severe hemorrhagic diathesis; high febrile reactions and inflammation induced by viral-cell damage, etc. An example of hemorrhagic lesions occurring in a vitally important organ, i.e., brain, caused by Marburg virus, is shown in Fig. I.12.

Transformation of host cells may also occur due to the presence of viral information with the oncogenic viruses. These transformed cells may lose contact inhibition, etc. and grow into malignant tumors. These transformed cells express specifically altered cell-surface components, and act as antigenic stimuli to the host's immune system. Activated T-cells altered by the viruses may lyze such cells; and thus prevent viral maturation so the viruses are destroyed. Therefore, this cytolysis plays a defensive role as well as causing damage to the host tissue. Other defensive mechanisms against viral infections include the action of interferon which interferes with intracellular viral multiplication and the neutralization of extracellular virus due to specific antibodies which prevents the spread of the viral infection in the involved organ or to other parts of the body. However, the humoral antibodies involved in viral neutralization may themselves damage the host tissues by forming immune complexes with viral antigens. These immune complexes may be deposited in organs, i.e., kidney glomeruli, and trigger the activation of the complement cascade to damage host tissues.

Viral infection in the mother may spread to the fetus to produce malformations of the newborn and/or cause abortion.

Due to the diversity in the pathogenesis of viral infections, the enumeration of concrete examples was intentionally avoided here and viruses are considered only briefly as the best examples of intracellular parasitism and host-dependent parasites.

Rickettsiae

The rickettsiae comprise a highly heterogeneous group of bacteria, but our presentation will be limited to the genus *Rickettsia*. The species of this genus are obligate intracellular parasites, dependent for survival in nature, on a vertebrate and invertebrate host. Some cause severe diseases in humans, such as epidemic typhus, Rocky Mountain spotted fever, and scrub typhus. These diseases, if promptly diagnosed, can be successfully treated with broad-range antibiotics. Other rickettsial species cause somewhat milder diseases, such as murine typhus, Siberian tick typhus, fievre boutonneuse, and rickettsialpox, while other species are not known to be human pathogens.

A discussion of the genus *Rickettsia* requires the recognition of three groups: the typhus, spotted fever, and scrub typhus groups. There is a small, but significant, difference in the % G+C of the DNA between the typhus (29–30%) and the spotted fever (32–33%) groups, but there is some DNA-DNA homology between some of the species of the two groups (30–50%). One of the species *R. canada*, has some properties that are typical of the typhus group and others that are typical of the spotted fever group. The base composition of the scrub typhus group is not known, but, judging from phenotypic characteristics, it is not closely related to the other two groups.

The basis for the evolutionary separation of the rickettsiae is probably the host of principal residence. *R. prowazekii*, the etiologic agent of epidemic typhus, resides principally in man and is transmitted from man to man by the body louse. *R. typhi,* the agent of murine typhus, resides principally in city rats and their fleas. *R. rickettsii,* the agent of Rocky Mountain spotted fever is transmitted by ticks from its rodent hosts. *R. tsutsugamushi,* the agent of scrub typhus, is a parasite of mites, which feed on rodents, usually rural rats, only once during their life cycle.

Despite the close dependence on the cells of their hosts, rickettsiae are rather typical Gram-negative bacteria. They do not have flagella or pili, but are surrounded by a conspicuous slime layer that can be demonstrated by the specific-antibody stabilization technique. The slime layer is readily lost during most laboratory procedures of separation of the rickettsiae from host-cell constituents. When placed in water or buffers of low ionic strength which do not contain Mg^{+2}, rickettsiae lose their microcapsular layer, a loss that is accompanied by greatly reduced metabolic activity and infectivity. The microcapsular protein that is lost is species-specific and highly immunogenic, eliciting protective immunity. Rickettsiae of the typhus and spotted fever groups, on the one hand, and rickettsiae of the scrub typhus group, on the other, differ considerably in the structure of the cell wall. The former have a thick inner leaflet and thin outer leaflet. The reverse is true in the scrub typhus rickettsiae, which, unlike the other rickettsiae, adhere tenaciously to host-cell constituents, from which they cannot be well separated without loss of activity.

Rickettsiae will grow in quite a number of primary and established eukaryotic cell lines. In the mammalian host the cells most commonly involved are the endothelial cells of capillaries. The infection of a eukaryotic cell by a rickettsiae requires the active participation of both. The infection can be foiled by metabolic inhibitors of either participant. The sequence of events leading to successful infection can be visualized as follows: (1) rickettsiae induce their own phagocytosis – a necessity when confronted with nonprofessional phagyocytes – by a mechanism that is not entirely clear; (2) the host cells internalize the rickettsiae in their phagosome; (3) if the rickettsiae are to survive, they must quickly escape into cytoplasm, a step which also prevents fusion of the phagosomes with the lysosomes. There is good evidence that escape involves phospholipase activity, although it is not known if the rickettsiae provides the enzyme or just stimulate a host enzyme. Rickettsiae vary in efficiency of escape from the phagosomes of professional phagocytes. The virulent Breinl strain of *R. prowazekii* is quite efficient, while the avirulent strain E usually succumbs to digestion by the lysosomes.

Once safely inside the cytoplasm, rickettsial multiplication proceeds by binary fission, but there are differences in some of the aspects of multiplication in the three groups. Typhus rickettsiae multiply until they reach a high concentration in the cytoplasm and the host cell finally disintegrates. Spotted fever rickettsiae, early in the cycle of multiplication, are found in the cytoplasm. However, occasionally they enter the nucleus of a host cell and in that location they multiply to high density. Scrub typhus rickettsiae multiply to high density in the perinuclear area of the cytoplasm, but have an unusual mechanism of release from the cell. They protrude from the surface of the cell, are released by a process akin to budding, and enter a new cell while still surrounded by some of the cytoplasm of the old host cell.

What is the contribution of the host cell to rickettsial multiplication? There is no evidence that rickettsiae have a glycolytic pathway, but they utilize glutamate or glutamine via the citric cycle. By this mechanism they maintain or restore their ATP level. They also have a carrier-mediated transport system for ATP and when the extracellular concentration is high, rickettsiae readily exchange their endogenous pool with extracellular nucleotides. However, there is no evidence that rickettsiae synthesize or degrade nucleotide monophosphates and their nucleotide metabolism is confined to the mono-, di-, and triphosphates. In appropriate rich media, rickettsiae were shown to be capable of synthesizing low levels of protein, but there is no real evidence that they synthesize RNA or DNA. Thus, for growth rickettsiae obviously depend on host cells for a variety of metabolites, including phosphorylated compounds, and, most likely, regulatory mechanisms. Enzyme regulation in rickettsiae remains largely unknown. In recent experiments it was shown that rickettsial citrate synthetase more closely resembles the enzymes of eukaryotic or Gram-positive cells, rather than those of other Gram-negative bacteria. When this enzyme is cloned in *E. coli* plasmids, it retains unusual characteristics.

What can we learn about rickettsial pathogenesis from their in vitro interaction with eukaryotic cells? As with other Gram-negative bacteria rickettsiae possess lipopolysaccharides which are released in the patient and undoubtedly contribute to the severity of the respective diseases. However, the most significant feature of rickettsial disease can be traced to the interaction with host cells. When rickettsiae are incubated with the erythrocytes of a number of species, including man, the cells are lyzed by a mechanism best described as abortive infection. A somewhat similar phenomenon occurs when large numbers of rickettsiae are incubated with susceptible nucleated cells. When the multiplicity of infection is very high the host cells are destroyed before they can be infected. Undoubtedly this happens in human infection when the rickettsiae are released from a first round of infection and attack neighboring cells in large numbers. Because of the predilection for endothelial cells, they cause petechial hemorrhages in the skin and in the brain, as in typhus fever, or cause sufficient damage to blood vessels to cause gangrene of the extremities, as in Rocky Mountain spotted fever.

In conclusion, there are three mechanisms by which obligate intracellular bacteria meet the challenge of the hostile environment of the eukaryotic cell. One is "grin and bear it". Some of the mycobacteria, the protozoan parasite *Leishmania donovani,* and the rickettsial agent of Q fever, *Coxiella burnetii,* thrive in the environment of the phagolysosome, with its digestive enzymes and low pH. Other parasites prevent hostile changes in their environment. *Chlamydia, Ehrlichia,* and the protozoan *Toxoplasma gondii,* prevent the fusion of the phagosome with the lysosomes and make their home in the cozy environment of the unmodified phagosome. Finally, venturesome parasites, such as the protozoan *Trypanosoma cruzi* and rickettsiae, escape their environment and make their home in the rich land of the cytoplasm.

Bacteria

Microorganisms, causing invasive (penetrating) infections employ many divergent virulence factors and four groups, spirochetes, mycobacteria, salmonellae, and listeria are discussed as "examples" in this subsection.

Example 1: Spirochetes

The main reason why spirochetes are selected for discussion, resides in the unique manner of their penetration of the host designated as "superficial parasitism" by Hayes et al. (1977). In addition, however, such properties as motility, antigen shifting, and even failure of stimulation of host-defensive mechanisms (Smith 1972) can be identified as virulence factors of spirochetes.

Spirochetes are highly motile, spiral microorganisms, whose shape is probably determined by the specific composition of the peptidoglycan in the cell wall. Beneath the outer envelope, they have characteristic axial fibrils which extend along the helicoid cylinderlike protoplasm. Using these fibrils, they move by flexing, undulating, and rotating to produce serpentine movements. This movement aids their penetration of the host's body through minute injuries of the skin. The organisms then readily move through the high-viscosity environments within the tissues, especially the intercellular spaces of connective tissues filled with lymph.

Among spirochetes, the causative spirochetal agent of recurrent fever, *Borrelia recurrentis,* a representative pathogen that uses the antigen shift (see later) phenomenon as a unique virulence factor, is transmitted to man via the bite of the human body louse. Subsequently, the spirochetes multiply and can be observed in the blood and tissues. Febrile attacks persist for several days and are followed by afebrile periods when the blood is not infectious – thus, the name, recurrent or relapsing fever (Baseman et al. 1979). The microorganisms disappear from the bloodstream, because of their sensitivity to specific antibodies produced during the period from infection through the febrile attack; complement action with the antibody produced eliminates the circulating microorganisms. However, *Borrelia* organisms surmount this humoral attack of the host by changing the antigenic specificities of their outer cell envelope (antigenic shift). This occurs by the development of certain mutants having new antigenic characteristics in a limited and predictable pattern (Felsenfeld et al. 1965). These new antigenic mutants then begin to multiply and the fever recurrs.

The causative agent of Weil's disease, *Leptospira icterohemorrhagiae* infects the human host percutaneously resulting in serious liver disorders with jaundice and febrile reactions which may result in death. But the same causative agents infect wild rats and mice without obvious symptoms even though they are circulating in the bloodstream and are excreted into urine. This is a typical example of a symbiotic host-parasite relationship based on *not*-stimulation of defensive mechanisms of the rodent animals by these microorganisms.

Another virulence factor of spirochetes has been convincingly demonstrated with *Treponema pallidum* using tissue cultures of various animal cell lines incubated together with *T. pallidum*. Hayes et al. (1977) observed by dark-field microscopy a specific orientation of virulent treponemes to host-cell surfaces. Treponemes attached to the host-cell membrane by their tapered ends and remained anchored in place yet actively motile. This association was temperature- and energy-dependent, with maximal attachment occurring at 37 °C. The initial surface parasitism by *T. pallidum* seems to saturate the putative receptor sites for *T. pallidum* on the host cell and make its surface refractory for further accession of the parasite. When the host-cell surfaces were pretreated with antibodies against the cell-surface receptors, a major decrease in *T. pallidum* attachment was demonstrated.

Similar observations were made by Baseman et al. (1979): Casein-activated peritoneal macrophages were obtained from normal rabbits and cultured on cover slips in Leighton tubes before exposure to freshly harvested *T. pallidum*. Treponemes readily adhered to macrophages and exhibited their unique polarity of attachment. Attached treponemes were actively motile and appeared totally resistant to the "phagocytizing" activity of host macrophages.

Based on these experimental observations, the pathogenesis of syphilis caused by *T. pallidum* can be explained as follows:

T. pallidum organisms infect the human host usually via sexual contact through mucous membranes and multiply at sites in the subcutaneous tissues. After 1 ~ 7 weeks of infection, these organisms induce the formation of a hard chancre, a painless proliferative lesion with induration and/or ulceration which disappears within 5 ~ 6 weeks. Probably the organisms penetrate deeper into the body not only by their active motility, but also by transportation on migrating macrophages whose surfaces are parasitized by *T. pallidum*. They reach various tissues, including skin and other organs, to produce the syphilis roseola and/or papulosa after about 3 months of infection. After passing through (or without) this period, chronic proliferative lesions containing fibroblasts, lymphocytes, plasma cells, histiocytes, and even necrotic foci, called gummas of various sizes are induced in various tissues and organs like skin, central nervous system, heart, large blood vessels, and liver. This third stage occurs after approximately 3 yrs. of infection. *T. pallidum* can exist until the fourth period, when degenerative changes in brain and spinal medulla tissues produce various serious disorders after more than 10 yrs. of infection. These serious outcomes of syphilitic disease result from the more or less unique virulence factors of spirochetes, i.e., active motility, not- or weak-stimulation of host-defensive mechanisms, and the "parasitism onto the host-cell surfaces". However, it has been demonstrated that spirochetes opsonized by specific antibodies and a complement component C3b are killed after phagocytosis (Baseman et al. 1979).

In summary, this example is characterized by unique virulence factors, e.g., not-stimulation of defensive mechanisms and parasitism onto the host-cell surfaces. Neither the nature of the virulence determinants coding for these factors nor their localizations have been elucidated because both of these factors are designated in functional terms and are not identified as related to specific materials produced by the parasite.

Example 2: Mycobacterium tuberculosis

Mycobacterium tuberculosis is selected for discussion, first, as one of the causative agents of *chronic* invasive (penetrating) infections versus those of acute or subacute character. Secondly, the long generation time of the human, the bovine, and the murine types of tubercle bacilli, e.g., approximately 18 h at shortest according to the measurement in artificial media by Dubos and Middlebrook (1947), plus their high resistance to the bactericidal mechanisms of the host animals, create a unique host-parasite relationship in which the defensive and adverse effects of the immune reactions (see p. 131) equilibrate delicately during the long period of infection. Under these equilibrating host-parasite relationships, the in vivo phase of *M. tuberculosis* can be differentiated from bacilli grown in cell-free medium by characteristic modifications of their surface structures that make the bacilli not-stimulating to host-defense mechanisms (p. 6). Third, *M. tuberculosis* (though a bacterial infectious agent) is a facultative intracellular parasite multiplying within macrophages as well as extracellularly following destruction of the host phagocytes. Therefore, this host-parasite relationship is suitable for a discussion of virulence factors that resist intracellular-destructive or bacteriostatic mechanisms.

A basic characteristic of this microorganism is its cell wall which has such a firm basal structure of murein (= peptidoglycan = mucopeptide) covalently linked to an arabinogalacton-mycolate that it causes the "acid fastness" of these bacilli.

Furthermore, it would be appropriate to select mycobacteria as a typical example of a parasite in which an unusual host-defense mechanism, "body-temperature" is found to be operating in the host-parasite relationship of mycobacteria because chickens, whose body temperature is about 42 °C, are susceptible to *Mycobacterium avium* serovars 1 and 2, which grow well in vitro at 43 °C (Feldman 1938), while they are very resistant to *M. tuberculosis* and *M. bovis*, for which the optimal growth temperature is 37 °C.

The pathogenesis of tuberculosis will be examined in detail with reference to the properties of the parasite- and host-resistance mechanisms.

Experimental tuberculous infection in rabbits, guinea pigs, and other susceptible animals begins with the recruitment of polymorphonuclear leukocytes (PMNs) from the bloodstream to the site of initial bacillary lodgment at the time of infection (Dienes and Mallory 1937). After phagocytosis of the bacilli by PMNs, it is not intracellular bacilli, but the bacteria-laden PMNs that die, and the leukocyte remnants together with their contents are engorged by macrophages (Rich 1951). During tuberculous infection, antibodies are formed (Middlebrook and Dubois 1948), and immunoglobulins reactive with *M. tuberculosis* are reported to be common among "normal" human subjects

(Bardana et al. 1973; Freedman et al. 1963). However, the passive transfer of either guinea pig (Raffel 1955) or rabbit (Reggiardo and Middlebrook 1974) sera containing antituberculous immunoglobulins demonstrated that these treatments were ineffective in producing protection of the recipient animals against challenge with virulent tubercle bacilli.

Irrespective of the presence or absence of antituberculous immunoglobulins, tubercle bacilli engulfed by macrophages are not killed by these cells. *M. tuberculosis* interferes with or antagonizes phagosome-lysosome fusion by the production of sulfatides, and thus avoids potential digestion by lysosomal enzymes (Goren 1977): i.e., virulent tubercle bacilli produce sulfur-containing glycolipids (Goren 1975; Goren et al. 1974, 1976 a) which produce this protection. Thus, certain intracellular parasites, including *M. tuberculosis,* may promote their survival within the host by acting from within phagosomes to prevent phagolysosome formation, and thus avoid exposure to lysosomal hydrolases (D'Arcy Hart and Armstrong 1974; Goren et al. 1976 b). When these sulfatides were added to cultured mouse peritoneal macrophages, they accumulated in lysosomes, which became incapable of fusion with phagosomes. Excellent experiments were performed by Goren (1977), in which small amounts of the sulfur-containing glycolipid were introduced directly into phagosomes by attachment to target yeast cells prior to their ingestion to demonstrate this "antifusion effect" of sulfatides. These sulfatides are families of structurally related sulfolipids and the chemical structure of the principal sulfatide from 40 strains of *M. tuberculosis* is shown Fig. I.13.

During this in vivo growth, tubercle bacilli reversibly acquire different characteristics from those of in vitro-grown bacilli. Segal and Bloch (1956) designated the in vivo- and the in vitro-grown virulent $H_{37}Rv$ organisms as Phe I and Phe II, respectively. He demonstrated the increase in Phe II in the quantity of hydrogen transfer and its capability to utilize exogenous energy sources detectable by a marked increase in Qo2. Evidence of reversible changes from Phe II to Phe I in the surface structures in these in vivo-grown tubercle bacilli were reported by Segal and Bloch (1956) and Kanai (1967 b, c). The quantity of lipid in the organism is different between Phe I and Phe II (Segal and Bloch 1965). Phe I bacilli of the *M. bovis* Ravenel strain were found to be devoid of the surface component, mycoside B which is produced by the bacilli grown in

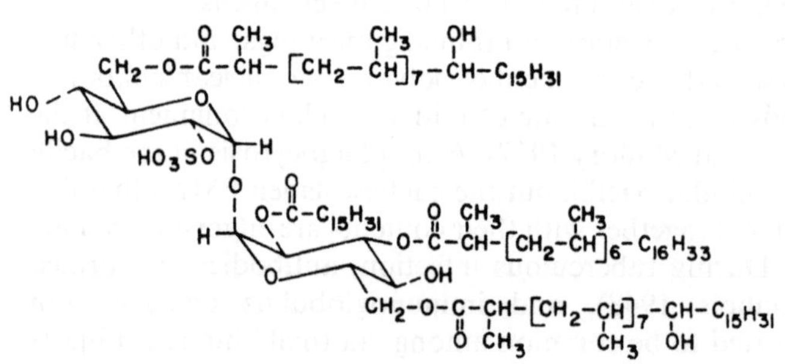

Fig. I.13. Chemical structure of sulfolipid I. [M. B. Goren, O. Brokl, and W. B. Schaefer (1974) Infect Immun 9:142–149]

vitro i.e. Phe II (Kanai et al. 1970). Some of the differences between Phe I and Phe II are derived from an intimate association of Phe I with macromolecules of the host, according to the data of Kanai (1967a, b) and Kanai and Kondo (1970). Kanai (1967a, b) showed that the virulent *M. tuberculosis* $H_{37}Rv$ and the attenuated, but infectious, *M. bovis* BCG organisms had associated with them an acid phosphatase (MAP) derived from the murine-host tissue during their growth in vivo in mice. When living in vitro-grown bacilli were exposed to MAP, they bound it. Furthermore, the presence of MAP on the surface of bacillary bodies seemed to block the activity of the acid phosphatase derived from these bacilli, i.e., the isozyme of parasitic origin. These results could be interpreted as the masquarading of the intracellular Phe I tubercle bacilli by the accumulation of host macromolecules on their surfaces that could render them somewhat indistinguishable from the host and therefore not-stimulating to host-defense mechanisms (p. 59).

Both of these two mechanisms confer on virulent tubercle bacilli resistance to intracellular destruction by macrophages. In detail, the phagocytized tubercle bacilli produce sulfatide and prevent the fusion of the phagosome in which the bacilli are included, with the lysosome, an organelle containing digestive enzyme-hydrolases (Goren et al. 1974). This mechanism gives intracellular bacilli time to induce the biosynthesis of the substance consisting of phospholipid and cholesterol esters called mycoside B, which confers on the bacilli resistance to the digestive action of hydrolases from lysosomes after the delayed phagosome-lysozome fusion (Kanai and Kondo 1970).

In addition, iron-transport mechanisms of mycobacteria serve to surmount the barrier of host-defense mechanisms, i.e., iron-chelating proteins act as one of the virulence factors. Mycobacterial siderophores (see Part II) are called mycobactin and were reviewed by Snow and White (1969). A short review (Wake 1980) deals with the divergence of opinions about the nature of mycobacterial siderophores between Kochan (1977) and Macham et al. (1975). Macham et al. (1975) have suggested that the lipophilic mycobactin molecule is located on the "boundary layers" of the mycobacterium and transports iron across the thick lipoidal layers of the mycobacterial cell wall as ferric ion, and that the ferrous form is then released from its carrier by the action of NADP-linked reductase (Brown and Ratledge 1975). Exochelin, a class of partly characterized water-soluble siderophores, on the other hand, diffuses freely from the bacillary surface in both desferric and ferric forms, and can steal iron from the ferritin of the host tissue, because it holds iron in solution at physiological pH values (Macham et al. 1975). The hypothesis is that the soluble ferric exochelines are taken onto mycobactin, which then carries the iron through the mycobacterial cell wall to the cytoplasmic membrane where the NAD-linked ferrimycobactin reductase makes the ferrous ion available to the cytoplasm (Fig. I.14). The iron assimilated by virtue of these siderophores, i.e., exochelin and mycobactin, is utilized to biosynthesize essential enzymatic systems containing iron as a vital element, e.g., cytochrome C, and to supply materials to promote bacillary multiplication.

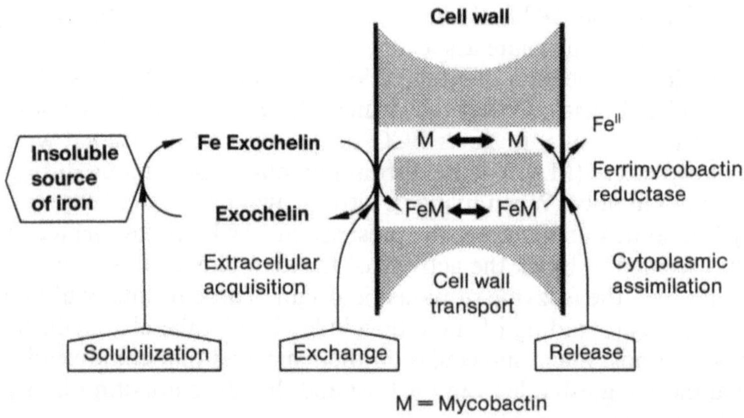

Fig. I.14. Proposed scheme for iron transport in the mycobacteria. (From Macham et al. 1975)

In addition to the above-stated *M. tuberculosis* virulence factors early investigators paid attention to the serpentine cord- or ropelike morphological appearance of aligned tubercle bacilli grown in liquid medium. Bloch (1950) considered this phenomenon as related to a virulence factor, and extracted a glycolipid from viable bacilli with petroleum ether and gave it the name of cord factor, regarding this as the responsible substance. This was identified as trehalose-6,6′-dimycolate by Noll et al. (1956), but its direct relationship with mycobacterial virulence has not been established. However, cord factor does inhibit the migration of leukocytes and is leukotoxic (Bloch 1950).

Kato (1968) reported that the succinic and malic dehydrogenase activity of mouse liver decreased approximately 24 h after the i.p. injection of cord factor and that this decreased enzymatic activity was restored by the addition of normal mouse liver homogenate supernatant. This enzymatic activity was found to decrease more remarkably, when the pathological changes in the infected guinea pigs and mice were more severe. By this indirect evidence, Kato et al. (1959) assumed that cord factor causes the metabolic lesions of host-enzyme systems which appear with an increase of bacillary damage. In addition, Bekierkunst et al. (1969) reported that injection of cord factor stimulated the formation of pulmonary granulomas in mice, but not rats. Kato (1971) showed that cord factor is not immunogenic, but it serves as a functional "hapten" when complexed with methylated bovine serum albumin (MBSA). When cord factor-MBSA emulsified in incomplete Freund's adjuvant was used as a vaccine, the mice were protected against both the toxic action of cord factor and infection with *M. tuberculosis* $H_{37}Rv$ (Kato 1971). In this context, cord factor may be included as one of the *M. tuberculosis* virulence factors.

However, the pathogenesis of tuberculosis should not be implied to be a result simply of the integrated action of various virulence factors of *M. tuberculosis* because it has evolved as a unique host-parasite relationship in which host-immune reactions are involved throughout.

From this point of view, the histopathological changes in infection will be considered.

At the local lesions, the formation of a tuberculous granuloma is a consequence of the recruitment of host leukocytes. A granuloma initially consists of plump, rounded mononuclear cells that vaguely resemble epithelial cells and are, therefore, designated as epitheloid cells. Each of these cells results from the ingestion of intact or fragmented tubercle bacilli by mononuclear phagocytes resulting in their transformation into epitheloid cells. At the margin of this epitheloid cell cluster and sometimes within the center of the cluster, cell membranes covering the individual cytoplasma of several cells disappear resulting in cell fusion to form large multinucleate giant cells. About this center of epitheloid cells and giant cells, there is a peripheral zone or collar of plump fibroblasts and lymphocytes. This whole structure is called a hard tubercle. Subsequently, the cells located at the center of tubercle lose their viability and become necrotic to produce "caseation necrosis" which is the most characteristic feature of the formation of a tubercle.

Tuberculosis is characterized by its chronic disease process. The rather exceptional acute lethal outcomes of infection are caused either by the formation of multiple disseminated tubercles or by the injury of vitally important organs or tissues, e.g., meningitis.

While virulent bacilli contained in the foci of disease resist the killing activities of host-cellular mechanisms and tend to multiply with a long generation time, the host-defense mechanisms recruiting mononuclear cells, fibroblasts, and lymphocytes lay siege to these foci enveloping them and calcifying the outer walls of the whole degenerated region containing the necrotic focus, preventing spread of infection and leading eventually in many cases to death of the bacilli. Thus, a long-term dynamic equilibrium is established between host and parasite in the individual host, until final outcome results either in the sequestration and eventual elimination of tubercle bacilli or the escape of the bacilli from the local tubercle to produce other new foci. In the latter case, large regions of some important organ, e.g., lung, intestine, kidney, brain, meninges, or vertebral column can be occupied by the masses of degenerated dead tissues or abscesses, characteristic of tuberculosis and result in the death of the host. Alternatively, both the tubercle bacilli and host can maintain an equilibrium state, in other words, coexistence, until the host dies due to some other cause.

It is to be noted that as long as viable bacilli remain in these local lesions in the body, the host remains tuberculin skin-test positive since when the organisms are eliminated by host reactions, drugs, or surgery, the host becomes tuberculin-negative.

Infection with tubercle bacilli influences the host body also by affecting the host's immune network. A typical expression of one of these changes is the well-known "tuberculin skin reaction" which is utilized to detect whether the host has experienced a sensitization with components of tubercle bacilli during infection with the bacilli. The bodies of tuberculin-positive individuals are in a state of delayed hypersensitivity to tubercle bacilli, i.e., a certain subset of their T-lymphocytes designated T_{DTH} is specifically activated by bacilli (refer to Part II). In other words, they have developed an "allergy" to these antigens. A great part of the pathological lesions of tuberculosis are brought about not by direct toxic action of the products of tubercle bacilli, but by these "suicide-like" allergic reactions which result in destruction of infected tissues. Thus, the virulence mechanisms of *M. tuberculosis* are greatly dependent upon the multifaceted defense mechanisms of its human host. However, the genetic control of virulence factors of *M. tuberculosis* remains to be elucidated.

Example 3: Salmonella

The genus *Salmonella,* consisting of *Sal. typhi, Sal. choleraesuis,* and *Sal. enteritidis,* is selected for discussion as one of the groups of Gram-negative bacteria causing acute invasive (penetrating) enteric infections. In spite of the extreme diversity in disease forms caused by *Salmonella,* their virulence factors conferring invasiveness are similar. Also, the pathogenicity of this genus is characterized by a wide host range, e.g., human, domestic and wild mammals, birds, and even insects. In addition, one species of this genus, *Sal. enteritidis* includes more than 1,400 serovars, e.g., *Sal. enteritidis enteritidis, Sal. enteritidis newport, Sal. enteritidis typhimurium, Sal. enteritidis paratyphi,* etc. that are for convenience abbreviated *Sal. enteritidis, Sal. newport, Sal. typhimurium, Sal. paratyphi,* respectively.

Because each serovar restricts itself more or less to a specific host animal using the latter as the ecological niche (p. 9), the genus *Salmonella* is a good example for the study of the concept, host-parasite relationships. It provides one of the best examples of the *specific* relationships between a particular host and its inherent parasites that have evolved during the evolutional history of ecosystems.

Considering the human host for an example, the diverse disease forms caused by *Salmonella* are gastroenteritis, the typhoidal forms, bacteremia with the severe form being called septicemia, and the healthy-carrier state. It is difficult to establish clear mutual discriminations among these forms because of possible overlapping and/or transitions from one disease form to another in a single patient.

The profound significance of salmonella in considering host-parasite relationships resides in the fact that any serovar of salmonella organisms has the possibility to induce any disease form depending on the specific relationships between its bacterial virulence factors and the multifaceted host-defense mechanisms. The virulence factor responsible for the induction of the typhoidal disease form is the ability of the organism to resist the intracellular-destructive mechanisms of the host's phagocytes. In this context, *Sal. typhi, Sal. paratyphi* (formerly *Sal. paratyphi*-A), *Sal. schottmuelleri* (*Sal. paratyphi*-B), *Sal. hirschfeldii* (*Sal. paratyphi*-C), and *Sal. sendai* are the causative serovars of the human form of typhoid, while the main disease form caused in man by *Sal. new-*

port and *Sal. enteritidis* is gastroenteritis and the most frequent causative serovar of human bacteremia and localized infected foci is *Sal. choleraesuis*. Infection with *Sal. anatum* induces mild gastroenteritis or the healthy-carrier state. In addition, *Sal. typhimurium* is isolated not only as the most common causative serovar of gastroenteritis, but also causes bacteremia and localized focal infections in rare cases (Rubin and Weinstein 1977).

From the point of view of experimental researchers, the specific host-parasite relationships in typhoid, have been difficult to investigate because infections of experimental animals with *Sal. typhi* and the other aforementioned human-specific serovars do not reproduce the same disease forms that occur in the human host and it is very difficult to infect most experimental animals with *Sal. typhi*.

Only the discovery of animal infection models with various *Salmonella* that will reproduce the same specific host-parasite relationships found in human infection with *Sal. typhi* can be expected to give the proper answers. Such selections are mice infected with *Sal. typhimurium* or *Sal. enteritidis* or rabbits given *Sal. hirschfeldii* (*Sal. paratyphi*-C) or *Sal. choleraesuis* (Kobayashi and Ushiba 1952). In these infections, the pathogens invade from the local sites of infection to spread via the lymph and blood and induce the formation of granulomatous "typhoid tubercles" in the lymph nodes, liver, spleen, and other organs after deposition and multiplication. If the pathological lesions continue to enlarge, the center of the granulomas becomes progressively necrotized, to release the organisms, and septicemias develop distributing the pathogens throughout the host's body. The host dies as a result of endotoxin (LPS) intoxication. A quite different disease process occurs if a large infecting dose is used, this results in an acute septicemic death designated as "primary sepsis" (Kobayashi 1940, 1942).

With the course of these experimental infections in mind, a discussion of the chemical components in the cell wall of *Sal. typhimurium* is relevant. This topic is important in the relationship between parasite and host, since the cell wall is the first structure to interact with the host cells.

Stocker et al. (1966) and Nikaido (1967) elucidated the fine biochemical structure of the cell wall lipopolysaccharide (LPS) by isolating a number of mutants derived from the wild-type strain LT2 of *Sal. typhimurium* and identifying the sequential sugar components of its polysaccharide portion. Each mutant was genetically defective in enzymes catalyzing a different synthetic step of polysaccharide biosynthesis by the bacteria. The results are shown in Table I.6. Nakano and Saito (1969) examined the virulence of each strain by infecting 6-week-old male outbred mice (DK1) intraperitoneally (i.p.) with 10^6 organisms. Although there are differences with regard to the susceptibility of various mouse strains, DK1 has been established to have high susceptibility to the LT2 (Ushiba et al. 1962). The results (Table I.6) showed that the strains possessing the O side chains consisting of abequosyl-mannosyl-rhamnosyl-galactose tetrasaccharide repeat units were virulent and killed almost all infected mice, while the strains lacking these O side chains, and therefore expos-

Table I.6. Relation between the sugar composition of the cell wall and virulence or immunogenicity of various strains of *Salmonella typhimurium* LT2 [modified from Nakano and Saito (1969)]

Strain	Sugar composition of cell wall									First infection[a]		Reinfection with wild-type strain[b]	
	Abe | Man-Rha-Gal	Abe | n-Man-Rha-Gal	GNAc-	Glc-	Gal | Gal-	Glc-	Hep-	Hep-	KDO-	Percent survival	Mean survival day (log) ± SE[c]	Percent survival	Mean survival day (log) ± SE[c]
Wild-type	+	+	+	+	+	+	+	+	+	0	0.55 ± 0.03		
SL 1027	+	+	+	+	+	+	+	+	+	0	0.58 ± 0.04		
SL 696	+	+	+	+	+	+	+	+	+	20	0.99 ± 0.04		
SL 1034	–	+	+	+	+	+	+	+	+	10	0.93 ± 0.04		
his 519	–	–	+	+	+	+	+	+	+	100		100	
his 695	–	–	+	+	+	+	+	+	+	100		100	
TV 119	–	–	+	+	+	+	+	+	+	100		100	
TV 149	–	–	+	+	+	+	+	+	+	100		100	
SL 1035	Trace	Trace	?	?	+	+	+	+	+	100		100	
SL 1036	Trace	Trace	?	?	+	+	+	+	+	100		100	
TV 148	–	–	–	?	+	+	+	+	+	100		80	0.72
SL 1069	–	–	–	–	–	+	+	+	+	100		90	0.85
SL 1004	–	–	–	–	–	–	+	+	+	100		100	
SL 1181	–	–	–	–	–	–	–	+	+	100		10	0.83 ± 0.06
SL 1102	–	–	–	–	–	–	–	–	+	100		0	0.79 ± 0.03

[a] First infection was by intraperitoneal injection of 10^4 organisms of each strain into a group of ten mice.
[b] Reinfection was by intraperitoneal infection with 10^4 wild-type organisms (100 MLD) on day 21 after the first infection. Percent survival and mean survival day of an untreated control group were 0 and 0.75 ± 0.02.
[c] The distribution of time of death in typhoid in DKl mice fits the logarithmic normal distribution.
Abbreviations: *Abe*, abequose; *Man*, D-mannose; *Rha*, L-rhamnose; *Gal*, D-galactose; *Glc*, D-glucose; *GNAc*, N-acetyl-D-glucosamine; *Hep*, L-glycero-D-manno-heptose; *KDO*, 3-deoxyoctulosonate.

ing a large area of the central portion of the antigen consisting of two L-glycero-D-manno-heptoses connected to the lipid portion KDO (called "R-core") showed intermediate virulence, although there were strains with no virulence in spite of their possession of nearly the same amount of O side chain material in their LPS. The presence of the latter exceptional avirulent strains indicates that the length of O side chains is not the sole virulence factor, but the action of the O side chain length and other virulence factors that will be described later separately, determine the expression of virulence by these pathogens. Thus, the longer the repeat units of the polysaccharide chains of O antigens, the greater the virulence of *Sal. typhimurium*. In other words, O side chains seemed to play the role of a virulence factor affecting the host-parasite relationship between *Sal. typhimurium* and the *mouse*.

Mice surviving infection in this experiment (Table I.6), were challenged with 10^4 organisms (100 MLD) intraperitoneally of the virulent wild-type LT2 organisms 3 weeks after the first infection. The capability of each mutant to confer protective immunity is also shown in Table I.6. With the exception of two mutants lacking one or both L-glycero-D-manno-heptoses, all mutants conferred a high protective resistance to the lethal challenge with the virulent LT2 organisms. This indicates that even O side chain deficient mutants (so-called rough mutants, e. g., strain SL 1004 in Table I.6) can serve as a living vaccine to induce protective immunity against infection with the wild-type strain. Later discussion about another virulence factor will be concerned with this point (p. 12). In other words, protective immunity against mouse typhoid could be induced without participation of most anti-O side chain antibodies, even though the O side chain itself is an important virulence factor as demonstrated in the experimental results shown Table I.6. Nakano et al. (1969) examined the correlation between the extent of multiplication of each mutant in the mice, and the immunogenicity of its LPS structure. After i.p. infection with 10^6 organisms of each strain, the average number of viable organisms in the spleens were counted at 6 h, and at 3, 7, and 21 days. All mutant strains that conferred protective immunity in the experiment shown in Table I.6 multiplied in the mice, peaking at different days. The mutant strains having a longer O side chain showed an earlier peak, though less than the wild-type which showed progressive multiplication and killed the mice within a few days.

The two mutants SL 1181 and SL 1102, the LPS of which contains one heptose plus 3-deoxyoctulosonate (KDO) and only KDO, respectively, did not multiply in the mice and, thus, produced no resistance to infection with the virulent strain LT2. These results suggest that multiplication of the vaccinating strain in mice is necessary for the establishment of protective immunity with the doses employed. As evidence to support this hypothesis, previous infections with SL 1181 or SL 1102 showed no influence upon the progressive multiplication of the wild-type strain given later. On the other hand, vaccination with mutant strains with two heptoses or longer side chains in LPS significantly inhibited the superinfecting wild-type organisms.

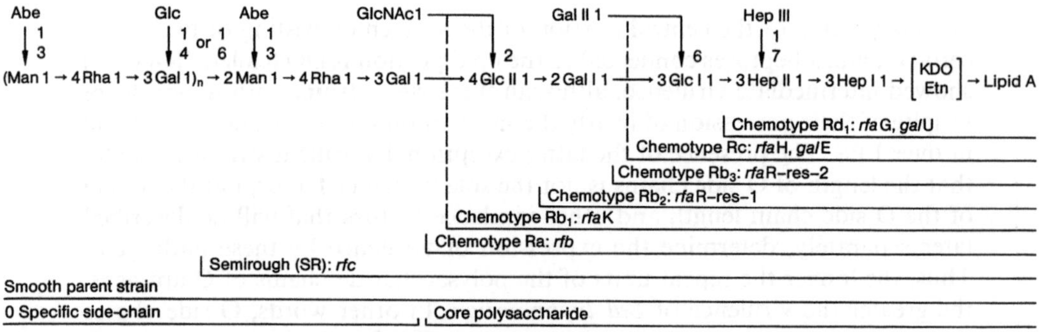

Phage attachment rate to smooth, semirough and rough *Salmonella* strains

Phage	Smooth	Semirough	Ra	Rb_1	Rb_2	Rb_3	Rc	Rd_1
P 22	297	<5	<5	<5	<5	<5	<5	<5
FO	<5	241	256	<5	<5	<5	<5	<5
Br 2	<5	<5	444	86	38	<5	67	<5
6 SR	<5	<5	264	24	<5	<5	33	<5
C 21	<5	<5	<5	<5	<5	<5	77	<5

Fig. I.15. General structure of the polysaccharide side chain of the *Sal. typhimurium* LPS in correlation with specific receptor sites for various bacteriophages. (From Lindberg 1973)

The *Sal. typhimurium*-mouse model employed in these studies could provide a clear explanation of the relationships of the O antigen to virulence and induction of immunity because the fine biochemical structure of the cell wall of *Sal. typhimurium* LT2 had been determined by Stocker et al. (1966) and Nikaido (1967) by isolating many mutants; each strain of which is deprived stepwise of sugar sequences of the O side chains until the deletion of two heptoses left KDO.

The differences in the length of O side chains among the *Sal. typhimurium* mutant strains can be determined by the use of phages absorbing specifically at unique biochemical structures on the bacterial surfaces which serve as receptors (Lindberg 1973). Again the general structure of the polysaccharide side chains of *Sal. typhimurium* LPS correlates with susceptibility to specific phages which utilize the receptors of the various mutants as shown in Fig. I.15. One can identify the chemotypes of given *Sal. typhimurium* strains by this method. As will be described later, this method was utilized also to demonstrate the existence of other additional virulence factors distinct from O side chains (Jones et al. 1982). Therefore, this evidence supports the relative significance of concrete sugar sequences of the LPS O side chains of *Sal. typhimurium* in the induction of typhoid-like disease in its specific murine host.

After the interaction of bacterial cell wall O side chains with the host mucosal epithelial cells, virulent *Sal. typhimurium,* e.g., strain LT2 with a mouse minimal lethal dose (MLD) of about 10^3 organisms when measured by i.p. injection, penetrates into the mucosal surfaces involving two steps: initial adhesion followed by endocytosis which results in invasion (Jones and Richardson

1981). The invasion of tissue cultured HeLa cells by virulent *Sal. typhimurium* is a useful model for the study of the adhesive and invasive properties of this pathogen. Utilizing this model, Jones and Richardson (1981) demonstrated that the form of adhesion which results in invasion is associated with the production of a mannose-insensitive (i.v. incapable of binding with mannose) hemagglutinin and mutants that fail to produce detectable quantities of this hemmagglutinin are both nonadhesive and noninvasive. Therefore, this hemagglutinin should be regarded as a virulence factor of *Sal. typhimurium* in addition to the aforementioned cell wall O side chains.

For the study of the genetic virulence determinant of this factor, Jones et al. (1982) examined the plasmid DNA content of six invasive and adhesive strains of *Sal. typhimurium:* all six strains were found to harbor at least one plasmid equivalent in size to the 60 Mdal plasmid designated pSLT which is normally resident in the *Sal. typhimurium* strain LT2. The derivatives of these strains that no longer harbored an autonomous 60 Mdal plasmid were significantly less adhesive and invasive in the HeLa-cell test. Additionally they had lost or reduced virulence measured by oral infection of mice ranging from 1,000- to 10,000-fold. Using the Tn 10 transposon carrying the tetracycline resistance gene Tc^r, and using phage transductions (see Chap. 3), Jones and Richardson (1981) localized the existence of this virulence determinant in both the automonous plasmid and in integrated states within the chromosome. However, only the bacteria with autonomous plasmid copies could express the virulent phenotype, thus, enabling such strains to survive the defense mechanisms of the host.

However, the possibility that the product(s) of this 60 Mdal plasmid is the specific antigen for the induction of protective cell-mediated immunity in mice remains to be determined. If the answer is yes, the vaccination of mice with an attenuated *Sal. typhimurium* lacking the O side chains (R-type), but harboring this plasmid, would protect the host as was demonstrated by the study of Nakano et al. with the R-strain 1004 (Table I.6).

Prior to these studies, Saito et al. (1960) investigated the relationship between tissue cultures of mouse monocytes and various strains of *Sal. enteritidis* including attenuated ones with a MLD of 10^6 organisms.

Virulent *Sal. enteritidis* No. 11 multiplied within the cultured monocytes with destruction of the host cells within 7 days. Being released, these extracellular *Sal. enteritidis* organisms continued extensive multiplication. In contrast, attenuated strains Jena and 11M hardly replicated, while strain 21 (purineless mutant of No. 11) multiplied somewhat within the monocytes. This observation on the cultured monocytes was confirmed by an in vivo experiment, when Nakano (1961) recognized the differences in the intraphagocytic multiplications of virulent No. 11 and attenuated Jena strains in the live Kupfer cells of infected mice under the electron microscope. Only slight multiplication of the Jena strain versus an extensive intracellular growth of No. 11 were seen.

In contrast, *Sal. typhi* which is specifically virulent for man, did not show any growth within mouse monocytes in tissue culture. Thus, the specificity of host-parasite relationships observed in the animal models was demonstrated at the cellular level, with monocytes.

This indicates that the intramonocytic interaction of the organisms is a critical factor for the determination of the specificities of host-parasite relationships in the induction of the typhoid-disease form of infection.

For the purpose of approaching the problems of immunization of humans against *Sal. typhi,* the knowledge of pathogenesis and bacterial virulence factors in this mouse-typhoid form of disease can provide the most logical experimental system to study immunization. In fact, the question of the effectiveness of killed or attenuated vaccines can be correlated with this critical factor, i.e., the intramonocytic interaction between *Salmonella* organisms and their specific hosts.

Ushiba et al. (1959) counted the number of the virulent *Sal. enteritidis* strain No. 11 organisms present i.p. in killed-vaccine-immunized mice or those immunized with the attenuated live-strain Jena which has been established to be deficient of O side chains (R-type) after infection. In the killed-vaccine-immunized mice, challenged with virulent organisms, the number of bacilli within the peritoneal cavity decreased promptly, but increased within the spleen and liver in the later period of infection. This resulted in the ultimate failure to protect mice in spite of the life-prolonging effect of the vaccination. In contrast, the number of intraperitoneal No. 11 organisms used as challenging bacteria did not decrease at the initial phase of infection in the mice immunized with the attenuated Jena strain, but virulent No. 11 organisms could not be recovered from their spleen and liver later in infection. As a result, the mice survived the challenge infection with No. 11 organisms. These contrasting results obtained by immunizing mice either by killed vaccine or by attenuated-live vaccine were explained by the authors as follows.

The mechanism of the initial prompt decrease of the No. 11 organisms in the killed-vaccine-immunized mice is attributable to the opsonization of bacteria by anti-O side chain antibodies of IgG class, which were demonstrated by tests of the serum. However, the virulent bacteria that survived after phagocytosis and were protected from the action of humoral antibodies were trapped by spleen and liver to initiate multiplication in the later period of infection.

In contrast, the virulent No. 11 organisms injected into attenuated-vaccine-immunized mice escaped from opsonization because the R-type Jena organism did not induce the production of anti-O side chain antibodies of IgG class. Therefore, the number of No. 11 organisms within the peritoneal cavitiy did not decrease at the initial phase of infection. However, those organisms which entered the bloodstream were trapped by the phagocytes present in the spleen and liver to be fixed there and destroyed. The fact that the mice immunized with the attenuated-living-vaccine strain Jena had survived the infection with the virulent No. 11 organisms indicates that immunization conferred on the mice the ability to destroy phagocytosed No. 11 organisms without participation of humoral antibodies.

Apart from these studies that evaluate the role of cellular immunity in protection against salmonellosis, Rowley et al. (1964) reported that cytophilic IgG antibody specific to the O side chain of an avirulent *Sal. typhimurium* strain M206 made mouse macrophages capable of destroying a virulent *Sal. typhimurium* strain C5 after binding with the phagocytic surface, although the failure of the passive transfer of anti-O side chain antibodies in the protection of mice against murine typhoid has been shown. This antibody-dependent bactericidal mechanism of macrophages must be discriminated from the intracellular bactericidal activity of activated macrophages, in which a lymphocyte subpopulation (subset) designated T_{DTH} participates by modifying the macrophages so that the latter acquire bactericidal activity (refer to Part II).

Turning the topic from the typhoidal disease form to the pathogenesis of *Salmonella* gastroenteritis, the whole process can be divided into the four steps: intraluminal colonization, limited invasion of the gastrointestinal mucosa, stimulation of intraluminal fluid accumulation, and dissemination from the intestine to the reticuloendothelial system (Giannella 1975). Thus, the four steps are analogous to the preceding example of *Sh. dysenteriae*, i.e., exudative infection (p. 42). The difference between the last step and the typhoid form of salmonellosis is determined by whether the pathogens are eliminated by phagocytic cells after local dissemination or not.

Another prerequisite for the multiplication of *Salmonella* organisms, irrespective of extra- or intracellular growth, is the transport of iron from the surrounding environment into the bacterial cells. The production of phenolic-type siderophores by *Sal. typhimurium* was described by Neilands (1957). Pollack and Neilands (1970) purified this siderophore and determined its structure as the cyclic trimer of 2,3-dihydrooxybenzoylserine to be called enterobactin, the same compound as enterochelin. Garibaldi (1972) and Puschmann and Ganzoni (1977) supported siderophore production by *Sal. typhimurium* as one of the virulence factors of this bacteria (see Part II, p. 120).

A question to be solved in future studies is the definition of virulence factor(s) that permit the intracellular multiplication of *Salmonella* organisms in macrophages, thus resisting their destructive actions. Enterobactin that might take over iron from the intralysosomal lactoferrin for bacterial multiplication could constitute one of them. Once this virulence factor is characterized as the main component of the virulence mechanisms responsible for the expression of the typhoidal disease form, it could be a leading candidate for the specific antigen for the induction of cell-mediated immunity. Another candidate is the mannose-insensitive hemagglutinin (Jones and Richardson 1981; Jones et al. 1982). This might be possessed by *Salmonella* organisms deficient in O-side-chain lipopolysaccharide which are, however, immunogenic, e.g., strain SL 1004 of *Sal. typhimurium* (Table I.6). In support of this hypothesis, alterations in the cell-wall lipopolysaccharides (LPS) while changing the virulence of *Salmonella* for mice, do not appear to influence their ability to invade rabbit ileum or HeLa cells (Giannella et al. 1973). On the other hand, Jones et al.

(1982) proposed that colicin plasmids, such as CF602 of *Sal. typhimurium* strain CR6600, may specify some virulence function. Therefore, they must be regarded as possibly specifying virulence factors. If this is the case, additional virulence factors, perhaps similar to those encoded on plasmids of other invasive, Gram-negative bacteria, also may be plasmid associated in *Sal. typhimurium* because of the plasmid nature of the bacteriocins.

In summary, experimental salmonellosis is an important model for the study of invasive (penetrating) infections as expressed in the typhoidal-disease form. However, the selection of appropriate experimental host-parasite models and the use of modern techniques of microbial genetics (Chap. 3) can be used to identify the virulence factor(s) and, furthermore, the virulence determinant(s) that are responsible for the invasiveness of salmonellae as have been described above. These virulence factor(s) might serve as immunizing agents for the induction of protective cell-mediated immunity against salmonellosis.

Example 4: Listeria monocytogenes

Listeria monocytogenes belongs to the Family *Corynebacteriaceae,* being a small Gram-positive rod. This species includes four serovars which are the causative agents of human and animal listeriosis with divergent disease manifestation, i.e., cerebral meningitis in pigs, goats and sheep, distemperlike symptoms in dogs and foxes; localized liver necrosis and generalized infections with monocytosis in rabbits and guinea pigs, and the abortion of necrotized fetuses in cows. Human listeriosis is expressed as meningitis or fetal septicemic granulomatosis with a high mortality among fetuses and newborns.

Mackaness (1962) selected the *L. monocytogenes*-mouse model as the best one for the study of cellular resistance to bacterial infection. The reasons for this selection were: (1) that nonhumural immunity could express itself in a pure form only against a true facultative intracellular parasite; (2) that the mouse develops a true acquired cellular resistance to *Listeria* (Osebold and Sawyer 1957); and (3) that *L. monocytogenes* immunity is not mediated by humoral factors (Julianelle 1941).

To examine the basis for the selection of *Listeria* for his studies, Mackaness (1962) assayed the capacity of the original strain (serovar I) of *L. monocytogenes,* a smooth phase derived from the National Collection of Type Cultures (NCTC 7939), to survive within the phagocytes of its natural host in suspensions of rabbit PMNs and macrophages. It is worthwhile to remember that the organism was readily ingested in the presence of normal serum, but that it survived poorly in both types of phagocytes. In particular, it was noted that 90% of the ingested bacteria were destroyed promptly upon ingestion by rabbit macrophages. To surmount this problem, Mackaness (1962) attempted to raise the virulence of *L. monocytogenes* employed, by animal passage and by "adapting" it to grow in macrophage cultures in vitro. A culture of the parent (wild) strain of *Listeria* was used to initiate a series of mouse passages in vivo,

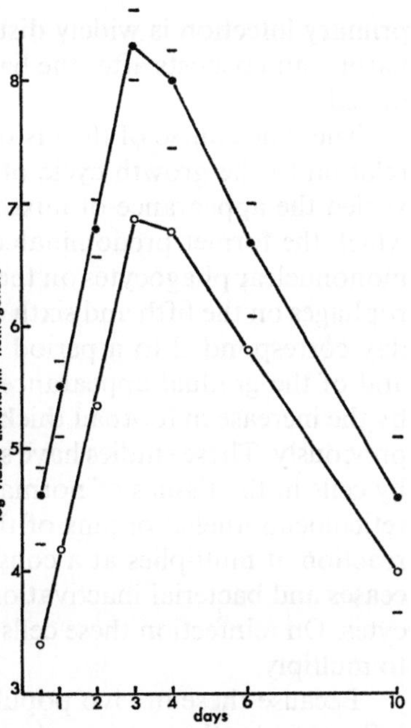

Fig. I.16. Mean numbers of viable *Listeria* recovered from the spleens (●—●) and livers (○—○) of normal mice at intervals following the i.v. injection of a 50% lethal dose virulent *Listeria*. The standard deviations of the means are shown for spleen counts only. The degree of variation in liver counts was somewhat less. (From Mackaness 1962)

and to infect macrophage cultures in vitro. Plaque-forming efficiency in vitro was used to select plaques on macrophage monolayers for the development of more virulent cultures (Mackaness 1962).

When the fate of wild and "macrophage-adapted" strains of *Listeria* in the mouse peritoneal cavities was studied, it became clear that only the latter strain maintained a high level of viability after entering peritoneal phagocytes (Mackaness 1962). The same author studied extensively the growth cycle of *L. monocytogenes* in vivo, by injecting 1.0 LD_{50} (1.1×10^5 organisms) of the pathogen intravenously (i.v.) into mice, with or without administration of anti-*Listeria* immune serum, and counting the viable bacteria within the spleen or liver.

As is shown in Fig. I.16, the viable number of *Listeria* increased logarithmically until they peaked on the third day of infection. Beyond this point, however, bacterial growth ceased and a rapid inactivation ensued. The data used in Fig. I.16 include counts made on mice which had died, because the mice were inoculated with a 1.0 LD_{50} dose.

In other experiments, the survivors of a primary infection with 0.5 LD_{50} were challenged i.v. on the 19th day with a lethal dose of 8.9×10^5 organisms that was equivalent to 9 LD_{50}. The challenge organisms failed to grow in the tissues of the convalescent mice, but survived long enough to permit an almost quantitative recovery of the inoculum on the fourth day after injection. It is apparent, therefore, that the antibacterial mechanism developed during the primary infection is widely distributed among organs, retains its bactericidal nature, and persists after the bacteria of the primary infection have been eliminated.

The time course of the histopathological changes in the liver and spleen in relation to the growth cycle of *Listeria* in vivo, was also followed which revealed the appearance of mixed PMNs and mononuclear phagocytes among which the former predominated at 24 h, to be followed by predomination of mononuclear phagocytes on the fourth day and by numbers of epitheloid macrophages on the fifth and sixth days. The interval between the third and fourth day corresponded to a period of concomitant growth and death of bacteria and of the gradual appearance of delayed-type hypersensitivity as measured by the increase in footpad thickness produced by culture filtrates injected 24 h previously. These studies have shown that *L. monocytogenes* is rapidly ingested by cells in the tissues of normal mice and becomes sequestered mainly in the reticuloendothelial organs of the body. Here, despite an acute inflammatory reaction, it multiplies at a constant rate for a period of 3 days. Growth then ceases and bacterial inactivation begins in lesions by the mononuclear phagocytes. On reinfection these cells alone respond, and in them the organism fails to multiply.

Because these in vivo population dynamics of *L. monocytogenes* were not influenced by the passive administration of mouse immune serum, the host's mononuclear phagocytes must have acquired antibacterial activity to be called activated macrophages (Mackaness 1962).

Thus, the nature of *L. monocytogenes* as a facultative intracellular parasite was confirmed as far as the macrophage-adapted strains were concerned. However, it is remarkable that even as large an infecting dose as 1.0 LD_{50} (10^5 organisms) of a virulent strain for the mouse shows a decrease in viable numbers after reaching the peak on the third day of infection (Fig. I.16).

On the other hand, all efforts to reveal an antibody that could account for the features of acquired resistance to *L. monocytogenes* were unsuccessful (Miki and Mackaness 1964). For example, the growth rates of *Listeria* in the spleens of mice infused repeatedly with normal or immune mouse serum, the peritoneal clearance of *Listeria* in the presence of normal or immune mouse serum, and the survival rate of *Listeria* in monolayers of mouse macrophages infected in the presence of normal or immune mouse serum showed no differences. Furthermore, Miki and Mackaness (1964) tried to solubilize possible cell-bound (cytophilic) IgM antibodies by extracting the spleens and peritoneal macrophages of normal or immune mice with urea solution. The injection of these extracts into recipient mice revealed no influence on the survival and growth of *Listeria* infecting these mice. Furthermore, mice survived a challenge with a lethal dose of virulent *Listeria* organisms only when intact, living spleen cells of the immunized donor mice were transferred passively to the challenged mice. The survival rates in lethally infected mice after the passive administration of the intact spleen cells of immune donors did not increase by the addition of the serum from either normal or immune donors.

In all these cases, it became clear that *L. monocytogenes* is a facultative intracellular parasite when its still unknown virulence factor(s) is being fully expressed, and that the protective immune mechanisms against this pathogen are cell-mediated.

Since then, the *Listeria*-mouse model has been employed almost exclusively for the studies of antibacterial cell-mediated immunity in the light of modern immunology. These studies and the comparative studies employing other intracellular parasites, e.g., *Brucella abortus,* and *Mycobacterium tuberculosis,* also, will be discussed in Part II. However, it must be kept in mind that the virulence factor(s) of *L. monocytogenes* responsible for its resistance to intracellular-destructive mechanisms have not yet been chemically identified.

Chapter 7 Summary

A particular contribution of this book is to revisit host-parasite relationships from the standpoint of the rapidly advancing knowledge of the biological picture of the living world, which has resulted from the establishment of molecular biology and revolution in immunology. The first discipline permits an understanding of biological phenomena as the modes of molecular movements, characterized by the replication of relatively constant nucleotide base sequences in the genetic code. The second discipline deals with one of the homeostatic regulatory mechanisms of higher animals, although it has developed as a result of the analysis of host responses to "foreign" invaders.

In contrast, the uniqueness of our discipline, host-parasite relationships resides in the fact that it deals with two discriminable biological systems which interact mutually with modification in each system as a result of this interplay. Therefore, we must understand the concept of host-parasite relationships as *fields,* the term derived from theoretical physics, where both components, i.e., the microorganisms and the hosts, mutually interact. The characteristic feature of host-parasite relationships is that both components are genetically controlled. In other words, they express the codes of genetic traits that are inherited, under the dynamic condition of the "field" at that moment.

The mechanisms which render microorganisms pathogenic have been considered using some typical examples in bacteria. However, nonbacterial microorganisms, such as viruses and rickettsiae, have been also described in a general way for the purpose of illustrating the whole concept of host-parasite relationships, which is a particular form of the ecological existence of microorganisms. The necessary background knowledge of pathogenicity and virulence have been delineated and the biology (classification, anatomy, physiology, etc.) of microorganisms was described (p. 3).

Diseases caused by microorganisms were divided into three categories of infections (p. 11). However, the classification into these categories of infection is no more than operational, and single species of microorganisms can cause infections belonging to different categories. Furthermore, host-parasite relationships are specific, if both genetically controlled components, host and parasite, are constant, but this is frequently not the case.

Taking such considerations into account, the effective mechanisms of virulence factors of typical pathogens in each category of infection were discussed. Generally, superficial infections exert only toxic and/or allergic effects;

exudative infections produce tissue damage in addition, and invasive infections produce septicemia, and widespread damage to organs.

On the other hand, a large number of bacteria constitute the normal flora. The existence of this normal flora itself can be either inert to the body or can result in beneficial effects for the host since disturbance of this equilibrium of flora and host due to dysfunction of some facet(s) of host-defense mechanisms can result (1) in the disease-opportunistic infection, (2) to an increase in susceptibility to infections by pathogenic organisms due to a lack of competition. These situations can be clarified only by the consideration of the host-defense mechanisms which constitutes Part II as well as the infecting microorganism per se.

It must be pointed out that recent advances in the technology of nucleotide base sequencing and identification of certain functionally significant sequences (e.g., Southern blotting method, p. 15) provide a powerful tool to clarify the mechanisms of host-parasite relationships at the molecular level. These sequences have been found to be distributed within the DNA chains of microbial chromosome, plasmid, and other replicons. Furthermore, these phenomena can be explained by the role "movable gene(s)" (McClintock 1951) that are currently designated transposons (Tn) and insertion units (IS), have played in the history of biological evolution. Such significance is not only important in connection with Part I of this book, dealing with parasitic components, but also might be so with Part II dealing with host animals. The transposons play roles in translocating bacterial functional genes, e.g., drug resistance genes, as well as the particular nucleotide base sequences that are virulence determinants. Thus, virulence determinant nucleotide base sequences can be incorporated into not only bacterial chromosomes, but also into phage DNA and plasmid DNA.

The latter two replicons can parasitize, i.e., "infect" bacteria and behave as informational sources for the expression of the virulence factors by the bacteria.

Therefore, microbial virulence is not fixed any more to a particular species or strain unlinked with specific characteristics, e.g., serological specificity of that microorganisms, since the virulence determinants are transferable from organism to organism. Therefore, the links described between serovar and virulence are not always absolute.

Thus, Part I proposes that host-parasite relationships of unicellular pathogenic microorganisms deal with a host parasitized by microorganisms parasitized by virulence determinants: triple biology.

However, it is also necessary to add the following consideration that some virulence factors mentioned in Part I are defined only in operational terms without identifying the responsible substances, e.g., "not-stimulation" of *M. tuberculosis*, because our knowledge is incomplete.

Including these operational definitions, virulence of a parasite is the integrated expression of all virulence factors inherent to that pathogenic microorganism. Therefore, considering toxin production as the sole virulence factor is an erroneous concept, since many virulence factors of the pathogen serve to circumvent one or more facets of host-defense mechanisms enabling it to cause disease which will be discussed in Part II.

Along with this consideration of host-parasite relationships at a more cellular level, studies at the epidemiological level have been discussed as well using some examples, e.g., *V. cholerae,* enteropathogenic *E. coli,* and *Shigella dysenteriae.* These examples enabled us to recognize the "transitionlike phenomena" in the history of epidemics.

With this long-term observation at the epidemiological level, also, the importance of the distribution of certain nucleotide sequences that are virulence determinants among various species and biotypes of microorganisms isolated at different periods of history has been discussed. This wide distribution of these virulence determinants might be mediated by their translocation among these microorganisms via transposons, i.e., conjugative plasmids or transducible (conversible) phages. After the virulence factors are distributed in this manner, the microorganisms might be subjected to natural selection by the host-defense mechanisms which are also evolving with evolutionary changes.

Thus, these studies provide an approach to the "origin of bacterial pathogenicity" and thus make it possible to explore "the future course of infectious diseases".

Part II
Defense Mechanisms of Host Against Infection

The purpose of Part II is to elucidate the mechanisms of host mammalian organisms that interact with microorganisms during their individual life processes (p. 2). This knowledge is important because the host mammalian organisms and microorganisms constitute themselves a series of ecosystems called host-parasite relationships which have developed during the 75 m. y. since the origin of life on the earth, during which time the host organisms have evolved into multicellular mammalian species (Hood et al. 1978). If the host organisms had failed to develop mechanisms for defense to maintain their viability against the selective pressure of the adverse effects brought about by microorganisms, these ecosystems themselves would have been eliminated, and the present variety of mammalian species would not have developed.

Needless to say, however, the individuals of a given multicellular host organism that start from one fertilized egg have completed a variety of differentiation processes that result in unique complex systems of tissues and organs which cooperate as a whole to maintain their viability and capacity to procreate. In other words, an individual host organism is an independent system mutually exchanging materials, energy, and information with its surrounding environment which involves the fundamental discrimination between the identity called "self" and the remaining part of the world called "nonself (or notself)". For example, death of an individual host is the irreversible loss of this identity of self for that member of the species. Therefore, it is the host-defense mechanisms that assure the maintenance of the continuity of self, and this is the reason why the word "self-defense mechanisms" has been used.

For example, the same self-defense mechanisms operate when transplanted tissues or organs are donated from another individual of the same biological species to another individual and are rejected by the recipient as "nonself", similar to the instance when the host resists infectious diseases caused by virulent microorganisms as was illustrated in Part I.

These two concepts: self-defense and host-defense, are closely related. For example, some endogenous tumor viruses can be transmitted from the parent to their progeny and may trigger the malignant transformation of cells so that they are recognized by the host-immune system as foreign or nonself and thus eliminated. Therefore, one way to treat neoplastic disease may be to enhance the activation of the host-immune system to recognize the nonself characteristics of malignant cells which are the subject of much current research. This same "immune system" plays an important role as one of the host-defense mechanisms against infections as will be discussed in Chap. 1, Part II.

It is possible to delineate the host mechanisms for defense that have developed during evolutionary history, if we concentrate our attention on the defense of a host against the infection with microorganisms.

After describing the mechanisms for the maintenance of self-integrity which establish the identity of multicellular host organisms – particularly mammalian – the defense mechanisms of the host will be described in the following order:

1. Anatomical barriers covering the host-body surface that play a role in limiting the easy invasion of foreign microorganisms.
2. Physiological mechanisms that play a role in defending the host body against infecting parasites in general, e.g., fever and reduced iron.
3. Competing normal flora which limits the survival and growth of new microorganisms which have the potential to produce adverse effects on the host.
4. Humoral and cellular components of the host which play a role in counteracting the parasite in nonspecific or specific ways. In these cases, nonspecific effects prevent the parasite from producing disease, irrespective of the species to which the microorganism belongs, while specific effects react with the particular products or components of the microorganism to prevent the development of disease.

Chapter 1 Recognition by the Immune System

The problem of establishing the mechanisms that discriminate between self and nonself has been solved with the rapid developments in immunology in the 1960s.

Phylogenetically, a host system specially differentiated for recognition of self and nonself appeared in the course of animal evolution at the primitive level of development occupied by the lamprey. Jerne (1973) has described this system and called it "the immune system", characterizing it as a diffuse organ having the assignment of monitoring the identity of the body. Basically it is constituted of lymphocytes and antibody molecules, which recognize both foreign molecules and one another. In man, it weighs ca. 1 kg, being comparable in the complexity of its functions to the nervous system sharing with it the ability to recognize various patterns and in being dispersed through most of the tissues of the body to form a network. For comparison, the nerve cells, or neurons are in fixed positions in the brain, the spinal cord, and the ganglia and their long processes, the axons, reach all parts of the body. In contrast, the immune system of man consists of about 10^{12} cells called lymphocytes and about 10^{20} molecules called antibodies that are produced and secreted by the lymphocytes, which reach most tissues through the bloodstream, and enter the tissues by penetrating the walls of the capillaries. After such movement they return to the vascular system via the lymphatic system, which collects lymphocytes and antibodies along with other cells and molecules and pours its contents back into the bloodstream via the thoracic duct by joining the venous system at the subclavian veins and these components of the immune system through the vascular system. Thus, in one instance in the nervous system, the cells remain fixed with processes extending to the various tissues; while in the immune systems, the cells are constantly moving in and out of the various tissues to maintain communication. Along the lymphatic vessels lymphocytes are pooled in lymph nodes that are located at anatomically determined positions in each region of the body. Their biological functions will be described later along with the functions of the bone marrow, the thymus, and the spleen.

Cohn et al. (1980) defined an antibody as an effector molecule with specific recognition capabilities, the expression of which is inducible or suppressible. The latter property, referred to as "self-nonself discrimination" distinguishes antibodies from other specific binding mechanisms, such as lectins, which are found in invertebrates and plants and are surmised to play protective roles. Within the partial structure of an antibody called "variable region", some

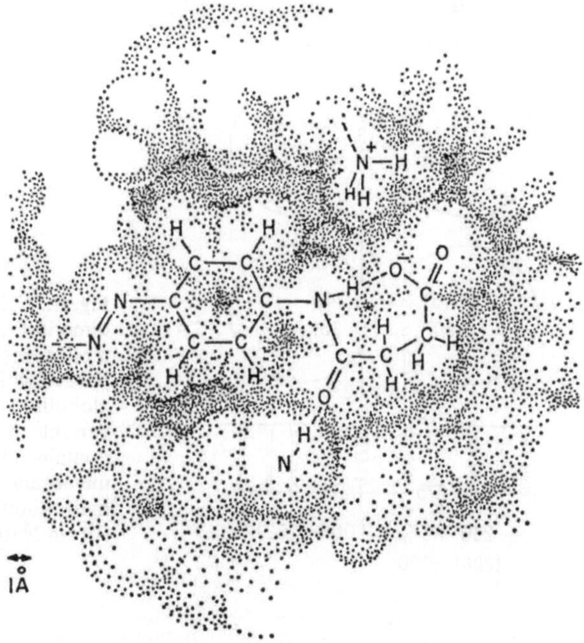

Fig. II.1. The shape of the combining site of a molecule of antibody for the hapten group p-azo-succinanilate and the shape of the hapten group itself [From L. Pauling (1948) Endeavour 7:43]

groups of limited lengths of amino acid sequences show tremendous diversity (designated CD = complementarily determining residues). The three-dimensional configurations of CD contain empty concave spaces, i.e., crevices, which can select a particular small molecular structure of any unique pattern via physicochemical binding forces exerted from the constituent radicals. The former molecular structure is designated "combining site" and the latter, "epitope" or "antigenic determinant". Their most appropriate metaphor is the relationship between a lock and a key (Fig. II.1). Recognition of the corresponding epitope is due to binding of the former by fitting it into the latter due to the mutual complementary patterns of both. The antibody secreted by lymphocytes contains another partial structure called the "constant region" as part of the immunoglobulin which will be described in Chap. 7 separately.

However, the antibodies are divisible into humoral antibody that is immunoglobulin secreted by induced bone marrow-derived B-cells, i.e., plasmocytes (p. 142) and cell-mediated antibody which is made up of two chains, α and β, each with two extracellular immunoglobulinlike domains, an amino terminal variable and a carboxy-terminal constant domain (Saito et al. 1984, Fig. II.2) and acts as the receptor on effector cells which are thymus-derived T-cells.

Thus, the pattern of recognition of the immune system is determined by the varieties of combining sites both in humoral and cell-mediated antibody. Therefore, a given molecular pattern of the antigenic epitope selects the appropriate three-dimentional molecular pattern of a combining site out of the large number of diverse ones (selective theory). In fact, experimental evidence has

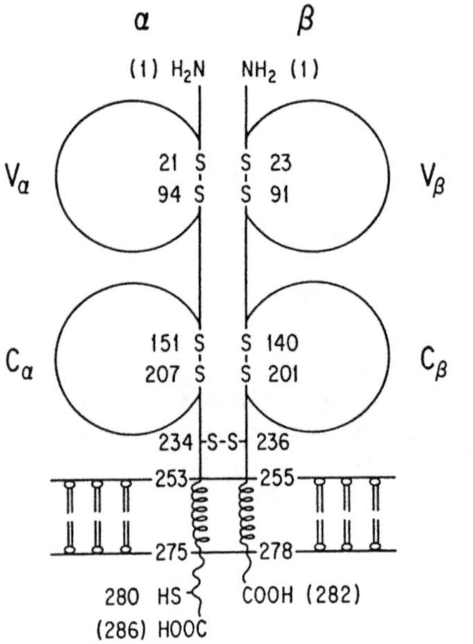

Fig. II.2. The overall structure of a heterodimeric T-cell (2C) receptor. Each receptor molecule is made up of two domains, α and β, each with two extracellular immunoglobulinlike domains, an amino-terminal variable and a carboxyl-terminal constant domain. The protein is anchored on the cell membrane by two (one for each chain) hydrophobic transmembrane peptides. (From Nature 309:757–762, 1984)

been accumulated to support firmly that the antibody molecules produced by a lymphocyte and by the cells of its clone all have identical combining sites. Therefore, theoretically, the numbers of combining sites in a single animal are equal to the number of its different lymphocyte clones.

The DNA of the cell controls the expression of combining sites. Each lymphocyte of a human or mouse is known to have 10^{-11} g of DNA equivalent to 10^{10} nucleotide base pairs which are contained in chromosomes. However, not all of these nucleotide base chains serve as the source of genetic information, because only short lengths of them designated as *exons* are active and are separated by nonfunctional base sequences designated as *introns*.

During the course of phylogenetic evolution, selective pressure from the pathogens belonging to the category of extracellular parasites (Part I) in the environment have selected those nucleotide base sequences (designated as V genes) that encode for the biosynthesis of the combining sites of humoral antibody versus these parasites. Furthermore, selective pressure to eliminate the nonself molecular and cellular species (designated as antispecies-R specificity) for the maintenance of self-identity by the multicellular organism (Chap. 2) and, in addition, selective pressure from the intracellular pathogenic parasites (Part I) in the environment, have selected those encoding for cellular antibody (Cohn et al. 1980).

In addition to the nonself epitopes, certain amino acid(s) occupying a position either in the variable or constant regions of the antibody molecule can be recognized by the combining site of another antibody of humoral or cell-mediated class when the former antibody is used as the antigenic stimulus. The epitope present in the variable region is called idiotype and the one in the constant region, allotype. Accordingly, the idiotype is unique to the specific antibody, while the allotype, is common to the animal group.

When antibodies produced by animal A are injected into animal B, animal B will produce antibodies against the idiotypic epitopes ("idiotopes") of the injected antibody molecules. That is also true when A and B belong to the same animal species and even when they are of the same inbred strain, that is, when they are genetically identical (p. 86). Furthermore, the idiotopes occurring on one antibody molecule are recognized by combining sites on a set of other antibody molecules and idiotypes on the receptor molecules of one lymphocyte and, thus, are recognized by the combining sites of the receptor molecules of a set of lymphocytes within one animal.

According to the "immune network theory (Jerne 1973)", the description of the immune system as a functional network of lymphocytes and antibody molecules is essential to its understanding, and the network as a whole functions in a way that is peculiar to and characteristic of the internal interactions of the elements of the immune system itself.

In conclusion, the immune system as a whole, irrespective of whether being communicated by the idiotype-antiidiotype network or by the associated recognition of nonself epitope (designated as X) by anti-X and species R (Chap. 2) by anti-R as is occurring in cell-mediated immune recognition (Cohn et al. 1980) recognizes nonself, including microorganisms by the use of antibodies of soluble-molecular nature and lymphocytes of cellular nature. The effector function of the respective antibody or lymphocyte is divergently expressed in a specific way, being determined by the "class" of antibody or by the classification of lymphocytes. It is a unique characteristic of the immune reaction that these effector functions can play not only a positive role in host-defense mechanisms, but also a negative role as will be described in the subsequent chapters. But another unique characteristic shared by all immune reactions is the specific interaction between epitope, i.e., antigenic determinant, and combining site. Because the host-defense mechanisms comprise many nonspecific interactions as will be mentioned in the subsequent chapters, the adjective "specific" will be used here to mean "via specific interaction between the epitope and the combining site".

Chapter 2 The Mechanisms for Maintenance of Self-Integrity

The discussion of host-defense mechanisms requires a parallel brief consideration of the mechanisms that integrate each constituent host cell to form a multicellular organism.

For the recognition of mutual self-identity of cells, the key role is played by the surface macromolecular substances exposed. This is similar in manner to the host-parasite interaction of microorganismic infection which starts from mutual contact between cell-surface substances of both host and microorganisms. The recognition of self-identity between cells resides in the exposed cell-surface substances that are expressed as R (Chap. 1), the products of the common "histocompatibility genes" unique to each individual host. If this meeting results in matching due to the recognition of the common molecular surface nature (like-like recognition), these cells are allowed to assemble to form multicellular tissues. This phenomenon of self-assembly has been confirmed by in vitro experiments (Moscona 1975).

Histocompatibility genes are located on a number of chromosomes of mammals, but the major histocompatibility complex (MHC) is a cluster of loci occupying a single chromosomal region in all mammals so far examined. The gene products of this region are involved in a variety of activities for defense mechanisms. For example, their ability to identify and destroy foreign cells may be important in protecting the developing fetus against invading maternal cells (or vice versa) and in immune surveillance for elimination of malignant cells which express new kinds of surface antigens. Certain gene products of this region mediate the cellular interactions of T-cells, B-cells, and macrophages. Immune response genes (Ir genes, p. 174) of the MHC regulate the magnitude of thymus-dependent immune responses to a variety of antigens. This regulation extends to cellular as well as humoral responses (p. 131). Ia molecules expressed by the IA region of mouse MHC play a fundamental role in the phenomena of associative recognition (p. 132). Genes in the MHC control the synthesis of C'4 protein of the complement system (Hood et al. 1978). Thus, the mechanisms for the maintenance of self-integrity cooperate with the host-defense mechanisms against infection by microorganisms, because the complement system constitutes itself a nonspecific humoral-defensive component as will be described later. Furthermore, detailed knowledge of the MHC region deserves consideration because one of its functions is to determine the grade of immune responses to stimulation by antigenic epitopes, including those of microorganisms.

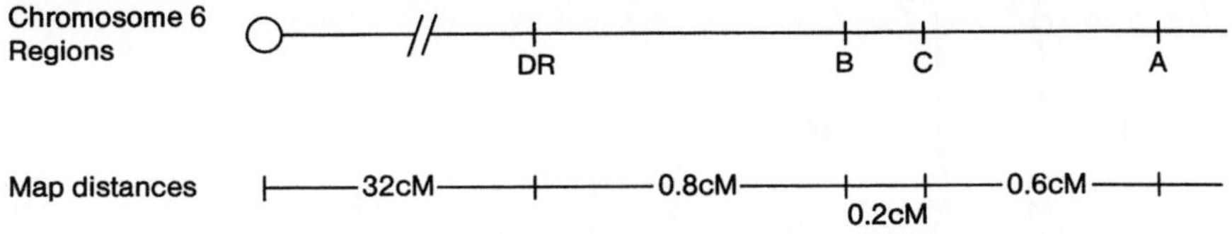

Fig. II.3. A genetic map of the human-HLA complex on human chromosome 6. Map distances (centimorgans) and regions are indicated. The *circle* represents the centromere

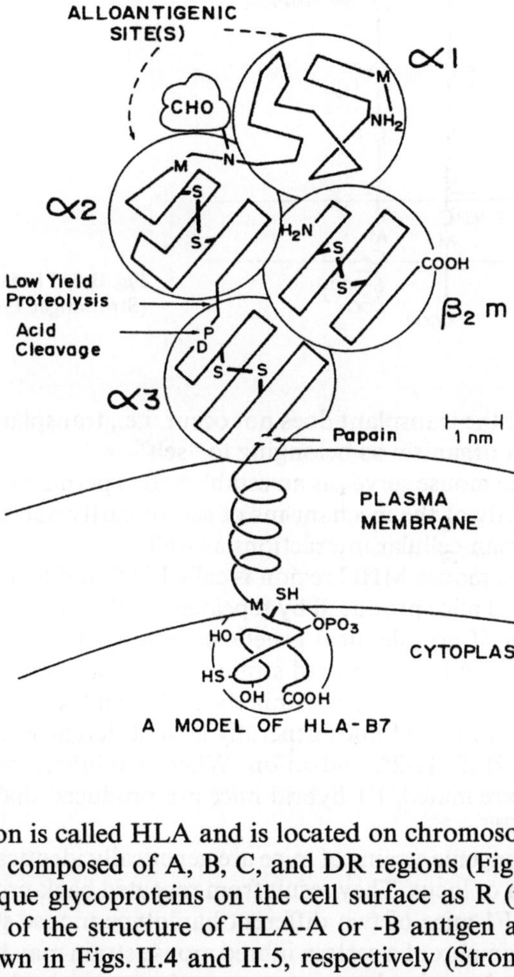

Fig. II.4. The structure of an HLA-A or -B antigen. (Strominger 1980)

The human MHC region is called HLA and is located on chromosome 6. The HLA gene complex is composed of A, B, C, and DR regions (Fig. II.3). Each region expresses unique glycoproteins on the cell surface as R (p. 86). Schematic representations of the structure of HLA-A or -B antigen and of HLA-DR antigen are shown in Figs. II.4 and II.5, respectively (Strominger 1980). These HLA types illustrated in these figures are selected examples of the diverse types that exist which require matching assays between donors and recipients, before organ or tissue transplantation can be successfully carried out. If MHC types coincide, the probability of "transplant take" is high, and rejec-

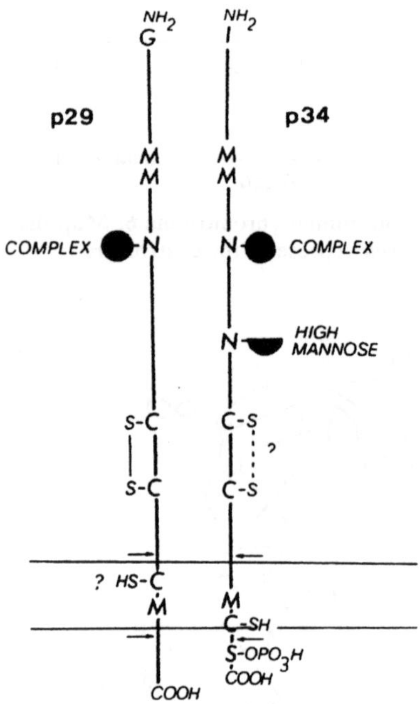

Fig. II.5. The structure of an HLA-DR antigen. (Strominger 1980)

tion of the transplant does not occur, i.e., transplant was recognized by the recipient organism as belonging to "self".

The mouse serves as an established experimental animal model feasible for the study of the mechanisms of self-integrity and the principles derived apply to human-cellular interactions as well.

The mouse MHC region is called H-2 and located on chromosome 17. Individual mice produced by repeated brother-sister-mating until they have very similar, if not identical genetic constitutions, are called inbred strains. More than 100 different inbred strains have now been developed. A unique combination of alleles at loci within the H-2 complex is termed a haplotype. Different inbred strains of mice generally have different H-2 haplotypes, which are denoted $H\text{-}2^a$, $H\text{-}2^b$, and so on. When two inbred mouse strains, e.g. $H\text{-}2^a$ and $H\text{-}2^b$ are mated, F1 hybrid mice are produced that have heterozygotic alleles of $H\text{-}2^{a/b}$.

Congenic strains of mice are genetically identical except for a single genetic region or locus. They result from repeated back-crosses and selections starting from F1 mice of two different haplotype mouse strains. By this method, the H-2 complex of a certain inbred mouse strain may be placed in the background of another inbred-strain genome of a particular haplotype. After production of certain congenic mouse strains, they can be maintained by self-crossing. Congenic mouse strains have been effectively employed to determine the genetic loci on the H-2 complex, which in turn, control the grade of immune re-

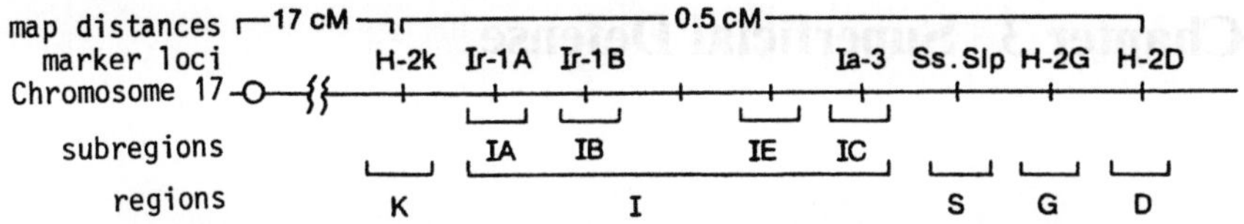

Fig. II.6. A genetic map of the mouse H-2 complex on chromosome 17. Map distances (centimorgans), marker loci, regions, and subregions are indicated. *Circle* represents the centromere

sponse to antigenic epitopes, including those of microorganisms, as will be seen subsequently.

The H-2 complex of the mouse can be divided into five major "regions" designated as K, I, S, G, and D associated with the H-2K, Ir-1, Ss-S1p, H-2G, and H-2D genes, respectively (Fig. II.6). The boundaries of each region are determined by intra-H-2 recombination between these traits. Whenever two traits are clearly separated by a cross-over and are unaltered by the recombinational events, the traits are assumed to be controlled by two separate genes. In a similar manner, the I region has been subdivided into I-A, I-B, and I-EC regions. In addition, there are other minor histocompatibility complex genes, e.g., H-Y gene(s), located on chromosome Y which is responsible for the determination of male sexuality.

In conclusion, it has been elucidated that multicellular mammalian host organisms consist of tissues and organs which share common cell-surface substances expressed on each component cell. These substances are unique to each individual and play an important role for the maintenance of self-integrity within the individual body. The main genetic traits controlling the expression of these substances on the cell surface are located on an MHC region of a particular chromosome unique to each mammalian species.

Since infection of host with microorganisms is initiated by mutual contact between surface substances of both microorganisms and host cells, an understanding of immunogenetics is an important prerequisite for the thorough understanding of the host-parasite relationships involving the recognition of nonself.

Chapter 3 Superficial Defense

Before birth, the developing embryo is defended against infection with microorganisms except those capable of vertical (transovarial or transplacental) infections like Rubella virus, the retroviruses, or *Treponema pallidum* by its mother's host-defense mechanisms throughout its intrauterine life. But emerging from the uterus, the infant comes into direct contact with microorganisms. Thus, the role of skin and mucous membranes covering the body surface as the host-defense mechanisms should be clearly considered as they are the first point of contact with the external world of microorganisms. Furthermore, there are deeply internalized tubes and cavities connected to the body surface, e.g., naso-pharynx, trachea, bronchi, lung alveoli; oral pharyngial cavities and the esophago-gastrointestinal tracts of digestive organs; urethra, and the vagina and uterus in female and prostate and seminal duct systems in male and auditory canals which reach to the tympanic membrane. In a strict sense, the inner surfaces of all of these ducts and cavities are, also, boundaries between the inner and outer world of the body, where the microorganisms make their first contact with the host. However, not only natural mechanical barriers, but also acquired resistance mechanisms which will be mentioned later, function on the surfaces of most internalized tubes and cavities, in contrast with the skin and the corneal surfaces. These mechanical barriers will be described as follows.

Skin

In the epidermis, the cells within the germinal layer repeat cell divisions to constantly generate new epidermal tissue. New cells transform into granular, and then into keratinous cells, destined to die and become discarded. Constant cell divisions at the germinal layer are supported by nourishment from blood capillary vessels penetrating the corium. In this way, a continuous flow of regenerated cells is established. These cells are also sources of nail and hair formations, composed of proteins and keratin.

Almost all microorganisms, except some fungi, such as *Trichophyton* and *Franscisella tularensis,* cannot penetrate through this surface barrier. If other microorganisms absorb onto the epidermal surface, the cell flow "washes" them off in time. Other defense mechanisms are performed by sweat and sebaceous glands. These secretions contain a basic protein, lysozyme (mol.wt.

about 17,500) which is a mucolytic enzyme with antibacterial properties. However, only sensitive microorganisms, like *Micrococcus lysodicticus,* are destroyed by this enzyme. Sebaceous glands located in close connection with hair roots produce lipids. Lipid extracted from the forearms of volunteers with acetone, showed killing activities for a variety of microorganisms. Data suggested that the active lipid might be oleic acid. Furthermore, many fatty acids, especially caproic and caprylic acids present in sebaceous secretions, showed bactericidal activity for several organisms. According to the review of Kanai and Kondo (1979), these studies were then extended to in vivo experiments in which staphylococci were inoculated on the standardized areas of the forearm skin after application of either 100% acetone, 74% ethanol, or soap to remove lipids and on control areas. Quantitative enumeration of the bacilli surviving 5 h later indicated that the treated skin favored the persistence of inoculated staphylococci as compared to untreated areas. This observation was confirmed with *Staphylococcus aureus, Staphylococcus pyogenes,* and *Candida albicans,* although the resulting data are divergent between individuals inoculated with *Staph. aureus,* i.e., the number of viable bacteria are greatly reduced in some individual volunteers, while persistence and multiplication of the bacteria were recorded in the other individuals. The quantitative reduction of experimentally inoculated *Staph. aureus* and *C. albicans* on the skin and the degree of their in vitro inhibition by acetone extracts of the skin surface was directly correlated. However, *Pseudomonas aeruginosa* was resistant to the antibacterial activity of the extracts. Sebaceous secretions of the scalp contain fatty substances mainly consisting of C_7, C_9, and C_{11} saturated fatty acids which are inhibitory for *Microsporom audouini,* a pathogenic fungus, causing ringworm. As a consequence of the increasing fungistatic action of sebaceous secretions of the scalp with maturity of the patient, ringworm of the scalp was found to be cured spontaneously in puberty. *Staphylococcus epidermidis,* which is more resistant to the bactericidal activity of long-chain fatty acids than *Staph. aureus,* constitutes itself as part of the resident flora of the skin, while the latter, is part of the transient flora. Because *Staph. epidermidis* has lipolytic activity which releases fatty acids from triglycerides, the growth of *Staph. aureus* may be inhibited by these fatty acids, and thus result in the predominance of the former.

The development of modern techniques of anaerobic cultivation has shown that the anaerobic bacterial genus *Propionibacterium* (formerly called *Corynebacterium*) occupies the first position among skin flora. These bacteria inhabit the deeper structures of the skin, e.g., the sebaceous gland ducts or hair roots, without stimulating host-defense mechanisms (p. 58) and are insensitive to the disinfecting action of fatty acids. Sebaceous substances, consisting of proteins and fatty acids, serve as their nutritional sources. Hormone imbalance at puberty, frequently disturbs this mutual compromise between host and parasite to cause acne or pimples, resulting in the stimulation of host-defense mechanisms, including the immune system. Significantly higher titers of anti-*Propionibacterium acnes* polysaccharide antibodies have been detected in pimple-

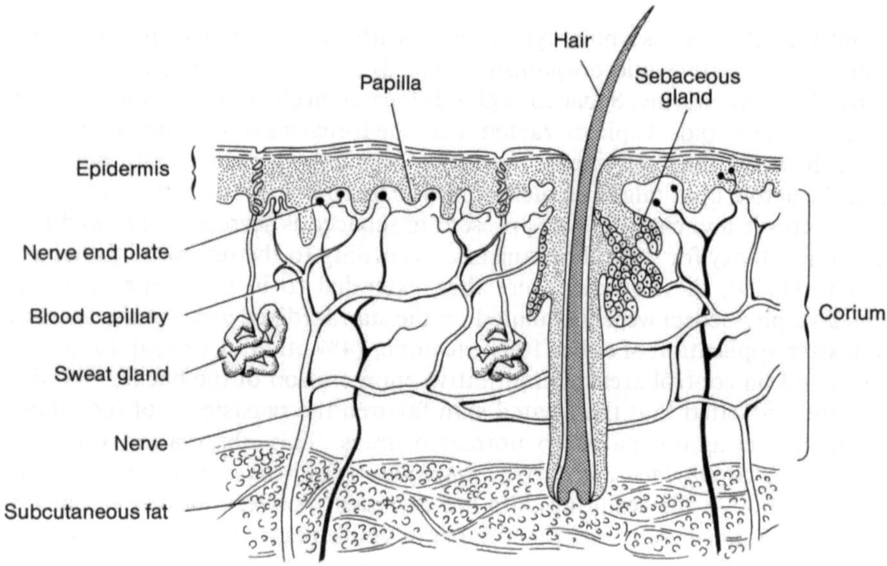

Fig. II.7. Longitudinal cut surface of human skin

bearing dermatological patients than in healthy controls or patients without pimples (Puhvel 1966).

When the skin is injured by trauma, or when the host-parasite compromise is unbalanced by iatrogenic causes, even the normal skin flora may invade to cause inflammation or bacteremia. This is one of the so-called opportunistic infections.

A diagrammatic representation of human skin cut surface is shown in Fig. II.7. This shows the structure of corium and subcutaneous tissue, including blood capillaries, end plates of nerve, fatty tissues, sebaceous glands and their drainage ducts, hair roots, and other connective tissues.

The skin plays not only a role as a mechanical barrier, but also is the site where humoral and cellular components of natural and acquired resistance function after recruitment from whole body response to local stimulation by the microorganisms present that are recognized as foreign, i.e, not-self or non-self.

The skin constitutes an ecosystem with its resident skin flora dominated by anaerobic bacteria. With regard to the fact that anaerobic bacteria predominate among the resident skin flora, a low partial pressure of oxygen in the vicinity of animal cells due to rapid O_2 consumption by their cytochrome oxidase has been hypothesized to encourage the growth of anaerobes. If exogenous O_2 is not supplied with enough speed within sebaceous glands, their drainage ducts and hair roots, limited areas of an anaerobic microenvironment are formed which are favorable for the growth of anaerobic bacteria which are resistant to the bactericidal action of the fatty acids present (Matsuyama 1982).

Cornea

The cornea of the eye is protected from injury due to reflex reaction resulting in closure of the eyelids. There are lacrimal (tear) glands within conjunctiva at its upper-outer portion, which secrete a continuous flow of tears over the corneal surface which stream into tear ducts producing a washing action and which contain lacrimase, the same enzyme as lysozyme which has antibacterial activity for a few microorganisms.

Mucous Membranes

Mucous membranes consist of mucosal epithelium and submucosa similar to the skin, except that mucosal epithelium is derived from endoderm embryologically, and it has a monolayer of cells which do not keratinize. Only small portions are exposed to the exterior at the mouth and anus.

These mucous membrane tissues cover the various internalized surfaces of organs. Each membrane is equipped with goblet cells secreting mucus and multicellular glands releasing secretions unique to the organs involved.

In addition, tracheal and broncheal mucous membrane cells are equipped with innumerable cilia that repeat synchronous rhythmical movements to move materials to the exterior. This is a mechanical natural resistance mechanism, because the inspired microorganisms dropped on these cilia are automatically transported in masses of mucus by their relay movements to be expectorated from the mouth. Some pathogenic microorganisms like phase 1 *Bordetella pertussis* are equipped with virulence factors to paralyze these movements which result in their successful colonization of these surfaces.

At the end of the bronchi and bronchili are alveoli, which do not have any mucous membrane, but consist of innumerable minute bags. The walls of these bags are covered by a monolayer of squamous epithelial cells, underneath which blood capillaries touch the basal membrane. As a natural resistance mechanism in the lung, alveolar macrophages are scattered among these alveolar spaces. They are ready to phagocytose foreign bodies that are inspired and reach this depth. If the inspired particles are virulent microorganisms, they may induce inflammation and lead to pneumonia, as will be described in a later chapter of Part II.

Various sorts of mucous membranes cover the digestive tract, 9 m long (in adult) from mouth to anus. Saliva secreted by salivary glands within the mouth has some antibacterial activity because it contains such enzymes as amylase, maltase, lipase, oxidase, and lysozyme which digest not only food components, but also cell-surface components of microorganisms. Iron-binding proteins, i.e., lactoferrin and the IgA class of immunoglobulin, are also contained in saliva, and play roles in the disinfection of the oral cavity.

The stomach glands secret hydrochloric acid, the pH of which is as low as 2.0. Many microorganisms are sensitive to this acidity and only a few may survive.

However, the quantity of hydrochloric acid secreted from the stomach glands varies depending on genetic and/or physiological conditions of the host. Therefore, the bactericidal effectiveness of this "stomach barrier" is not always constant. Furthermore, the acidity of stomach juice can be diluted by water taken together with a meal, neutralized by alkaline foods or drugs, e.g., sodium bicarbonate, to a nonmicrobicidal level. Besides hydrochloric acid, β_2-glycoprotein within stomach juice may play the role of a host-defense mechanism by activating certain complement pathways.

Microorganisms circumventing the stomach barrier, enter the small and large intestine, sequentially where the activity of stomach juice is neutralized by the alkaline intestinal juice. These digestive organs themselves constitute a unique ecosystem where microorganisms find their niche and multiply after being liberated from the bactericidal environment of the stomach. The gut has a much larger net area than the appearance of intestine. In fact, small projections called villi, bordered by monolayers of cylindrical cells, with about 600 brushlike microvilli on the surface of each cylindrical cell form the net surface area. Microfiber-structures, i.e., the glycocalyx, consisting of glycoprotein and polysaccharides covering the microvilli surfaces are called the brush borders. These surfaces are electronically negatively-charged because of their acidic mucopolysaccharide content, e.g., sialic acid and uronic acid. This characteristic is important for the adsorption and transportation of intraluminal bacteria, digestive enzymes, and various substances produced by them onto or into the host cells. There are glands secreting alkaline mucus, which contains chymotrypsin within the crypts among these villi. The basal cells of the crypt are dividing constantly, and the generated cells transfer sequentially to become the epithelial cells of the villi, so that a constant flow of epithelial cells analogous to the constant desquamation turn-over of epidermal cells of the skin, is established. This mechanism plays a role in natural defense, because microorganisms without special virulence factors for successful colonization, are washed off from the intestinal epithelium by this mechanism.

Another mechanical defense mechanism is the active peristalsis of intestinal muscles which forcefully transports luminal contents to the exterior to defecate from the anus. Due to this physical force, microorganisms without special virulence factors do not adhere to the luminal wall for colonization or penetration. Another remarkable natural defense mechanism is the normal intestinal flora which will be discussed in a separate section because of its importance and complexity.

Many small lymph nodes called Peyer's patches are scattered along the intestinal wall. These are places from which macrophages migrate, as well as the locations where immune reactions occur, e.g., mutual interactions among T- and B-lymphocytes and macrophages, produce acquired-specific immune-resistance mechanisms, including antibodies. Among the immunoglobulins, immunoglobulin A (Ig A) is secreted to become coproantibody. Ig A is secreted in a monomeric form from the plasma cells within Peyer's patches, and circulates in the blood and lymph in the same form. However, as it crosses the intestinal walls, secretory pieces (SC pieces) bind two monomers so, coproantibody Ig A is of dimer configuration.

Also, exocrine organs like the liver and pancreas pour their respective digestive juices containing enzymes into the intestine. The liver also produces bile which enters the intestine at the duodenum. This bile contains Ig A and bile salts which affect the persistence and growth of certain microorganisms.

In addition to these specific immune mechanisms, nonspecific defensive factors, e.g., lysozyme and lactoferrin (p. 120), are secreted into the intestinal lumen to play an inhibiting role against the intraluminal multiplication of microorganisms.

The anatomical structure of the ileocecal junction with the cecum and the *appendix vermiformis,* favor invasion by pathogenic bacteria, e.g., *Salmonella.* Therefore, ulcer formations due to inflammatory reactions (p. 159) are frequent in the cecum, and the upper part of large intestine. The host-defense mechanisms, i.e., phagocytosis by polymorphonuclear leukocytes (PMNs) and monocytes, activation of reticulo-endothelial macrophages, production of specific antibodies, and induction of delayed-type hypersensitivity (DTH, pp. 148, 161), which normally affect the host-parasite relationships of these ecosystems, may be altered when the host homeostasis is destroyed by the influence of other endogenous causes, e.g., malnutrition and/or mental stress.

Chapter 4 The Role of Body Temperature: Fever and Reduced Iron

Both host and parasite organisms have their optimal temperature range for life. Therefore, the body temperature of the host animal constitutes an important factor for the maintenance of the relatively constant relationships inherent in the healthy host, but it also affects the expression of virulence factors by pathogenic microorganisms which determine the disease processes.

The body temperature of mammals and birds is regulated to keep a narrow range being unique to each species. Temperature is controlled by neural and hormonal adjustments whose center is located in the anterior hypothalamus. The exception is the lower body temperature of some mammals during their hibernating periods when the normal flora as well as infectious microorganisms are kept in a static state in the body. These hibernating animals, e.g., tarabagans in Mongolia, are reported to be killed when their hibernation is over by the multiplication of *Yersinia pestis* which infected them before hibernation (Pollitzer 1954). This is evidence that lower body temperature plays a role in the defense mechanisms of the hosts.

On the other hand, fever, a state brought about when the thermoregulatory center of the anterior hypothalamus is stimulated in such a way that the body temperature is elevated to levels higher than normal, is prone to be erroneously regarded as hazardous for the host organisms, because of its adverse effect on the feeling and consciousness of the hosts. However, many studies have indicated that fever is a host-defense mechanism during bacterial and viral infections. Kluger and Rothenburg (1979) attributed the beneficial effects of an elevated body temperature to such mechanisms as (1) an increase in leukocyte mobility; (2) increased leukocyte killing of ingested microorganisms; (3) an increased production of interferon; and (4) altered levels of the blood concentrations of many nutrients. With regard to the last mechanism, Kluger and Rothenburg (1979) demonstrated that not only warm-blooded animals, but also reptiles (*Dinosaurus dorsalis*) develop a fever during infection with *Aeromonas hydrophilia* (body temperature goes from about 38 °C to 41 °C) and this fever increases their survival rate. During these infections the plasma-iron concentration decreases. This example indicates that reducing the levels of plasma iron, which is essential for the multiplication of bacteria, plays the role of a defense mechanism as will be described later.

The mechanisms of fever generation in warm-blooded animals are diagrammatically illustrated in Fig. II.8.

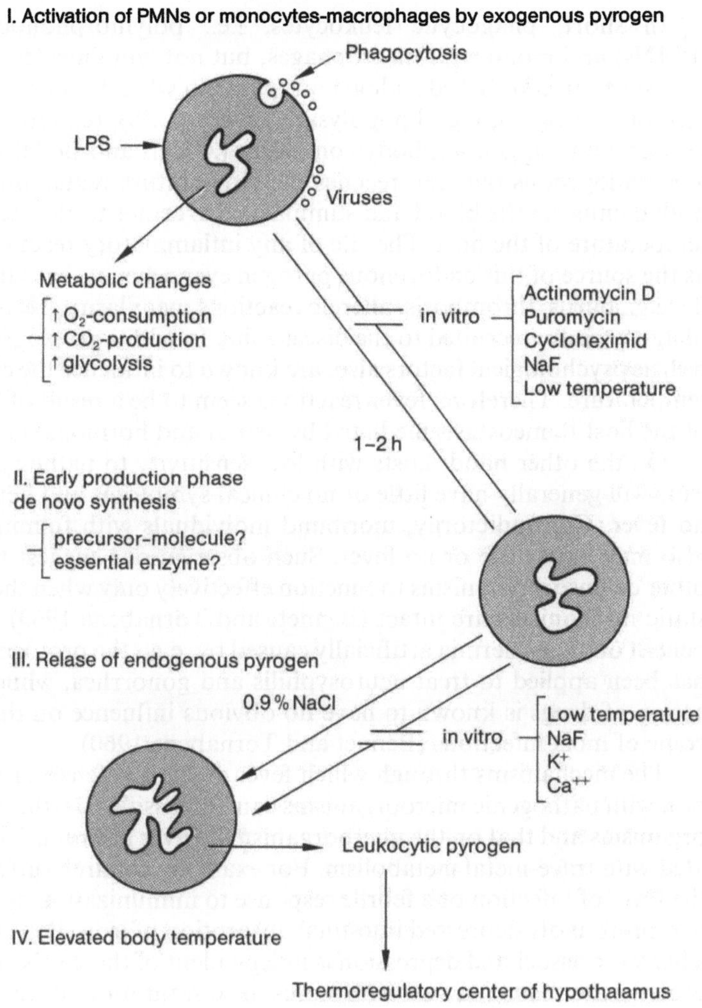

Fig. II.8. Mechanisms of fever genesis [modified from H. Hahn (1974) Immunität und Infektion, 2:69–72].
I Normal PMNs or monocytes-macrophages are activated by the contact with phagocytosis of exogenous pyrogens (LPS, bacteria, viruses, etc.). This activation process can be inhibited by the inhibitors of protein synthesis or transcription (actinomycin D) in vitro. **II** It is not yet clear whether precursor molecules of endogenous pyrogen are newly synthesized or some enzyme transforms a hypothetical propyrogen into an active endogenous pyrogen during the activation process. **III** The inhibitors of protein synthesis are not effective in blocking the release of endogenous pyrogens, while the inhibitors of glycolysis (NaF, jodacetate) are effective. These of aerobic metabolism (e.g., cyanide) are not effective, demonstrating that the energy for endogenous pyrogen release is supplied by glycolysis, not by aerobic metabolism. **IV** Endogenous pyrogen might change the sodium-calcium ionic ratio in the thermoregulatory center of the anterior hypothalamus to result in the elevation of body temperature or the release of 5-hydroxy-tryptamin, noradrenalin, and adrenalin by the hypothalamus. Bacterial lipopolysaccharide could also affect it directly according to the latter hypothesis

In short, phagocytic leukocytes, i.e., polymorphonuclear leukocytes (PMNs) and monocytes-macrophages, but not lymphocytes, liberate a mediator substance(s) named endogenous pyrogen when they are stimulated by exogenous pyrogens, e.g., lipopolysaccharides (LPS), (endotoxins of bacteria), viruses, and antigen-antibody complexes (Atkins and Bodel 1972). The liberated endogenous pyrogen reaches the temperature-regulating neurons in hypothalamus via the blood and stimulates this center to elevate the whole body temperature of the host. The site of any inflammatory reaction can also serve as the source of this endogenous pyrogen even when its causative agent is sterile, e.g., burns, thrombosis, allergic reactions, neoplasms, because a number of phagocytes are recruited to the disease site. In addition, exogenous physical as well as psychological factors also, are known to influence the elevation of body temperature. Therefore, fever reactions seem to be a result of labile deviations of the host homeostasis mediated by neural and hormonal networks.

On the other hand, hosts with low sensitivity to pathogenic microorganisms will generally have little or no clinical symptoms and hence have little or no fever. Contradictorily, moribund individuals with fulminating infections also may have little or no fever. Such observations suggest that fever assists other defense mechanisms to function effectively only when the body's homeostatic mechanisms are intact (Bennett and Tornabene 1960). In practice, the benefit of hyperthermia artificially caused by, e.g., the protozoan *Plasmodium*, has been applied to treat neurosyphilis and gonorrhea, while the antipyretic action of drugs is known to have no obvious influence on the course or outcome of most infections (Bennet and Tornabene 1960).

The mechanisms through which fever exerts a defensive role against infection with pathogenic microorganisms can be classified as: the effect on the host organisms and that on the microorganisms. Fever has been found to be associated with trace-metal metabolism. For example, children suffering from either the fever of infection or a febrile response to immunization have a consistently and profoundly depressed intestinal absorption of iron (Beresford et al. 1971). This fever-associated depression is independent of the host's state of nutrition, growth rate, bone marrow iron stores, iron saturation of transferrin, plasma-iron level, or the nonfebrile rate of iron absorption. In a different study using adults, an average of 7.2% of an oral dose of radioiron was absorbed into the blood of normal persons, whereas only 1% of the dose was absorbed by febrile patients (Dubach et al. 1948).

In young adult volunteers, following exposure to *Francisella tularensis*, plasma-iron concentration was decreased in two phases (Pekarek et al. 1969). In the first, or exposure phase, a modest decline of plasma iron occurs during the incubation period and is independent of the development or absence of subsequent illness. In the second, or febrile phase, a greater decline is seen. It begins at the peak of neutrophile leukocytosis. Endotoxin was found to cause hypoferremia and to depress the intestinal absorption of iron.

When either endotoxin, polyriboinosinic polyribocytidilic acid (poly-I:C), *Diplococcus pneumoniae*, or the attenuated vaccine strain of *F. tularensis* was

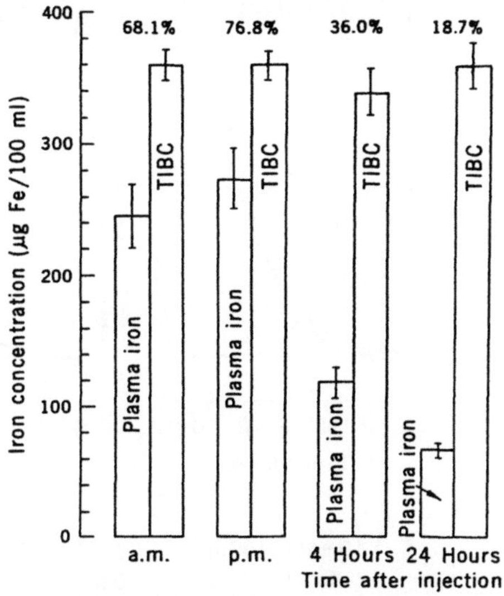

Fig. II.9. Plasma iron (PI) and total iron-binding capacity (*TIBC*) in eight male New Zealand white rabbits. Prior to infection with live pathogenic bacteria (*Pasteurella multocida*), the rabbits' PI concentrations averaged 247 µg per 100 ml during the morning and 275 µg per 100 ml during the afternoon hours. After infection with bacteria, the PI concentrations fell to 118 µg per 100 ml by 4 h and to 66 µg per 100 ml by 24 h. The TIBC did not change during this period. The *percentages* above each set of bar graphs indicate the percentage saturation of the plasma protein, transferrin. (From Kluger and Rothenburg 1979)

administered intraperitoneally into rats, plasma-iron and zinc levels were decreased. All these in vivo reactions result from the action of a small-molecular-sized protein called endogenous pyrogen released from PMNs. This substance has been isolated from peritoneal exudates of animals and shown to induce fever, hypoferremia, and hypozincemia.

The evidence to support the hypothesis that one of the mechanisms behind the adaptive (or benefical) role of fever is the reduced ability of pathogenic bacteria to grow well at elevated temperatures in an iron-poor medium was presented by Kluger and Rothenburg (1979). In their in vivo experiments, the blood plasma-iron concentration (PI) and the percentage iron saturation of the plasma protein, and transferrin (TIBC) levels of New Zealand white rabbits were measured prior to and at intervals after i.v. injection of five pathogenic *Pasteurella multocida* organisms. Figure II.9 shows that PI concentrations decreased, while TIBC did not change appreciably. The average rectal temperature of eight rabbits was from 39.7 °C ± 0.17 °C prior to injection of bacteria and 40.85 °C ± 0.29 °C by 4 h after injection and 40.90 °C ± 0.30 °C by 24 h after injection. In their in vitro experiments, the growth of *Pasteurella multocida* at various incubation temperatures in media containing graded concentrations of iron were measured by absorbance on a spectrophotometer. The effect of the added iron chelator deferoxamine mesylate was also examined in parallel experiments. The results are shown in Fig. II.10, indicating the synergistic effects of temperature and reduced iron on the growth of *Pasteurella multocida*. The effect of the added iron chelator was greater at the febrile temperatures of 41 °C and 42 °C. Thus, the results of the in vivo and in vitro ex-

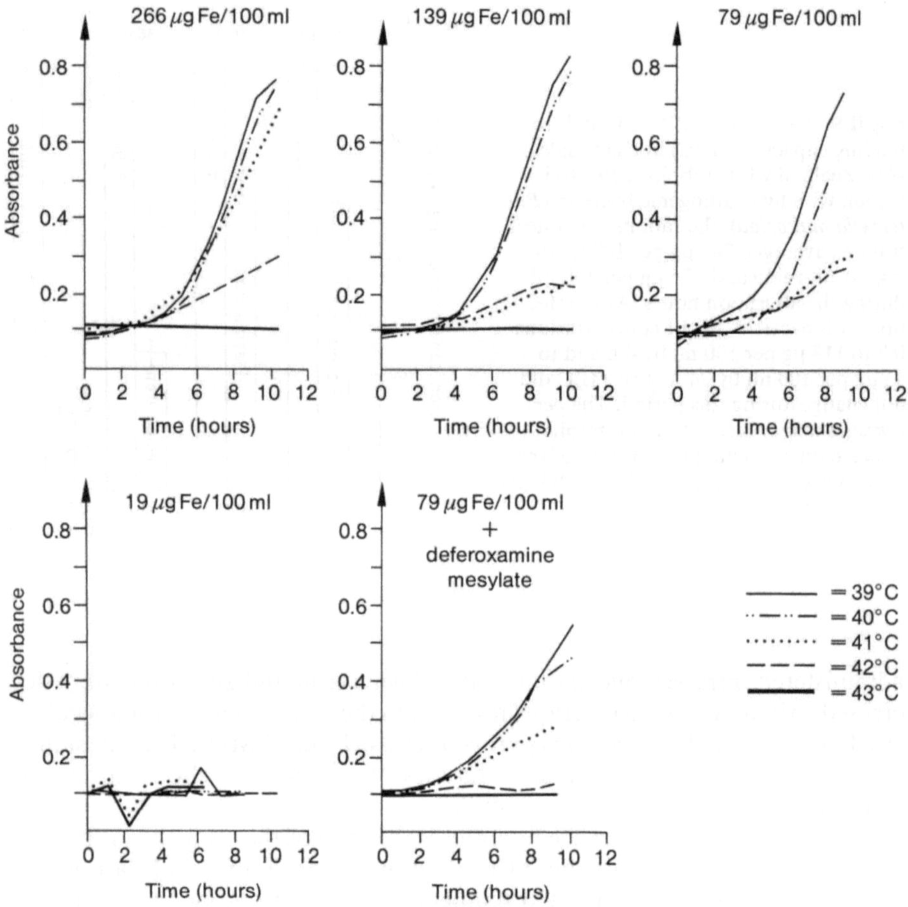

Fig. II.10. Growth of *Pasteurella multocida* (measured as absorbance on a spectrophotometer) at temperatures of 39 ° to 43 °C and in iron concentrations of 19 to 266 µg of iron per 100 ml of growth medium. At an iron concentration corresponding to the normal plasma concentrations of iron in uninfected rabbits (266 µg per 100 ml), the bacteria grew well at 39 °, 40 °, and 41 °C. When the bacteria were grown in a medium containing concentrations of iron corresponding to that which occurs in rabbits during infection (139 or 79 µg per 100 ml), the growth of the bacteria was not inhibited at the afebrile temperatures of 39 ° or 40 °C, but was depressed at the febrile temperature of 41 °C. When the iron concentration was reduced to subphysiological levels (19 µg per 100 ml) the bacteria did not grow at any temperature. The addition of the iron chelator deferoxamine mesylate to the growth medium depressed the growth of the bacteria at all temperatures, but clearly had a greater effect at the febrile temperatures of 41 ° and 42 °C. (From Kluger and Rothenburg 1979)

periments were sufficient to explain the mechanisms of their earlier finding that infection of rabbits with live *Pasteurella multocida* (a common rabbit pathogen) gave higher survival rates in correlation with the fever reaction of the animals (Vaughn and Kluger 1977; Kluger and Vaughn 1978).

In the case of viral infections, it has been known that small variations in temperature can result in major alterations of the course of virus infections in cell cultures, embryonated eggs, and infected animals. The influence of supraoptimal body temperature on the course of viral diseases was studied experimentally by A. Lwoff and M. Lwoff (1958) using poliomyelitis virus in association with virulence studies of temperature-sensitive mutants. He showed that higher temperatures reduced the effects of infection with polio virus.

In viral infections, plasma-iron and zinc concentrations fall early even in those subjects who remain asymptomatic, e.g., when adult human volunteers are immunized with live-attenuated Venezuelan equine encephalomyelitis virus. The decrease is intensified just before the temperature elevation and is maximal at the height of the febrile phase. Plasma iron and zinc levels return to normal as the body temperature decreases to normal.

The sequence of febrile phases and of declines in the blood concentration of iron and zinc seem to indicate that fever might occur as a part of the fail-safe defense when the early phase of hypoferremia and hypozincemia are not enough to defend host organisms against infection and allow the microorganisms to produce enough exogenous pyrogen to stimulate leukocytes so that endogenous pyrogen is produced and fever develops. Simultaneously, enhanced declines in plasma iron and zinc would also occur and iron absorption through the intestine would be blocked.

Secondarily, fever might provide a defensive role by directly interferring with the ability of the invading prokaryotes or eukaryotes to synthesize substances such as exotoxins, antibiotics, or siderophores. The production of these substances is not essential for the normal life cycle of the producer. However, the beneficial roles of these substances for the survival of these microorganisms are exhibited when the latter relate with the other microorganisms or live within the body of host animals. The mechanism of such interference by fever is as follows:

Generally speaking, all metabolites biosynthesized for normal life cycles are termed primary metabolites versus those biochemically bizarre substances produced by the cell equipped with unique synthetases that catalyze transformation of surplus primary metabolites along the pathway to the formation of those substances termed "secondary metabolites". With some exceptions, the optimal conditions, e.g., the concentrations of trace metals, inorganic ions, pH, redox potential, and temperature, required for the biosynthesis of the secondary metabolites are more restricted within narrower ranges than those conditions required for the biosynthesis of the primary metabolites which are a prerequisite for cell division, i.e., vegetative growth.

Figure II.11 is a diagrammatic presentation of the temperature ranges that microorganisms require for their vegetative growth and the narrower temper-

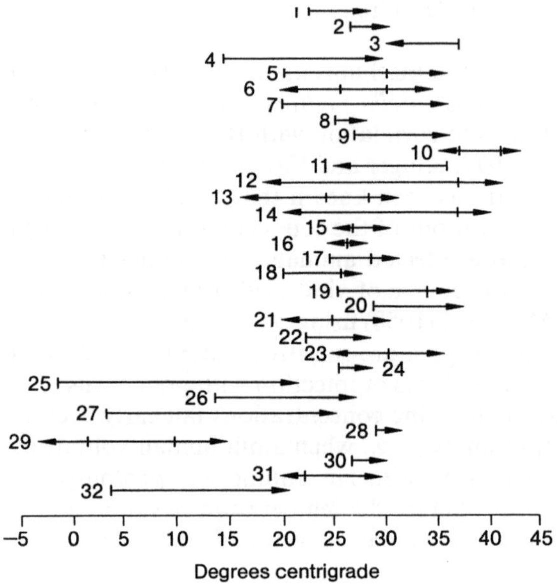

Prokaryotes	Arrow no.	Secondary metabolite
1.	*Anabena flos-aquae*	Toxin
2.	*Aphanizomeron flos-aquae*	Toxin
3.	*Bacillus cereus*	Toxin
4.	*Bacillus cereus* var. *alesti*	Pigment
5.	*Bacillus polymyxa*	Polymyxin
6.	*Clostridium botulinum* types A, B, E	Toxin
7.	*Listeria monocytogenes*	Hemolysin
8.	*Microcystis aeruginosa*	Fast-death factor
9.	Pasteurella pestis	Murine toxin
10.	*Pasteurella pestis*	V & W antigens
11.	*Pseudomonas aeruginosa*	Pyocyanine
12.	*Salmonella typhi*	Vi antigen
13.	*Serratia marcescens*	Prodigiosin
14.	*Staphylococcus aureus*	Enterotoxin B
15.	*Streptomyces griseus*	Streptomycin
16.	*Streptomyces jamaicensis*	Monamycin
17.	*Streptomyces niveus*	Novobiocin
18.	*Streptomyces rifamyces*	Rifamycin
19.	*Vibrio cholerae*	Vascular permeability factor
Eukaryotes		
20.	*Ashbya gossypii*	Riboflavin
21.	*Aspergillus flavus*	Aflatoxin
22.	*Aspergillus niger*	Citric acid
23.	*Aspergillus niger*	Malformin
24.	*Claviceps purpura*	Ergoline alkaloids
25.	*Fusarium poae*	Alimentary aleukia toxin
26.	*Fusarium* sp.	Zearalenone
27.	*Fusarium tricintum*	Scirpene toxin
28.	*Gibberella fujikuroi*	Gibberellin
29.	*Penicillicum martensi*	Penicillic acid
30.	*Penicillium notatum*	Penicillin
31.	*Penicillium rubrum*	Rubratoxin
32.	*Penicillium* sp.	Tremortin

Fig. II.11. Suppression of secondary metabolism by temperatures that permit vegetative growth. *Vertical bars:* temperature at which maximum yield of metabolites is obtained. *Arrow heads:* temperature at which minimum yield of metabolites is obtained. (From Weinberg 1974)

ature ranges required for the biosynthesis of the secondary metabolites. It is apparent that vegetative growth of 6 of all 32 species of microorganisms is possible through ranges of temperature that are greater than that required for the secondary metabolites. This means that even a slight elevation of body temperature can suppress the biochemical pathways for the production of secondary metabolites without interferring with those of primary ones, i.e., vegetative growth. As was mentioned in Part I, a number of the virulence factors of pathogenic microorganisms belong to the category of secondary metabolites as seen in Fig. II.11, their production is often controlled by the gene(s) locating on plasmids (Okanishi 1977).

Generally, the influence of temperature on secondary metabolism has been observed to occur during some portion of the transitional period between cessation of vegetative growth and the actual synthesis of the secondary substances. This period is comprised of the sequence of events that include (1) repression of that portion of the genome responsible for cell growth; (2) derepression of the portion required for transcription of secondary metabolite synthetases; and (3) the transition from the cessation of vegetative growth to the actual synthesis of the secondary metabolites. In some cases, the temperature optima for the transitional period in which synthetase formation occurs differ from that needed for activity of the synthetases. In other secondary-metabolic processes, the yield of the metabolites is good at both the temperature optimum for vegetative growth and that best for production of the synthetases. Thus, it is conceivable that infecting pathogenic microorganisms cease the biosynthesis of secondary metabolic virulence factors when the host's body temperature is elevated beyond the optimal range (Bennett and Nicastri 1960). The experimental evidence for this possibility will be included in the next discussion.

During the interaction between host and infecting microorganisms the influence of the host body temperature is increased when both host and parasite compete for common nutrients, e.g., essential trace elements or some organic substances. The host inhibits the multiplication of invading microorganisms by depriving them of these nutrients; therefore, Kochan (1973) used the term "nutritional immunity" for such a defense. For example, mammalian species whose seminal vesicular and placental tissue are devoid of erythritol are considerably more nutritionally immune to brucellosis than cattle, sheep, goats, and hogs which have the substance, since erythritol stimulates the growth of the organisms. Likewise, nonrenal tissues that contain lower amounts of urea than does the kidney would be somewhat nutritionally immune to those pathogens that produce urease and, thus, use this substrate for growth. Earlier studies concerning host-tissue environments that might be deficient in organic nutrients essential for growth of potential pathogens have been summarized by Garber (1956). Such eukaryotes as *Ochromonas, Euglena, Saccharomyces,* and *Candida* have elevated trace-metal requirements at elevated temperatures, although only the last is pathogenic. Likewise, such prokaryotes as *Pseudomonas* and *Salmonella* have increased iron requirements at high temperatures.

Table II.1. Hematological data[a] and average survival[b] of the three groups of mice fed with low-iron diet, low-iron diet plus parenteral iron, or normal diet (summary of four repeated experiments). (Modified from Puschmann and Ganzoni 1977)

Mice (No. infected)	Hemoglobin (g/100 ml)	Serum iron (μg/100 ml)	Total iron-binding capacity (μg/100 ml)	Whole-body iron (μg/animal)[c]	Surviving animals (%)
Group A					
Low-iron diet (15,15,15,26)	10.0±2.0 ($P<0.001$)[d]	116±76 ($P<0.001$)[d]	601±101 ($P<0.001$)[d]	1.180±310 ($P<0.001$)[d]	58.5 ($P<0.01$)[d]
Group B					
Low-iron diet plus parenteral iron (17,17,11,14)	11.7±0.9 ($P<0.001$)[e]	314±77 ($P<0.001$)[e]	571±113 (NS)[e]	1.714±568 ($P<0.001$)[e]	19.7 ($P<0.05$)[e]
Group C					
Normal diet 10,10,15,20)	12.9±1.0 (NS)[f]	330±152 (NS)[f]	409±31 (NS)[f]	1.639±449 (NS)[f]	10.8 (NS)[f]

[a] Statistical evaluation between groups was done by Student's test.
[b] Statistical evaluation by the Chi-square test.
[c] Average body weights of groups A, B, and C were not significantly different.
[d] Group A to group C.
[e] Group A to group B; NS, not significant.
[f] Group B to group C; NS, not significant.

The suppressing effect of high temperature on siderophore production by bacteria was convincingly demonstrated by Garibaldi (1972). Normal in vitro growth and production of siderophores occurs at 36.9 °C in the presence of 0.5 μM iron; however, at 40.3 °C, 50 μM iron is required.

Among those trace metals, such as V, Cr, Mn, Fe, Co, Ni, Cu, Zn, and Mo that are biologically of interest, the concentration of iron is most critical for efficient secondary metabolism in bacteria, as is that of zinc in yeasts and molds. Although at optimum temperatures bacteria grow equally well at concentrations of available iron ranging from 1.0 to 1000 μM, their formation of secondary metabolites generally is stimulated or inhibited by quantities between 3.0 and 100 μM. It can be predicted that temperature control of prokaryote secondary metabolism will be found to be based, at least in part, on the temperature sensitivity of the synthesis of either siderophores or their corresponding repressors (Garibaldi 1972). In this connection, Kitos et al. (1953) demonstrated that the ability of trace elements to interfere with citric acid synthesis by *Aspergillus niger* was more efficient at elevated temperatures. It is quite probable that, as with prokaryotes, trace-element transport systems in eukaryotes will be found to be temperature-sensitive.

The level of plasma copper has been known to rise in acute and chronic infections. Normal concentrations are approximately 17–20 μM, whereas patients with infections have values of 25–38.5 μM. The high levels return to normal in severely ill patients who become "metabolically exhausted" as well as in cases in which the infection disappears. The hypercupremia might possibly serve to inhibit microbial growth and/or secondary metabolism, and temperature might affect these roles of copper as indicated. In an in vitro study of the toxicity of copper on the growth of *Escherichia coli* in a synthetic medium at various temperatures, 2 μM copper permitted immediate growth at 35 °C, but delayed multiplication for 5, 8, and 12 h, respectively, at 37 °, 40 ° and 42 °C. The corresponding delays for 3 μM copper at the four temperatures were 8, 15, 18, and 25 h, respectively.

These results support the hypothesis proposed by Weinberg (1966) that the body temperature of the host influences greatly the behavior of parasitic microorganisms in vivo through trace-metal metabolisms in the host body as well as by the temperature-dependent requirement for the growth of the microorganisms.

However, the defensive role of reduced iron, i.e., hypoferremia, is obvious even in the absence of an elevated body temperature of the host. As evidence for this, Puschmann and Ganzoni (1977) fed two groups of weanling NMR1 mice with a commercial diet containing less than 5 mg of iron per kg. One of these groups received an intraperitoneal injection of 1 to 1.5 mg of iron as iron dextrin. These groups and a third one, receiving a stock diet containing 360 mg of iron per kg, were infected with 90 to 150 organisms of *Sal. typhimurium* intraperitoneally 10 to 14 days thereafter. Five mice of each group were sacrificed at the time of infection for hemoglobin, serum iron, and total iron-binding capacity measurements. For whole-body iron assay, all mice were dissolved in nitric acid at the end of the study. The results of these measurements are shown in Table II.1. The percentage of survival mice was largest when the amount of free iron available for the pathogen was minimal, and smallest when the mice were fed with a diet containing iron.

In contrast, mammalian sensitivity to microbial infection is increased at the time of hyperferremic episodes. The examples are: (1) destruction of liver cells containing ferritin (as in viral hepatitis or louse-borne relapsing fever); (2) silent or overt hemolytic anemia (as in malaria, bartonellosis, sickle cell disease, leukemia, lymphomas, or Hodgkin's disease); or (3) an overload of iron from exogenous sources (as when neonates are born with excess iron that had been obtained from maternal plasma, or when excess dietary iron causes hemochromatosis).

In conclusion, the aforementioned facts indicate the importance of the physical ionic microenvironments surrounding the infecting microorganisms in the host for the understanding of host-parasite relationships, although this field of study has not been recognized nor considered in a satisfactory manner.

Chapter 5 The Defensive Role of Normal Microbial Flora

Self-defense mechanisms are already becoming effective during the intrauterine developmental stage of the host embryo, but the establishment of host-defense mechanisms is completed only after the release of the neonatal body from the confinement within the uterus, i.e., birth. Before this moment, the mammalian embryo is sterile due to the host-defense mechanisms of the pregnant mother, although a few exceptions exist. These exceptions are the transmission from the parents to the fertilized eggs of particular viral genomes that have been incorporated into the chromosomes of the ovary and/or sperm cells or the penetration of some particular microorganisms to the fetus, e.g., *Treponema pallidum, Toxoplasma,* rubella virus, etc. by breaking through the placental defensive barriers. In addition, the newborn host can receive certain microorganisms from the mother who secretes contaminated milk just after birth. These exceptional modes of transmission are called "vertical infections". However, microscopic observations of the green-brown or black-colored sticky stool which is excreted within 1–2 days of birth, reveals no bacteria or a very small number, demonstrating the sterility of the neonatal body.

The germ-free neonatal body can exist under natural conditions only for a very short period, because various microorganisms that are either resident in the mother's body, e.g., gynecological organs or gut, or resident in the surrounding outer world, initiate multiplication after colonizing the skin, the mucous membranes, and body cavities of the newborn. In fact, the appearance of *Streptococcus, E. coli, Clostridium,* and yeast organisms within the newborn feces were recognized 3–4 h after defecation of the initial sterile stools. The numbers of bacteria increase rapidly after suckling; and the total number of bacteria contained in the feces of a 1-day-old newborn exceeds a 10^{11} g^{-1} concentration in almost all human babies. Usually, *Bifidobacterium* begins to appear on the third to fourth postnatal day, and becomes most predominant around the fifth postnatal day, substituting for the predecessor bacteria, *E. coli, Streptococcus, Lactobacillus,* and *Clostridium* which are reduced in number during this short period. This transitional process of the newborn fecal microbial flora is diagrammatically illustrated in Fig. II.12 until the seventh postnatal day when the equilibrated ratios of six bacterial species are stabilized (Mitsuoka and Hayakawa 1972).

However, the "normal intestinal flora" is no more than an example of various normal microbial floras which have been established in or on various locations of the host body via similar processes of microbial colonization. The

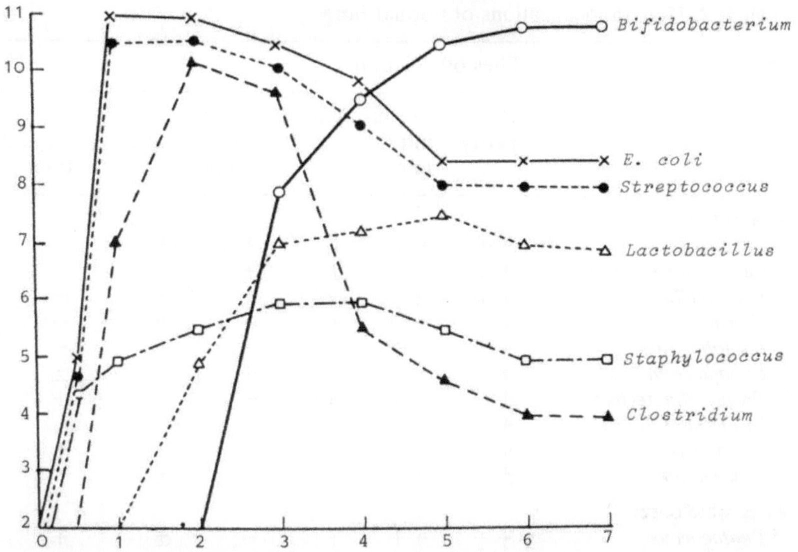

Fig. II.12. The transition phenomenon of normal fecal bacterial flora of neonatal human baby until seventh postnatal day. *Ordinate:* Log number of bacteria per 1 g of stool; *abscissa:* day after birth. [Modified from Mitsuoka and Hayakawa (1972)]

various genera of bacteria that are resident in the various locations of the human body as the normal microbial flora are shown in Table II.2.

From the viewpoint of microbial ecology, the aforementioned process of colonization by the microorganisms of a wide range of taxonomic species of the initially germ-free newborn body can be regarded as the distribution of these microorganisms to their respective niches on or in the host where the most appropriate living conditions are offered to them by the host.

At the first stage of the establishment of the microbial flora, only those microorganisms that require certain nutrients, growth temperatures, pH ranges, reduction-oxidation (redox) potentials, oxygen concentrations, etc. are selected for colonization by the respective niche where all these requirements are satisfied. This is the reason why *Streptococcus, E. coli, Clostridium,* and yeast colonize at first in the newborn intestine. However, the environmental conditions of these niches are altered by the colonization of the pioneer microorganisms as the result of interactions between the host and microorganisms and among microorganisms themselves. These detailed interrelationships will be discussed at the end of this chapter with regard to normal *intestinal* microbial flora in particular. Such transition phenomena at the various niches proceed until a steady state of relatively constant constitution of microorganismic species are established as determined by certain mutual reactions among the various species of resident microorganisms and the physiological conditions of the niche.

Table II.2. Human localizations of normal flora

Bacteria	Sites of isolation						
	Oral cavity	Lower intestine	Vagina	Outer sexual organ	Urinary tract	Upper respiratory tract	Skin
Anaerobic rods							
Bacteroides	++	++	+	+	+	+	−
Fusobacterium	++	+	±	+	+	+	−
Leptotrichia	++	±	−	±	−	+	−
Vibrio	+	+	+	−	±	+	−
Lactobacillus	+	++	++	−	±	−	−
Eubacterium	+	±	+	+	+	±	±
Propionibacterium	±	++	++	−	−	+	−
Bifidobacterium	+	++	++	−	−	−	−
Actinomyces	+	++	−	−	−	+	−
Clostridium	±	+	±	−	±	−	−
Anaerobic cocci							
Peptococcus	++	++	+	−	±	+	+
Peptostreptococcus	++	++	++	++	++	++	−
Veillonella	++	+	+	−	±	+	−
Spirochaeta	−	±	−	+	−	+	−
Aerobes							
Staphylococcus (coag. −)	+	+	+	+	++	++	+
Staphylococcus (coag. +)	+	+	±	−	±	++	+
Streptococcus	++	++	±	+	±	++	±
Diplococcus (S. pneumoniae)	±	−	−	−	−	+	−
Neisseria	++	−	+	−	+	+	−
Corynebacterium	+	+	+	+	+	+	+
Mycobacterium	−	+	±	+	+	−	±
Coli group	+	++	±	+	+	+	−
Klebsiella	+	+	−	−	+	+	−
Proteus	−	+	+	−	+	+	−
Pseudomonas	−	±	−	−	+	±	−
Hemophilus	+	−	+	−	±	+	−
Mycoplasma	+	+	+	+	±	+	−

Thus, in addition to the self-defense mechanisms inherited from the parents, the host-defense mechanisms of the integrated ecosystem of the hosts together with normal microbial flora, in general, are effective for the maintenance of the homeostatic state of the host. Furthermore, the normal microbial flora is not without effect in enhancing the other host-defense mechanisms by activating self-defense mechanisms, i.e., the immune system, as will be described later. Also some host animals are dependent on the nutrients, e.g., vitamins produced or digested by the normal microbial flora, as another instance of its important functions.

The normal intestinal flora will be discussed as a representative example of such flora. The mouth is the main entrance route for these intestinal bacteria. Pediatricians observe that mother-milk-fed babies are more resistant to microbial infections than formula-fed babies. One of the mechanisms for this is that mother's milk contains colostrum, which includes large amounts of lactoferrin (see Chaps. 6 and 7). Another reason for this is the difference in the intestinal flora between mother-milk-fed babies and others. Even the recent development of improved dry-milk formulas has not yet completely overcome these discrepancies. Mother-milk-fed babies evacuate stools containing 10^{10}–10^{11} g^{-1} *Bifidobacterium* group, but only around 10^8 g^{-1} of *E. coli* or enterococci. Few anaerobic organisms like *Bacteriodes* or putreficative bacteria can be detected within this stool, the pH of which is as low as 4.5 to 5.5. In most cases of formula-fed babies, the *Bifidobacterium* group is predominant also but, occasionally, some babies who have no detectable *Bifidobacterium* organisms in their stool have been found among the formula-fed infants. The remarkable differences between these two groups are: first, with formula-fed infants there is more than a ten fold increase in facultative anaerobic bacteria, such as *E. coli* and enterococci, and increased numbers of highly anaerobic *Bacteroides* bacteria; second, a higher stool pH 5.7 to 6.7 is seen and; third, lower oxidation-reduction potentials exist in the formula-fed groups. The last means a lower content of oxygen within the stool of formula-fed babies that favors the increase in numbers of *Bacteroides*. On the other hand, the production by the *Bifidobacterium* group of lactic acid and acetic acid and the weaker buffering action of human milk in comparison with cow's milk, results in the lower pH of the stool of mother-milk-fed babies. This condition is favorable for the continuous growth of *Bifidobacterium* group and unfavorable for the growth of *E. coli* and putrefactive bacteria.

When these babies stop breast-feeding and begin taking foods similar to adults, step by step, the intestinal flora changes: the numbers of *Bacteroides, Eubacteria,* and of anaerobic streptococci become predominant as is the case in the adult. Within the *Bifidobacterium* group itself, the bacterial species change from the neonatal pattern into the adult pattern. The intestinal flora of adults shows a fairly stable and similar pattern in the state of health.

This adult normal flora contains anaerobic bacteria, such as *Bacteroides, Eubacterium,* and anaerobic staphylococci and the most important human lactic acid bacteria, *Bifidobacterium* are predominant in these stools. The total number of these predominant bacteria is 10^9–10^{11} per 1 g of stools, and they maintain a mutual balance in the intestine. The numbers of *E. coli,* enterococci, lactobacilli, and *Veillonnela* groups are usually not more than 10^5–10^8 g^{-1}. The *Clostridium welchii,* staphylococci, and yeasts are less than 10^5 g^{-1}. However, the isolation rate and number of *Bifidobacterium* groups decrease with human aging. When stools of 65–85-year-old persons were compared with younger adults the sampling frequencies of *Clostridium welchii* had increased remarkably even up to 92% of the organisms isolated. These relationships of the respective bacterial numbers per 1 g of stool to the approximate aging process is shown in Table II.3.

Table II.3. Age differences of human fecal normal bacterial flora. (From Mitsuoka and Hayakawa 1972)

Age group	Total bacterial number	Bacteroid	Bifido-bacteria	Catena-bacteria	Strepto-cocci	Entero-bacteria	Lacto-bacilli	Clos-tridia	Veillo-nella	Pepto-strepto-cocci	Staphy-lococci	Yeast
I Suckling under 4 days old; 17 cases	10.1±0.5[a]	8.6±1.7 (41)[b]	9.3±2.5 (47)	0 (0)	8.5+1.3 (100)	9.3±1.3 (100)	6.4±1.9 (53)	5.9±3.0 (35)	5.6±2.2 (29)	0 (0)	6.2±1.7 (100)	3.5±1.3 (41)
II Suckling between 4 and 90 days old; 60 cases	10.5±0.6	8.2±2.3 (57)	9.9±1.5 (90)	9.7±0.5 (7)	8.1±1.9 (100)	8.8±2.0 (100)	7.3±2.4 (75)	6.9±2.0 (32)	6.3±2.7 (55)	9.0±0.6 (5)	6.8±1.7 (100)	4.0±1.6 (28)
III Children between 4 and 6 years old; 29 cases	10.8±0.3	10.4±0.6 (100)	10.1±0.6 (97)	9.9±0.4 (24)	7.8±1.3 (100)	8.0±1.4 (100)	7.0±1.8 (93)	5.7±2.0 (79)	5.2±1.9 (86)	8.1±0.9 (14)	4.0±1.3 (86)	4.2±1.3 (86)
IV Adult between 20 and 64 years old; 29 cases	10.8±0.4	10.3±0.6 (100)	9.8±0.7 (100)	9.9±0.8 (93)	7.7±1.3 (100)	8.2±1.3 (100)	6.7±1.8 (100)	4.8±1.7 (52)	4.8±2.3 (55)	8.9±1.8 (52)	4.4±1.8 (90)	3.7±1.2 (69)
V Old people, more than 65 years old, 72 cases	10.5±0.5	10.0±0.8 (100)	9.4±0.8 (85)	9.5±0.9 (76)	8.2±1.3 (100)	7.8±1.3 (100)	8.0±1.5 (99)	6.6±1.9 (92)	6.1±2.0 (64)	7.7±2.2 (35)	4.3±2.1 (60)	4.6±1.5 (71)

Significance of bacterial number and the sampling frequencies

I: II	*	●(●)	●(**)	**(●)	●(●)	●(●)	●(●)	●(●)	●(●)	●(●)	●(●)	●(●)
I: III	**	**(**)	●(**)	**(●)	●(●)	**(●)	●(**)	●(**)	●(**)	**(●)	**(●)	●(**)
I: IV	**	**(**)	●(**)	**(**)	●(●)	**(●)	●(**)	●(●)	●(●)	**(**)	**(●)	●(●)
I: V	**	**(**)	●(**)	**(**)	●(●)	**(●)	**(**)	●(**)	●(*)	**(*)	**(**)	●(*)
II: III	*	**(**)	●(●)	●(*)	●(●)	●(●)	●(●)	●(**)	●(**)	●(●)	**(*)	●(●)
II: IV	*	**(**)	●(●)	●(**)	●(●)	●(●)	●(**)	**(●)	●(●)	●(**)	**(●)	●(**)
II: V	●	**(**)	*(●)	●(**)	●(●)	**(●)	*(**)	●(**)	●(●)	●(**)	**(**)	●(**)
III: IV	●	●(●)	●(●)	●(**)	●(●)	●(●)	●(●)	●(*)	●(*)	●(**)	●(●)	●(**)
III: V	**	*(●)	**(●)	●(**)	●(●)	●(●)	**(●)	●(●)	●(*)	●(●)	●(*)	●(●)
IV: V	**	●(●)	*(●)	*(●)	●(●)	●(●)	**(●)	**(**)	●(●)	●(●)	●(**)	*(●)

The mark in the parenthesis: Significance of sampling frequencies. ● = not significant; * = s(0.05); ** = s(0.01).

[a] Log mean average value ± standard deviation of the number of bacteria per 1 g of stool taken from the positive individuals.

[b] The ratio of detected individuals (%).

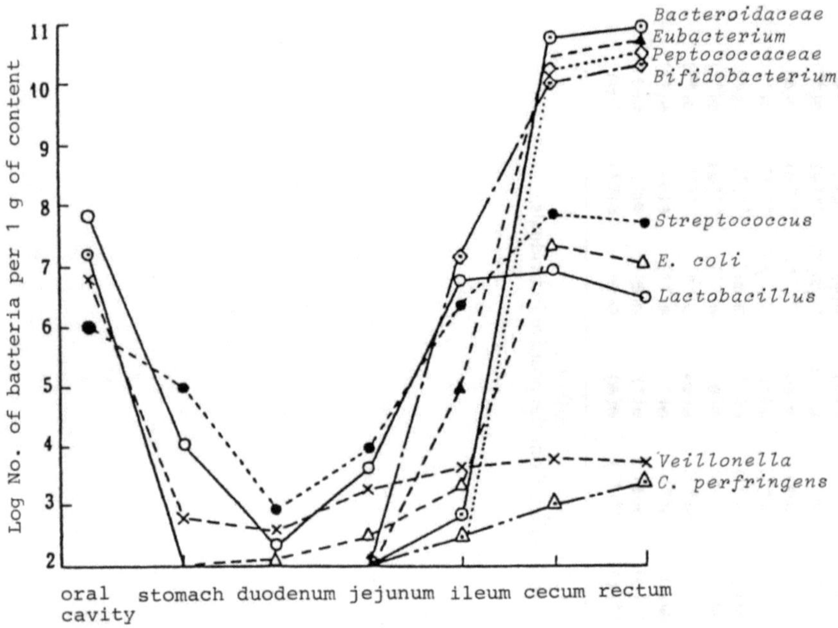

Fig. II.13. The normal bacterial flora distributed along the digestive tract of healthy human individuals. *Ordinate:* Log number of bacteria per 1 g of content; *abscissa:* local parts of human digestive tract. [From Fig. 4 Mitsuoka (1982)]

Changes in the pattern of various bacteria are seen in the different parts of the digestive tract in the numbers of organisms per gram of feces. A diagrammatic representation of the approximate number of respective bacteria per 1 g of content, taken from various parts of the digestive tract is shown in Fig. II.13.

As is shown in Fig. II.13, about 10^7 of various aerobic and anaerobic bacteria except *E. coli* are contained within 1 g saliva. These numbers decrease in the stomach where the pH fluctuates between less than 2.0 and more than 4.0 depending upon the presence of food, saliva, and gastric juice components.

At the time of the transfer of the gastric contents from the stomach to jejunum, the sterilizing effect of gastric juice is at its peak as was mentioned before. Therefore, the upper intestine contains less than 10^4 organisms per gram. Lactobacilli, yeasts, or *Veillonnella* are the organisms found. In addition to the effect of gastric juice, bile is poured into the duodenum through the bile duct, and the intestinal mucous membrane secretes lysozyme which acts as an antimicrobial factor. Bacterial numbers increase rapidly in the lower intestine, because the alkaline intestinal juice neutralizes acidity of the contents, which move downwards at a slower speed than in the upper part of the intestine. The bacterial number in the lower ileum reaches 10^7 organisms per gram, composed of *Bacteroides, Bifidobacterium, Eubacterium,* and anaerobic streptococci. The content of the cecum, the beginning of the large intestine, contains as many as 10^{10} bacteria per gram, being composed predominantly of anaerobic organisms, such as *Bacteroides, Bifidobacterium, Eubacterium,* and *Clostridium* and about 10^5–10^8 per gram of *E. coli,* streptococci and enterococci, lactobacilli, *Veillonnella,* and staphylococci. Thus, the human intestinal flora has different patterns of bacteria depending upon the age of the person and intestinal location, but is roughly stable at any one age or location. However, these average patterns can be more or less divergent due to the uniqueness of the individual physiology of the digestive organs, e.g., peristalsis of intestine, secretion of gastric juice; and due to differences in the diet of individuals. In short, a unique ecosystem becomes established within each individual's intestine which is a relatively constant form of a host-parasite relationship. This is a result of mutual adaptation and acclimatization between enterobacteria and the intestinal mucous membrane, in a typical ecological balance.

As the result of the continuous growth and turnover of intestinal bacteria, healthy mammalian intestines establish a unique normal intestinal flora that differs among various animal species and respective individuals. The predominant residential bacterial species belonging to this normal intestinal flora are: *Bifidobacterium, Bacteroides, Eubacterium,* and anaerobic streptococci groups. The less common ones are *E. coli* and enterococci. Although these patterns of residential normal flora are unique to each animal species or even to each individual, these flora have resulted from the developmental history of each individual since birth due to mutual interactions between the host and parasite to result in a condition of relative stability.

The *Bacillus* group, staphylococci and pathogenic organisms like *Shigella,* the cholera group, and *Salmonellae,* are brought into the intestine as contaminants of food, etc. and are destined to be excluded sooner or later from this conservative intraintestinal ecosystem. They are called "transient passengers" to be differentiated from resident organisms. Even naturally acquired *Vibrio cholerae* is spontaneously excluded from the patients' intestine generally within 1–2 weeks (Wake, unpublished data).

After a bout of acute diarrheal disease, the patient receiving appropriate treatment acquires immune resistance to transient passengers which eliminates them as the normal flora organisms recolonize the gut. Thus, the relative stability of normal flora unique to each individual host and, more loosely, to each host species, constitutes itself a restorable mechanism that plays a role in excluding transient passengers that can be pathogenic for the host. This can be seen when a broad spectrum antibiotic, like tetracycline, is used for treatment of a patient and the susceptible normal flora is reduced. In such an instance a tetracycline-resistant *Streptococcus* may colonize the gut and remain until treatment with an appropriate antibiotic or the inhibiting normal flora is restored.

Along the whole length of the gut, the duodenum has little bacterial flora. In contrast, the jejunum and ileum have microbial flora. Here, the villous epithelial cells are vigorously generated and from the depth of crypts translocate to the tops of villi and are discarded into the lumen after their death to carry bacteria away. However, multiplication of the intestinal bacteria synchronizes with this cell turnover and thus maintains the stable steady state.

In the large intestine, numerous varieties of bacteria cover the intestinal wall which is covered with mucin sticking to the epithelial mucous membrane. It was demonstrated that a simple protein, lectin, bridges between the polysaccharide chains of the *E. coli* surface glycocalyx and the villous glycocalyx (Mitsuoka 1978) to serve as an attachment and sterilizing mechanism.

From the viewpoint of microbial ecology, on the other hand, mutual symbiotic, antagonistic, or competitive interrelationships among intestinal microorganisms have been elucidated step by step. Recent advances in the techniques for anaerobic cultivation have shown that the predominant bacterial species of intestinal flora are anaerobes. Mechanisms similar to those involved in the establishment of skin anaerobic flora (p. 92) explain certain of the anaerobic microenvironments in the vicinity of the intestinal mucous epithelial cells. Furthermore, organisms that grow in the upper parts of the intestinal tract consume oxygen to produce anaerobic conditions within its lower parts. Therefore, the lower the part of the intestine, the more predominant are the anaerobic organisms.

As an example of the mutual antagonisms among these microorganisms a variety of substances, collectively called bacteriocins, are produced by some bacteria which show inhibitory effects on other bacteria which have appropriate receptors, e.g., *E. coli* produces colicin, *Pseudomonas aeruginosa,* pyocin and *Bacillus megatherium,* megacin. Another example of such mutual antagonism is the production of volatile fatty acids by the fermentation activities of some bacteria which inhibit growth of the other bacterial species (Mitsuoka 1978).

Also competitive interrelationships of bacteria for acquiring common essential nutrients exist, and siderophore production by some microorganisms which permits them to sequester iron from their competitors is also well known.

Change of pH, of oxidation-reduction potential, production of peroxide, or hydrogen sulfide due to the activities of some microorganisms play a role in either inhibiting or stimulating the growth of other microorganisms.

Thus, the significance of the normal intestinal flora as a host-defense mechanism resides in the complexities of interactions among the microorganisms that are themselves utilizing the intestinal lumen as their ecological niche. Once these interactions reach a compromising state through the process of the mutual antagonistic and symbiotic influences, the normal flora plays the role of a buffer zone that absorbs new organisms; and the disturbances resulting from their presence are soon restored without causing disease.

Chapter 6 The Nonspecific Humoral Components

Within the individual body, all cells in various organs and tissues are bathed in the interstitial fluid which is collected via the lymphatic system to be poured into the bloodstream (Chap. 1). The source of the body fluids of the interstitial space is the water absorbed from the environment by the energy-requiring transport system of the intestinal epithelial mucosal cells plus plasma components of the blood which infiltrate beyond the endothelial barrier of the blood capillaries. Blood cells with migrating activity, accompany this infiltration under normal conditions. Some of them are fixed to the tissue, but others circulate along with the flow and the body fluids and reenter the closed circulation system of the bloodstream.

Along with these cellular components, the body fluid thus constitutes the «mileau intérieur» (Bernard 1857) for each cell.

The difference between the outer environment and mileau intérieur resides in the fact that the latter maintains a constancy of temperature, O_2 partial pressure, pH, and other physicochemical conditions within narrow limits as long as the organism lives. All defense mechanisms, but one, i.e., normal microbial flora, so far mentioned in Part II, contribute to the maintenance of this relative constancy. In addition, the selective permeability of the intestinal mucous epithelium and of the kidney glomeruli, respiratory functions of the lung and skin, biochemical turn-overs (detoxifications) occurring in the liver, and secretions of hormonal substances by various cells, are so regulated that the biochemical composition of the body fluids fluctuates within narrow ranges. This phenomenon which results in the relative constancy of the milieu intérieur is called "homeostasis" (i.e., maintaining the same state) (Cannon 1929). Therefore, a disease caused by endogenous or exogenous etiology may disturb this homeostasis and this disturbance itself, also, can be regarded as a disease.

From another point of view, however, the body fluid, including plasma of an animal species, can be regarded as a tissue of liquid form that has evolved phylogenetically. Therefore, the molecular species in the composition of the body fluid have been enriched during the course of evolution, as the cells secreting them have been diversified into more complexity. Among these molecular species, some should play defensive roles in parallel with other functions, e.g., transporting metabolic substances or directing cell differentiation. It is well-known that even monocellular organisms secrete defensive molecules, e.g., antibiotics or bacteriocins.

The animal lectins secreted by multicellular organisms express a binding activity capable of agglutinating microorganisms and cells. An interesting parallel between the function of hemagglutinins of the Australian crayfish, an arthropod, and vertebrate immunoglobulins has been observed. These agglutinins serve as opsonins and increase the rate at which circulating phagocytes take up erythrocytes or bacteria (Marchalonis 1974). The body fluid of the genus *Limulus* is coagulated by minute amounts of bacterial endotoxin and this potentiality seems to defend such an ancient organism against bacterial invasion.

Thus, all examples support the hypothesis proposed in this book that the evolution of the "host-parasite relationships as a *field"* originated from the ancient interaction between the two ancestor organisms that recognized the counterpart organism as nonself (not-self).

According to this hypothesis, evolution has resulted in a diverse range of humoral components that play roles for host defense in higher animal bodies. These components themselves have diversified into nonspecific and specific types when the immune system (described in Chap. 1) appeared in the individual animal organizations. Therefore, the nonspecific humoral components are devoid of a "combining site", but are equipped with some ligand molecule that binds to a particular exogenous chemical configuration. Any substance(s) or cell equipped with the latter produces reaction. Specific cellular components, such as a variety of lymphokines (LKs), interferons, and interleukin 2 (IL2) as will be discussed later, belong to *non*specific humoral components according to this definition. On the other hand, all classes of immunoglobulins belong to the specific humoral components since they possess unique combining sites.

The nonspecific humoral components as factors in the basic knowledge for the understanding of host-parasite relationships will be discussed in this chapter, in the following order:

1. The complement system, which is a system characterized by relative independence from cellular components in spite of its important role in host-defense mechanisms.
2. Transferrin and lactoferrin, which are also humoral components, but can be localized intracellularly and play a role as host-defense mechanisms.
3. Cytokines, which are a multiplicity of hormonelike factors, regulating the activities of a large variety of cells that participate both in immunologically-mediated and in nonimmunologically-based inflammatory reactions, e.g., lymphokines, monokines, prostaglandins, and leukotrienes.
4. Interferons, in particular, characterized by their antiviral effect, although belonging to cytokines.

Complement System

Mammalian blood plasma though appearing as a fluid substance like the other body fluids, such as lymph and cerebrospinal fluid, is not a simple medium

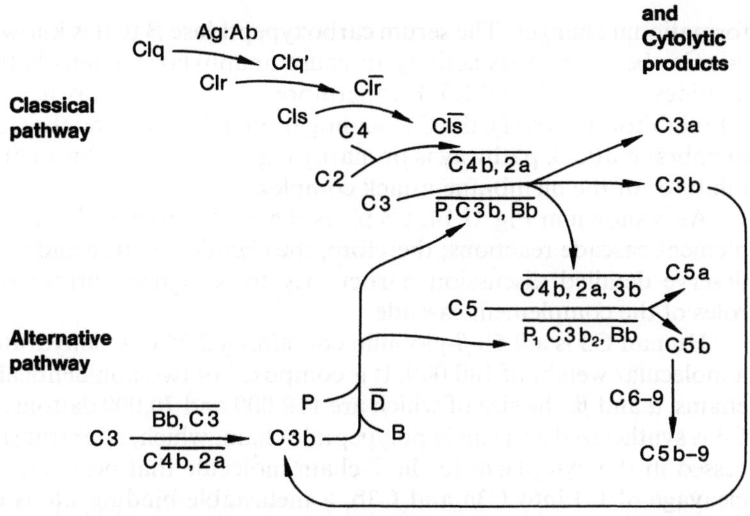

Fig. II.14. The pathways (cascade) of sequential interactions of 20 complement proteins

only for suspending and nourishing cellular components or filling interstitial spaces. A nonspecific humoral system comprising 20 plasma proteins has been found as a factor to "complement" the interaction between specific antibodies and antigens and to result in hemolysis or bacteriolysis in certain test systems which makes it an integral part of the immune system. Each component of these 20 complement proteins exercises its biological function in cooperation with other components sequentially in three principal modes. (1) It forms ligand molecules, such as C3b, that can bind to biological particles through its capability to translocate binding sites and then link these particles to phagocytic cells by binding to their C3b receptors in a way to be called metastable binding. (2) Complement elaborates hormonelike molecules. C3a, C4a, C5a, and Bb elicit responses from a variety of cells that relate to inflammation. For instance, C5a causes directed migration of monocytes, while Bb causes cessation of migration and cell spreading. And, (3) complement generates a unique membranolytic, multimolecular complex which is capable of weakening and dissembling lipid bilayers of cell membranes and of forming transmembrane channels. A most simplified diagrammatic representation of these sequential interactions called the "complement cascade", is depicted in Fig. II.14. Thus, six proteins constitute the classical pathway of the activation of complement: C1q, C1r, C1s, C2, C3, and C4, and six the alternative pathway: C3, B, D, B1H, C3b inactivator (C3b INA), and properdin. Whereas the latter three proteins are considered regulatory proteins, two of these, B1H and C3b INA, are essential for activation of the alternative pathway. Five proteins participate in the membrane-attack pathway: C5, C6, C7, C8, and C9. All these pathways involve protein-protein, protein-carbohydrate, and protein-lipid interactions. They involve protein activation by highly restricted proteolysis or by conformational changes. The serum carboxypeptidase B that is known as anaphylatoxin, because of its activity to cause anaphylaxis, controls the activation peptides of C3, C4, and C5. Furthermore, classical pathway regulators are the C1 inhibitor (C1 INH), the C4-binding protein (C4bp), and the C3b LNA. The membrane-attack pathway is primarily regulated by the S-protein which is the inhibitor of the membrane-attack complex.

As is shown in Fig. II.14, C3 plays the central role in all pathways of complement cascade reactions; therefore, the chemical nature and behavior of C3 deserve detailed discussion particularly to recognize further the biological roles of the complement cascade.

Human C3 is a 9.5S β-globulin containing 2.7% carbohydrate and having a molecular weight of 180,000. It is composed of two nonidentical polypeptide chains, α and β, the size of which are 110,000 and 70,000 daltons, respectively. C3 is synthesized as a single polypeptide chain, which, after translation, is processed in the cytoplasm to the 2-chain molecule that occurs in serum. After cleavage of C3 into C3a and C3b, a metastable-binding site is expressed on C3b through which C3b can firmly attach to a large variety of biological particles. One of the functions of particle-bound C3b is so-called opsonization, i.e., it marks particles for ingestion by phagocytic cells and serves as a ligand to C3b specific cell-surface receptors.

The complement cascade reaction can involve cellular defensive components as follows: polymorphonuclear leukocytes (PMN) and mast cells have cell-surface receptors for both C3a and C3b, and monocytes and macrophages, for C3b. C3b molecules on the surface of target particles promote the adherence phase of phagocytosis. Engagement of particle-bound C3b with C3b receptors on phagocytes causes increased oxidative metabolic activity and the production of superoxide anions, hydrogen peroxide, hydroxyl radicals, and singlet oxygen. Thus, antibody-dependent cellular cytotoxicity (ADCC, p. 118) is markedly enhanced by the presence of C3 fragments on the target cells. All C3 fragments can enhance the ADCC reaction, in which iC3b is the strongest, C3d is strong and C3b moderately active.

On the other hand, the flow of the "cascade" called the alternative pathway can be activated by certain bacteria, viruses, virus-infected cells, or parasites in the total absence of specific antibody (Fig. II.14). Some clinical observations on congenital deficiencies in the complement system provide evidence for the elucidation of the function of each element of complement. Individuals with homozygous deficiencies of C3 or of the C3bINA have been described and found to suffer from life-threatening, recurrent infections. The sophisticated process of alternative pathway activation is initiated from random deposition of C3b on the surface of biological particles and is amplified by a discriminatory interaction of bound C3b with surface structures. At the initial step occurring in the fluid phase, C3 is hydrolyzed by the initial enzyme and changes into C3b-like C3. The latter deposits on the potential target particles via its metastable-binding site. After the binding is realized by the loss of a thioester bond, the particles can activate factor B with the help of factor D.

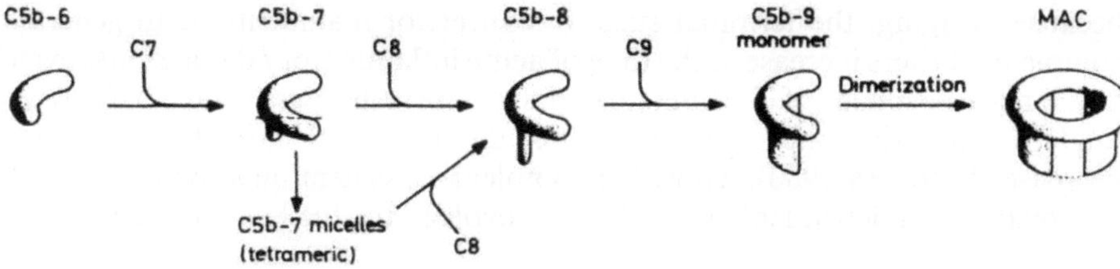

Fig. II.15. Structural aspects of MAC formation. (From Podack et al. 1980)

Schreiber et al. (1979) isolated 11 protein systems of the complement complex and showed that certain Gram-negative bacteria do activate the system in the absence of immunoglobulin and are killed in the process. Lysis of bacteria required, in addition to the 11 proteins, lysozyme. Properdin was not essential, but killing or lytic activity of the system was reduced by 50% when properdin was deleted from the system. All other proteins including C9 were needed for bacterial killing and lysis. It appears, therefore, that maximal aggregation of the membrane-attack complex is necessary to achieve the degree of reorganization of the outer lipid layer of the bacteria that causes killing.

In addition to the alternative pathway leading to cytolysis and bacteriolysis, the elucidation of the molecular dynamics of the membrane-attack pathway has been one of the new developments during the past several years leading to an understanding of it as a constituent of defense mechanisms mediated by nonspecific humoral components (Müller-Eberhard 1980).

The membrane-attack complex (MAC) is exclusively responsible for the membranolytic activity of complement. The structural aspects of MAC formation are diagrammatically represented in Fig. II.15. Not only is the antibacterial activity of complement in vitro entirely a function of MAC, but also enveloped viruses, including RNA tumor viruses, are lyzed by the action of MAC on their lipid envelopes so that their infectivity is destroyed. The significance of MAC formation for human-host defense is indicated by the fact that individuals with a homozygous deficiency of one of the MAC precursor proteins often suffer from repeated bacterial infections, particularly meningococcal meningitis and disseminated gonococcal infections.

In conclusion, the complement system among nonspecific humoral factors plays a defensive role against invading microorganisms by executing three "cascade reactions" involving C3 as its central element. Although this system cooperates with the immune system intimately, it constitutes a more primitive and basic defense mechanism than the latter system (Nishioka 1980). First, the initial defensive factor to react with foreign microorganisms after their breaking through surface-defense mechanisms (Chap. 3) is the complement system, because C1, C3, and B are present adjacent to the mucous epidermis (Ogawa and Yamazaki 1978). Second, C3 can be recognized in the 4-week-old human fetus when its immune mechanisms are still immature. Furthermore, the complement system is still active even after delayed hypersensitivity is diminished because of aging, the terminal stage of cancer, or malnutrition. In general, complement titers increase at the time of acute inflammatory diseases and even at the negative phase of T-cell-mediated immune responses (Chap. 7). Only at the terminal period approaching death or the moribund state, do these titers decrease (Nishioka 1980). Thus, the complement system must be recognized as a relatively independent entity that has evolved for the host defense.

Transferrin and Lactoferrin

Due to the mechanisms described in Chap. 4, the free-iron concentration in the blood plasma of host animals is decreased at the time of fever and this phenomenon itself is effective in limiting the multiplication of the infecting microorganisms (Bullen et al. 1978; Weinberg 1971, 1974; Payne and Finkelstein 1978b).

The subject of this section in Chap. 7 is a consideration of the two nonspecific molecular species, i.e., transferrin of blood plasma and lactoferrin, another protein, both characterized by their strong binding affinity for iron. Therefore, the answer to the question "How does the reduced level of iron play a defensive role against microbial infection?" will be found in this section.

The key point regarding iron in microbial infections resides in the fact that iron is essential for most living things, including microorganisms and mammalian hosts. Within the latter organisms, this metal is locked up in ferritin, hemosiderin, myoglobin, and in the hemoglobin in red cells. The origin of this iron is, of course, absorption of food constituents through the intestinal wall. The detailed mechanisms of this iron absorption is schematically presented in Fig. II.16.

Transportation of the absorbed iron to the respective tissues of the body is the function of a single-chain glycoprotein (transferrin) having a molecular weight of about 76,000, that belongs to the β-globulins. This transferrin is synthesized in the liver and its human serum concentration is less than 9 mg ml^{-1}, (30 μM), although it increases during pregnancy and decreases in liver diseases. In the normal state of the body, transferrin is only partly saturated with Fe and has an exceptionally high association constant of about 10^{36} for the metal. This means that the amount of free iron in equilibrium with this protein is only about 10^{-18} M, which is far too low for normal microbial growth. To obtain Fe from normal tissues, microorganisms must therefore possess iron-chelating agents, called siderophores, with association constants similar to transferrin (Chap. 4).

After absorption of Fe^{2+} from the intestinal lumen through the wall, it is oxidized by means of ferrioxidase to become Fe^{3+} and is bound to transferrin (Fig. II.16). Transferrin functions to transport Fe^{3+} to the bone marrow, where about 200×10^9 erythrocytes containing 6 g of hemoglobin are synthesized per day in the adult. The receptors of erythroblasts receive the Fe^{3+} bound to transferrin with an exceptionally high association constant of about 10^{36} under enzymatic reduction as Fe^{2+}. The remaining part of the iron after

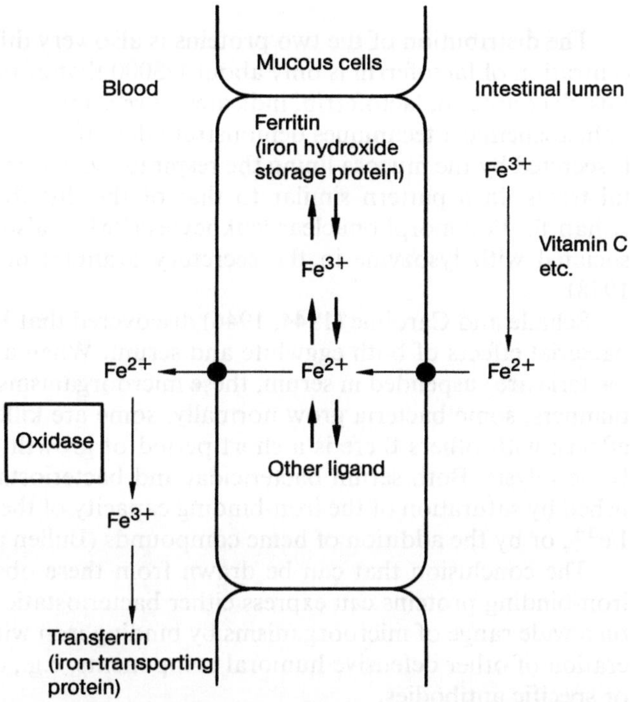

Fig. II.16. Absorption of iron from the intestine (a role of transferrin). (From Abb. 4.23, Biochimie, p 147)

being utilized for the synthesis of heme and other iron-containing proteins and enzymes, is distributed into the liver parenchymal cells, bone marrow reticuloendothelial cells, and spleen and liver cells in the form of ferritin and of hemosiderin for storage. This deposited iron is mobilized again into the bloodstream only at the time of urgent need, such as blood loss or during pregnancy, at a time when the fetus requires iron (Jungermann and Möhler 1980).

Transferrin binds $2Fe^{3+}$ mol^{-1}, but the other ions, like Cu^{2+}, Cr^{3+}, Co^{3+}, Cd^{3+}, Mn^{2+}, Ni^{2+}, Ga^{3+}, In^{3+}, trivalent lanthanides, Sc^{3+}, Pu^{4+}, Vo^{2+}, Al^{3+}, or Zn^{2+} also bind to it. Zn^{2+} transferrin supplies Zn^{2+} to the lymphocyte (Bullen et al. 1978). Circulating plasma normally contains sufficient transferrin to bind 60 μEq of Fe. This capacity of plasma is designated "total iron-binding capacity", abbreviated TIBC. Actually, however, only 30% of this capacity is occupied by the bound iron in a healthy individual. Such a percentage of actual occupation is designated "saturated iron-binding capacity", abbreviated SIBC (Chap. 4).

Lactoferrin (referred to above) is another iron-binding protein that is able to exert a powerful inhibitory effect on microbial growth which behaves like transferrin. The chemical nature of lactoferrin, including its molecular weight, is similar to transferrin, but its amino acid composition and antigenic properties are different.

The distribution of the two proteins is also very different. The serum concentration of lactoferrin is only about 1/3000 that of transferrin. The name of this glycoprotein, lactoferrin, indicates its relationship to human milk. Immunohistochemical techniques demonstrate that after local synthesis, lactoferrin is secreted by the mucosa lining the respiratory, gastrointestinal, and urogenital tracts, in a pattern similar to that of the distribution of secretory IgA (Chap. 8). Polymorphonuclear leukocytes (PMN) also contain lactoferrin associated with lysozyme in the secretory granules of the cell (Bullen et al. 1978).

Schade and Caroline (1944, 1946) discovered that Fe^{3+} abolishes the antibacterial effects of both eggwhite and serum. When a number of pathogenic bacteria are suspended in serum, these microorganisms react in three different manners; some bacteria grow normally, some are killed rapidly (bactericidal effect); with others there is a short period of growth followed by prolonged bacteriolysis. Both serum bactericidal and bacteriostatic effects can be abolished by saturation of the iron-binding capacity of the serum transferrin with Fe^{3+}, or by the addition of heme compounds (Bullen and Rogers 1969).

The conclusion that can be drawn from these observations is that these iron-binding proteins can express either bacteriostatic or bactericidal activity on a wide range of microorganisms by binding iron with or without the cooperation of other defensive humoral components, e.g., complement, lysozyme, or specific antibodies.

For example, the bacteria that can grow in serum must possess a specific mechanism for acquiring iron from partly saturated transferrin. One possible mechanism is a direct interaction between receptors on the bacterial cell and the Fe-transferrin complex. This is analogous to the reaction occurring between transferrin and the reticulocyte. A second possibility is the secretion by the organisms of a low molecular weight iron chelator capable of removing Fe^{3+} from the transferrin molecule, the resulting iron chelate then being taken up by the bacterial cell. This bacterial ability to utilize transferrin-bound Fe^{3+} is included in the category of virulence factors, because the bacteria cannot multiply within the host without the iron.

The mechanism by which humoral antibodies neutralize such an iron-transport system and the restoration of the iron transport by the presence of excess free iron or hemin are diagrammatically illustrated in Fig. II.17.

These situations are explicable as follows:

When the iron-binding capacity of either transferrin or lactoferrin was saturated by the addition of enough iron, their ability to function in these antibacterial systems was lost. In addition, antibody or complement, was also

Fig. II.17. The interaction between *E. coli*, iron-binding protein, and *E. coli* antibody. **1** Situation in normal serum or milk without antibody; **2** inhibition of Fe-transport by antibody; **3** saturation of iron-binding proteins or addition of heme makes Fe available to bacteria despite antibody. [Reproduced from Ciba Symposium No. 42 (new series) "Acute diarrhoea in childhood" Elsevier, Amsterdam 1976]

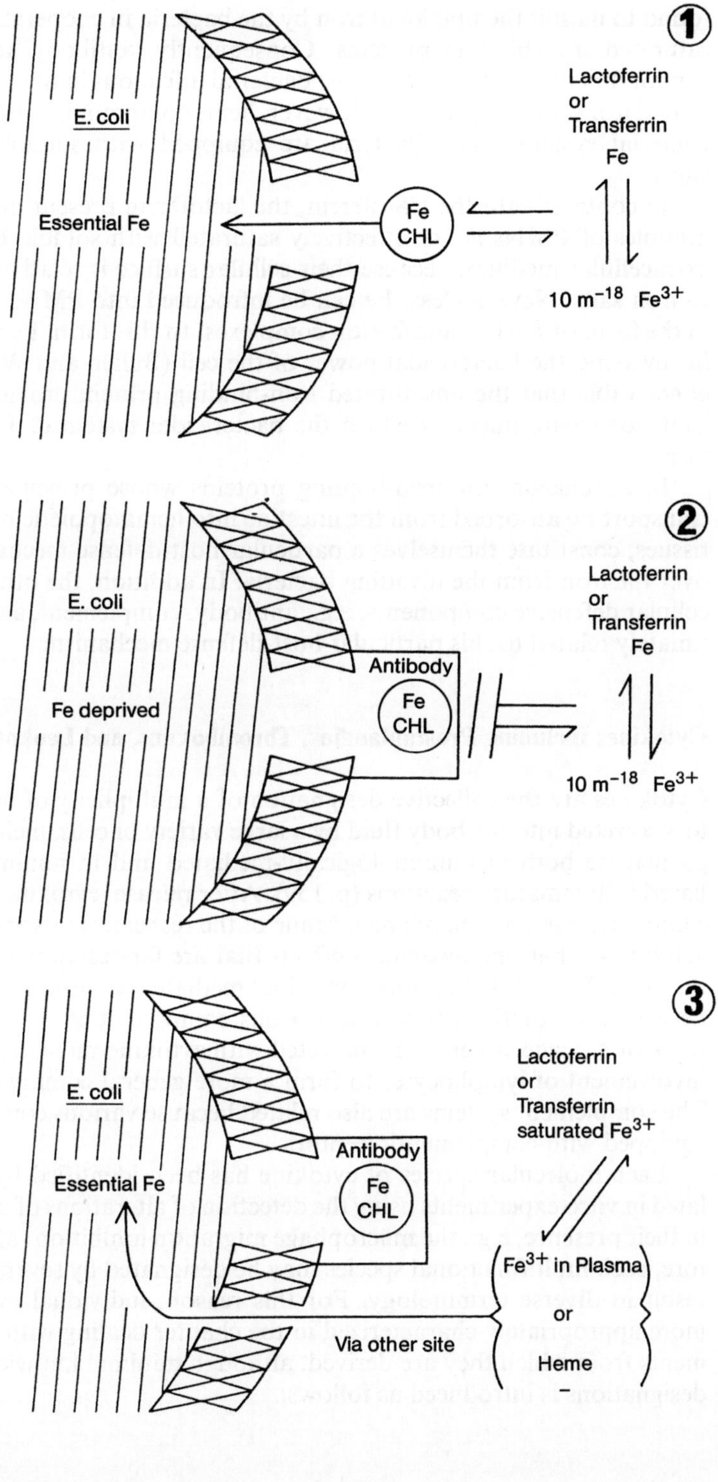
1
E. coli
Essential Fe
Fe
CHL
Lactoferrin
or
Transferrin
Fe
10 m−18 Fe3+
2
E. coli
Fe deprived
Antibody
Fe
CHL
Lactoferrin
or
Transferrin
Fe
10 m−18 Fe3+
3
E. coli
Essential Fe
Antibody
Fe
CHL
Lactoferrin
or
Transferrin
saturated Fe3+
Fe3+ in Plasma
or
Heme
Via other site

found to inhibit the uptake of iron by the bacteria in cooperation with the unsaturated iron-binding proteins. Consequently, antibody and complement cannot function effectively against bacterial infections in the absence of unsaturated iron-binding proteins. However, heme compounds can bypass this antibacterial system, when bacteria are equipped with suitable heme-binding sites.

In contrast with the transferrin, the lactoferrin present inside the specific granules of PMNs is not effectively saturated with soluble Fe added to the extracellular medium, because their cellular surface is relatively impermeable to iron salts. Nevertheless, Fe can be introduced into PMNs by presenting it in the form of ferritin-antiferritin complexes. In this form, Fe is quite effective in reversing the bactericidal power of the cell (Bullen and Wallis 1977). It is conceivable that the unsaturated iron-binding protein provides the essential iron-free environment in which the bactericidal system of PMNs can function.

In conclusion, the iron-binding proteins whose principal function is to transport Fe absorbed from the intestine into hematopoietic organs and other tissues, constitute themselves a particular host-defense mechanism by taking over the iron from the invading bacteria. In addition, the other humoral and cellular defensive components, e.g., antibody, complement, and PMNs are intimately related to this particular host-defense mechanism.

Cytokines Including Prostaglandins, Thromboxans, and Leukotrienes

Cytokines are the collective designation of a multiplicity of hormonelike factors secreted into the body fluid by a large variety of cells, including those that participate both in immunologically-mediated and in nonimmunologically-based inflammatory reactions (p. 159). After release, cytokines bind to the receptors exposed on the cell membrane of the respective target cell and alter its activity, so that one recognizes effects that are far-reaching over distance via the body fluids. Furthermore, cytokines mediate the formation of cascades or networks of reactions involving various component cells. The immune network or cascade system are connected with cytokine-mediated systems by the involvement of lymphocytes to form a more general communication system. The complement systems are also related because various component cells are equipped with complement receptors.

Each molecular species of cytokine has been identified by the use of isolated in vitro experiments using the detection of alterations of cellular activities in their presence, e.g., the macrophage migration inhibition (MIF) test. Therefore, each multifunctional species may be designated by several tests and thus result in diverse terminology. For this reason, individual cytokines will be more appropriately characterized in the chapter dealing with the cellular elements from which they are derived, although minimal knowledge about their designations is introduced as follows:

Cytokines that are produced by lymphocytes have been called lymphokines, and those produced by monocytes, monokines. This does not exclude the possibility that other cells produce cytokines. In addition to this main classification, a list of individual cytokine species with its abbreviated designation deserves elucidation for the convenience of following discussions on the diversely related interacting reaction systems.

Thus, macrophage-activating factor (MAF), migration-inhibition factor (MIF), lymphocyte-derived chemotactic factor (OAF), colony-stimulating factor (CSF), and interleukin 2 (IL2) belong to such lymphokines with the functions their designations indicated. A few lymphokines, e.g., lymphotoxin and tumor migration factor, have been demonstrated to exert direct protective effects against exogenous microorganisms. Oppenheim et al. (1982) identified macrophage (Ia-positive)-recruiting factor (MIRF) with the known function of recruiting Ia-positive macrophages into the peritoneal cavity. With regard to molecular function, it is capable of converting Ia-negative macrophages into Ia-positive ones. The IL2 that stimulates the growth of T-lymphocytes is the same molecular species as the T-cell-growth factor (TCGF). In addition to this function, the same molecule, i.e., IL2, stimulates the secretion of T-cell replacing factor (TRF) which promotes the differentiation of B-cells to produce specific immunoglobulin G (IgG, p. 136). Another distinct lymphokine, killer-helper factor(s) (KHF) is also secreted by IL2-stimulated T-cells and is involved in the generation of cytotoxic T-cells.

Among a number of the monokines, lymphocyte-activating factor (LAF) or interleukin 1 (IL1), have been most extensively studied. In fact, Cohn and Scott (1982) disclosed the secretory nature of the macrophage in addition to the classic expression of macrophage function as professional phagocytes or "scavenger cells". Table II.4 lists the secretory products of mononuclear phagocytes and presents data which revolutionizes the concept of macrophages from professional cells responsible for endocytosis and disposition of the exogenous agent to include secretory functions.

Table II.4. Secretory products of mononuclear phagocytes

Complement components	*Hydrolases-neutral pH optima*
Classical pathway	
C1	Muramidase
C4	Neutral proteinases
C2	Plasminogen activator
C3	Collagenase
$C3_b$ Inactivation	Elastase
B_1H	Angiotensin-convertase
C5	
Alternative pathway	*Hydrolases-acid pH optima*
Factor B	Proteinases
Factor D	Peptidases
Properdin	Lipases
	Phospholipases

Table II.4 (continued)

	Ribonuclease
	Deoxyribonuclease
Coagulation factors	Phosphatases
V	Glycosidases
VII	Hexosaminidases
IX	Sulfatases
X	
Thromboplastin	
Prothrombin	Arginase
Peptides	Lipoprotein lipase
Glutathione	*Arachidonate metabolites*
Enzyme inhibitors	*Cyclooxygenase products*
Plasmin inhibitors	Prostaglandin E_2
α-2 macroglobulin	Prostaglandin $F_{2\alpha}$
	Prostacyclin
	Thromboxane
Binding proteins	*Lipoxygenase products*
Transferrin	
Transcobalamin II	*Leukotriene* B?
Fibronectin	Leukotriene C
	Leukotriene D
Nucleosides	Leukotriene E
Thymidine	
Uridine	HETE's
Adenosine	5
Guanosine	12
	15?
Platelet activating factor	other dihydroxy Cmpds
Reactive metabolites of oxygen	Factors inhibiting growth at:
Superoxide anion	Tumor cells
Hydrogen peroxide	Lymphocytes
Hydroxyl radicals	Myeloid precursors
Singlet oxygen	
Factors promoting growth of:	Other Hormone-like factors
	Endogenous pyrogen 1 & 2
Lymphocytes (B & T cells)	Interferon α, β
Myeloid progenitors	Insulin-like activity
Mononuclear phagocytes	
Erythroid progenitors	
Fibroblasts	
Capillary endothelium	
Neutrophil chemotactic factors	
Factors regulating protein synthesis of:	
Hepatocytes	
Serum amyloid	
Haptoglobin	
Synovial cells	
Collagenase	
Plasminogen activator	

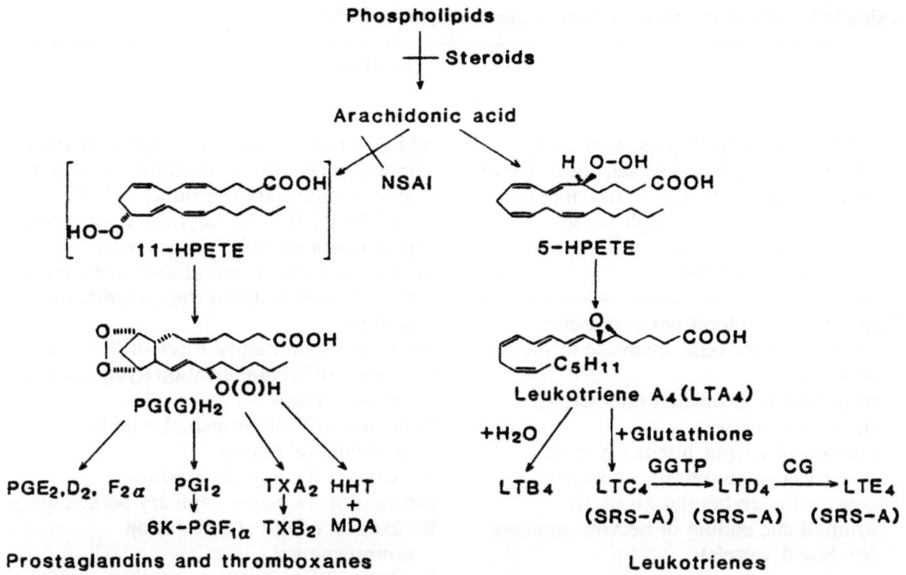

Fig. II.18. Formation of prostaglandins, thromboxanes, and leukotrienes. The intermediate 11-HPETE is bracketed because there is only partial evidence of its occurrence. *NSAI:* nonsteroidal anti-inflammatory compounds. (From Samuelsson 1983)

However, cytokines are secreted by other cell species, also. For example, fibroblast cells produce CSF, interferon (INFβ), MAF, and thymocyte-activating factors (TAF). Polymorphonuclear leukocytes (PMNs) produce IFN. Therefore, a given cytokine can be the product of multiple-cell types.

Furthermore, virtually all mammalian tissues thus far examined release free fatty acids, generally arachidonic acid from esterified lipid sources resulting in the biosynthesis of prostaglandins (PGs) and thromboxane A_2 by the enzymatic action of cyclooxygenase on the one hand, and of leukotrienes by the action of lipoxygenase (Fig. II.18), on the other. A wide variety of hormones, inflammatory or immunological stimuli, calcium ionophores, ultraviolet light, the membrane active component of bee venom, tumor promoting agents, and even mechanical agitation can bring about the release of arachidonic acid; and each of these arachidonate metabolites can be potent mediators of far-ranging biological effects of divergent characteristics (Table II.5). Those of the PG pathway, particularly PGE_2, have been implicated as inflammatory mediators. The biological activities of thromboxane A_2 and prostacyclin as well as a radical with destructive activity deriving from oxygen occupying the central position as additional products of this biosynthetic pathway are to be considered as potential inflammatory mediators. On the other hand, the products of the lipoxygenase pathway, i.e., leukotrienes, also appear to be important inflammatory mediators (Samuelsson 1983). Unlike the PGs, some of which play important roles as biological regulators, the actions of the lipoxy-

Table II.5. Important effects of prostaglandins (Losert 1975)

Typical effect of PGA	Typical effect of PGE
(PGA(c) > PGE ≫ PGF) Inhibition of stomach juice secretion, relaxation of arterial vessels, decrease of blood pressure (PGC > PGA), increase of sodium excretion through kidney	(PGE > PGA ≫ PGF) Inhibition of allergic phenomena (in vitro model), increase of hormonal secretions: anterior pituitary hormones (STH, ACTH, TSH, LH), thyroid, progestrone glucocorticosteroid, progestrone
Typical action of PGF	Inhibition of overstimulation in the nervous system, especially on the sympathetic ending
Contraction of blood vessel muscles, constriction of veins, increase of blood pressure Contraction of bronchial muscles, bronchial stenosis Inhibition of corpus luteus function (progesteron secretion), luteolysis Increase of overstimulation on the sympathetic ending of nervous systems (on blood vessels)	Increase of contractive force of the heart Inhibition of fat mobilization (lympolysis) in the fatty tissue Relaxation of the bronchial muscle, bronchial dilatation Increase of the eyeball pressure Increase of the blood capillary permeability Relaxation of the uterus during nonpregnancy
Mixed actions of PGE-PGF	Inhibition of platelet aggregation ($PGE_1 > PGE_2$)
Increase of platelet aggregation (not PGE_1) Contraction of intestinal muscles Contraction of uterus muscles during pregnancy	

genase products, leukotrienes, appear to be exclusively of a pathological nature (Kuehl and Egan 1980). Among them, leukotriene B_4 causes adhesion and chemotactic movement of leukocytes and stimulates aggregation, enzyme release, and generation of superoxide in polymorphonuclear cells (PMNs).

In conclusion, host-defense mechanisms involve a more general biological control system than the immune network and the nonspecific reaction cascade. This system involves arachidonic acid as the precursor molecule. This example indicates that any ubiquitous cell-membrane active component can play a defensive role by being involved in pathogenic changes, e.g., inflammation.

Interferon (IFN)

Among the numerous humoral components with defensive functions, interferon (IFN) deserves to be characterized in some detail because of its significance as a defense mechanism present in virtually all vertebrate species. It is included in the "disease formula" (p. 3) of Part 1. IFN is no longer a series of loosely defined factors, but rather a group of full-fledged glycoprotein hormones, each one of which, within a given animal species, has a somewhat different structure.

Table II.6. List of main interferon inducers

1. Double-strand RNA type	Synthesized or natural double-strand RNA; replicative intermediate (RI) of virus; replicative form of virus; polyanionic macromolecules, such as pyran copolymer
2. Microbial product	Lipopolysaccharides of *E. coli*, etc
3. Low molecular form	Tilorone and its derivatives; antibiotics, e.g., cycloheximide
4. Immune form	Antigens responsible for the sensitization of donor lymphocytes after separation; mitogens responsible for the blastogenesis of lymphocytes, e.g., lectin

IFN is secreted by a variety of cells in response to the triggering action of a variety of exogenous compounds called *interferon (IFN) inducers*. A list of the main IFN inducers is given in Table II.6.

Originally, the demonstration of interferon was derived from the recognition of its antiviral activity. Nagano and Kojima (1954) reported a "virus inhibitory factor" (IF) using rabbit skin inoculated with vaccinia virus and Isaacs and Lindenman (1957) described "interferon" using chick embryos inoculated with influenza virus. They found that intracellular multiplication of these viruses was inhibited after treatment of the host tissues with the active or inactivated viruses. IFN synthesis began 2–3 h after the stimulation of animal cells with the inducer, and the inhibition of viral multiplication peaks 6 h after contact of virus-infected cells with IFN. Thus, IFN is effective before specific immune reactions are detected and can explain the phenomenon of "interference" in which a viral infection of the animal cell with the same or different viruses prevents their superinfection by a second virus. The quantitative assay method of IFN is based upon the addition of serially diluted test samples to the virus (usually VSV), vesicular stomatitis virus-infected cell cultures and is expressed as units in comparison with an international standard. This method can detect a 10^{-11} g ml^{-1} concentration of IFN. One IFN molecule is effective for more than one virus-infected cell. The specific activity of IFN is about 10^9 biological units mg^{-1} protein and it acts on cells in vitro at concentrations in the range of 3×10^{-14} M or in other words, the equivalent to 1.0 pg ml^{-1}.

IFN inhibits the intracellular multiplication of both DNA- and RNA-containing viruses and the effect is mediated by the cell. However, IFN itself functions as an antigen, i.e., it is equipped with epitopes and its antiviral effect can be neutralized outside the cell with a specific serum. Furthermore, IFN has been shown to act not merely as an antiviral humoral component, but also as an important regulatory entity in the multi-tiered host-defense mechanisms. These effects of IFN on host eukaryotic cells detectable in vitro and in host-animal tissues in vivo involve a variety of nonspecific and specific cellular components of the host-defense mechanisms.

To summarize concisely without discussing the individual target cell species, IFN inhibits cell multiplication in vitro and in vivo; inhibits or enhances antibody formation in vitro and in vivo; enhances lymphocyte cytotoxicity in vitro and in vivo; enhances macrophage activity in vitro and in vivo; enhances expression of surface antigens in vitro and in vivo; enhances production or release of specific substances, e.g., cyclic GMP and histamine in vitro and inhibits or enhances delayed-type hypersensitivity in mice depending on the time of its administration. The detailed effects involving the target cellular component and in vivo phenomena are discussed throughout Part II.

Another characteristic distinguishing IFN from specific antibody is that it acts, i.e., its effect is host-species specific, only on cells of the same species from which it was produced to inhibit virus multiplication. On the other hand, it is nonspecific in the context that the multiplication of all DNA and RNA viruses is inhibited in cells when the IFN-secreting and virus-infected cells are derived from the same species.

At the early stage of investigation of IFN, IFN was classified into type 1 which is secreted by many types of cells, including lymphoid cells after induction with viruses or double-stranded RNA; and type 2 which is secreted by lymphoid cells by the induction with some antigens and some mitogens. As the biochemistry, i.e., isolation, purification, and structural determinations of IFN progressed, type 1 was classified into type α (IFN α) which is secreted by blood leukocytes by induction with viruses or RNA and type β (IFN β) which is secreted by fibroblast cells by induction with synthetic double-stranded RNA and viruses. Both are stable at pH 2.0 and with molecular weights (mol.wt.) of 20,000–40,000. Correspondingly, type 2 was called immune interferon because of its secretion by antigen- or mitogen-induced lymphocytes and its designated type γ (IFN γ). This is acid-unstable and with the mol.wt. of 40,000–70,000. All these interferons are antigenically distinct.

It can be concluded that the IFN system constitutes itself an aspect of the host-defense mechanisms as a more primitive fundamental nonspecific humoral component than the components constituting the specific immune system. In addition to its unique defensive activities against viral infections, the IFN system plays important roles to coordinate various activities of specific as well as nonspecific cellular components of the immune network.

Chapter 7 Specific Humoral Components: Immunoglobulins

Chapter 1 delineated immunoglobulins as protein molecules diversified to function for specific recognition of a tremendous number (10^6) of unique epitopes present in nature. The same molecules, however, exert a variety of biological functions called efferent arms of the immune system versus afferent arms: recognition. To summarize, the characteristic of the specific humoral components resides in the dual functions of immunoglobulins that are classified into IgM, IgG, IgA, IgE, and IgD, depending on their biological effects in the host organisms and molecular structure.

These effects were first recognized as "antitoxin" because of the neutralization of bacterial toxins (Behring and Kitasato 1890). Next, Pfeiffer and Issaeff (1984) found "bacteriolysis" of *Vibrio cholerae* in the peritoneal cavity of guinea pigs immunized with the same microorganisms. The proposal of "agglutinins" (1896) and "precipitins" (1897), and "opsonins" (Wright and Douglass 1903) were successively described until Zinsser (1921) proposed a unitarian theory of antibodies and included these serum factors in a single concept, antibody. This term was derived from the host-defensive role of this factor.

On the other hand, however, Pirquet (1906, 1910) established the concept "allergy" as the adverse effects of immunization to the host animals in such phenomena as "Koch's phenomenon" and anaphylaxis (Richet 1902; Arthus and Breton 1903). Of these adverse effects of immunization, "the immediate type allergic reactions" was found to be the effect of "skin-sensitizing antibody or reagin" (Prausnitz and Küstner 1921).

In this context, the biological functions of the specific humoral components, which can play a defensive role as well as showing adverse effects for the host animals, will be discussed.

Structures of Immunoglobulins

After immunizing with a specific antigen, the proportion of serum globulins occupied by any specific antibody is increased. Since Tiselius (1937) dissociated serum proteins into several fractions using electrophoresis and found that all of the specific activities associated with antibodies fall into the fraction of γ-globulin, the antibodies acquired the chemical designation, irrespective of specificity, as "immunoglobulins".

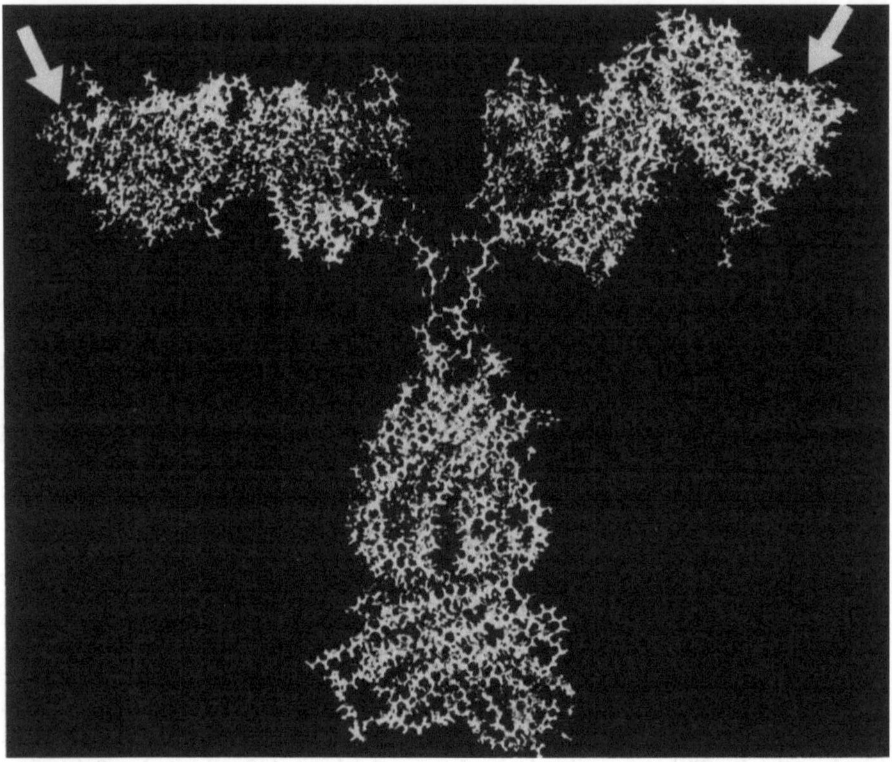

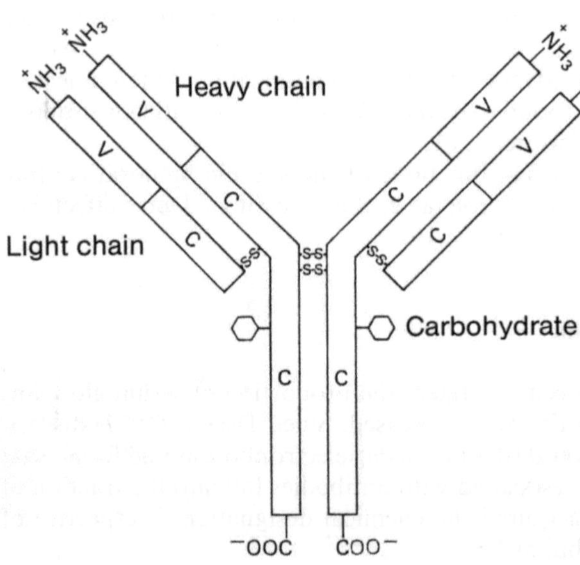

Fig. II.19. The structure of antibody (IgG). *Above:* molecular model reconstructed on the basis of Roentgen-structural analysis. *Arrows* indicate antigen-combining sites (From Marx 1975). *Left:* scheme of the antibody structure; *v*, variable region, *c*, constant region. The inner disulfide bridges of polypeptide chains are not indicated

The fine structure of each class of immunoglobulin has been disclosed using enzymatic treatment, amino acid sequencing, X-ray crystallography, and electron microscopy.

Recent advances in biological techniques, such as cell fusion (hybridoma) and genetic engineering have enabled us to prepare enough immunoglobulins or their molecular fragments with a single specificity for detailed study.

To summarize the results of these investigations, immunoglobulin molecules of all classes were found to share common basic structures consisting of light and heavy chains. Along the amino-terminal half of two heavy chains, two light chains are located parallel with the former, being bridged by disulfide bonds. Treatment of immunoglobulins with pepsin under pH 4.0, produced a fragment containing combining site, the responsible amino acid sequences for antigen-recognition, designated the $F(ab')_2$ domain. On the other hand, another portion of the basic structure, which is responsible for exerting biological effect and discriminating classes and designated Fc, is produced by the treatment with papain under pH 7.0. Then two pairs of light-chain amino-(NH_2-)-terminal halves of heavy chains are partner products of Fc and called Fab. Before enzymatic splitting, the amino-terminal half, the carboxyl-half of two heavy chains are connected to form two whole heavy chains, respectively. The Fc part consists of two heavy-chain carboxyl-terminal parts which are also bound to each other by bridging with disulfide bonds. Therefore, the molecule

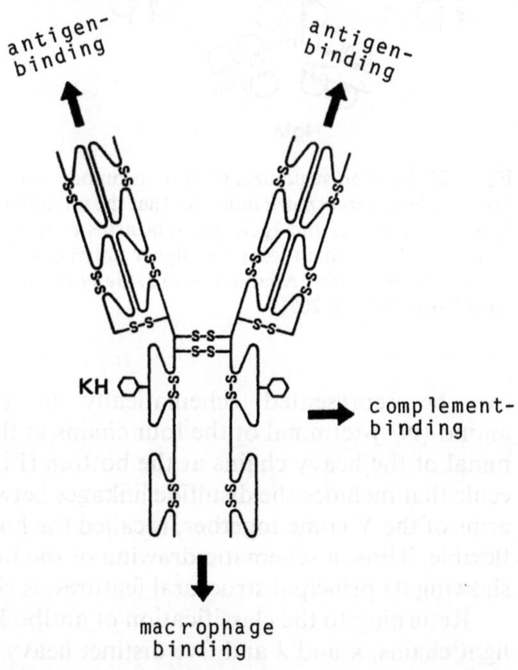

Fig. II.20. A basic structure of an antibody with its sites responsible for dual functions

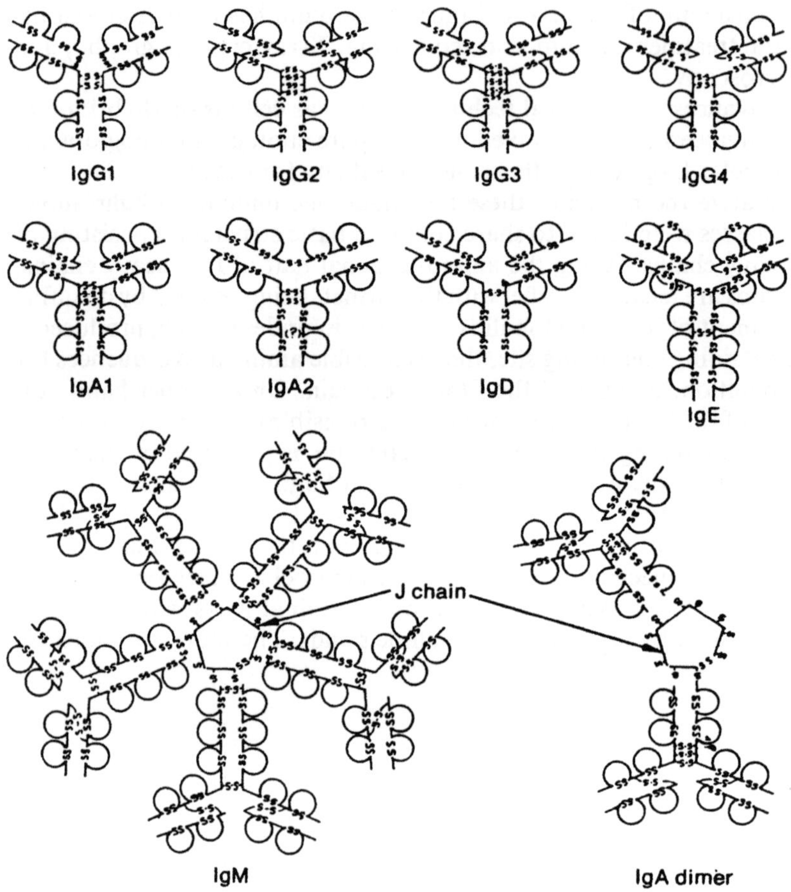

Fig. II.21. Subunit structures of various human immunoglobulins. -S-S- indicates a disulfide bridge. The *question mark* indicates that the interchain disulfide bridge structure is unknown. IgG_1, IgG_2, IgG_3, and IgG_4 are subclasses of IgG. IgA_1 is the monomeric subclass of the IgA class. The J-chain joins the higher polymeric forms of IgA as well as the five subunits of the IgM molecule. [Adapted from J. Gally in The Antigens, M. Sela (ed), Academic Press, New York, 1973, p 209]

can be represented schematically in the form of a Y, with the amino-(N^-)-terminal of the four chains at the top and the carboxyl-(C^-)-terminal of the heavy chains at the bottom (Fig. II.19). The portion of the molecule that includes the disulfide linkages between heavy chains where the three arms of the Y come together, is called the hinge region. The arms of the Y are flexible. Thus, a schematic drawing of the human immunoglobulin molecule, showing its principal structural features, is shown in Figs. II.19 and II.20.

Returning to the classification of antibodies, again, there exist two distinct light chains, κ and λ and five distinct heavy chains, μ, γ, α, ε, and δ. One can prepare antiserums for each type of chain by immunizing another species of animal (heterologous immunization). Using these antiserums as reagents, one can classify immunoglobulins into each class and subclass of heavy chain and each type and subtype of light chain. This is possible because these different chains are characterized by antigenic determinants unique to respective isotypes, but common among all individuals of the same species.

The combinations of light and heavy chains in the assembly of these proteinaceous fragments to form the basic structure of immunoglobulins, determine the antibody classes, IgM, IgG, IgA, IgE, or IgD, corresponding to the class or subclass of heavy chains that the complete immunoglobulin molecules take up at the time of their biosynthesis within antibody-forming cells.

On the other hand, each class of human immunoglobulin, for example, possesses a different grade of polymerization of the basic structure as follows: IgG, IgD, and IgE are monomers; IgA can exist as monomer, dimer, or trimer, being connected at the C-terminal through the J-chain in the latter two polymers. Serum IgAs are mostly monomer, but at the secretion into the outer world of its polymeric forms, by passing through the epithelium, another molecule called a secretory component (sc), a product of epithelial cells, is added. IgM are pentamers equipped with ten antigen-binding sites possessing a J-chain at the center to join five Fc regions of respective basic immunoglobulin structure together. With each class of immunoglobulin, further classifications into several subtypes have been made, based upon the positions and number of the disulfide bonds between heavy chains and in the number of basic subunits in the molecule. These structural differences can be reflected in the biological function of the immunoglobulin molecules. For example, human subclass IgG_4 does not bind with complement to differ from IgG_1, IgG_2, and IgG_3 subclasses. Subclass IgA_1 is the monomeric subclass of the IgA class. Diagrammatic representations of these immunoglobulin molecules are shown in Fig. II.21.

Biological Characteristic and Time Course of Each Class of Antibody Production

The five classes of antibody are very different in their concentrations within normal human serum, in normal serum half-life, and in relative amount produced in the primary and secondary immune responses.

For IgM, the mean adult serum level and half-life are 1.0 mg ml^{-1} and 5 days, respectively. IgM rapidly increases in its blood concentration, i.e., after 2–3 days of antigenic stimulation; reaching its peak after 1 week, and then decreases to be replaced by the slower increasing IgG. The patterns of concentration changes of both immunoglobulins is characteristic for the particular antigen, a representative antibody response is outlined in Fig. II.22.

IgM is about 1000 times more effective on a molar basis in agglutinating cells by cross-linking than are monomeric divalent antibodies. Its opsonizing

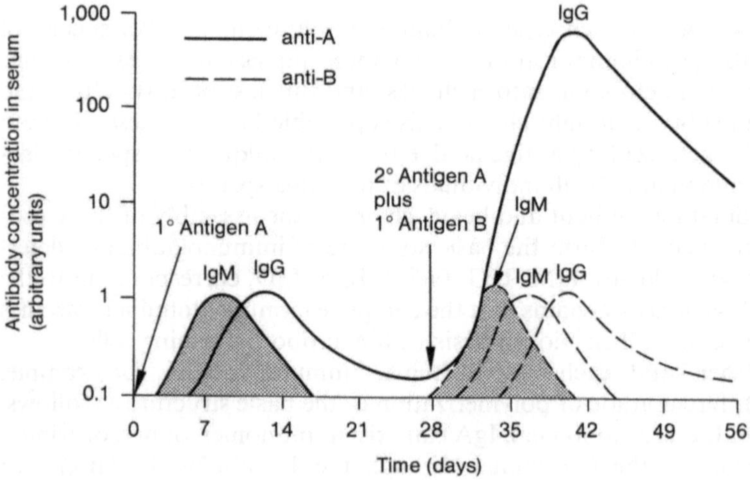

Fig. II.22. Kinetics of IgM and IgG appearance in the serum following primary and secondary immunizations. The secondary response to antigen A demonstrates that the IgM response is slightly more rapid and intense than the primary IgG response. However, most of the secondary response is of the IgG class

effect, i.e., stimulating effect for phagocytosis through binding to an antigenic target cell, as well as its destructive efficacy for these cells by complement fixation are remarkable. Because of its large size, it enters the interstitial fluids slowly. It does not cross the placenta to enter the fetal circulation. IgM also is displayed as a monomer on the surfaces of B-cells to be regarded as a receptor immunoglobulin (cytophilic antibody).

IgG, the mean adult serum level and half-life of which are 12 mg ml^{-1} and 25 days, respectively, increases its blood concentration after high-dose antigenic stimulation. IgG is the most prevalent antibody in the blood, and is a major antibody in tissue spaces. IgG triggers complement fixation except for the IgG_4 subclass and shows opsonizing effects for antigenic particles by coating them with antibody. IgG is the only class of antibody that can cross the placenta to provide immunity for a developing fetus.

IgG production after antigenic immunization continues for long periods peaking at about 2 weeks in its blood concentration. A second and subsequent immunization with the same antigen brings about a more rapid and higher increase of IgG concentration with less production of IgM than the first time. In this way, one can prepare hyperimmune serum for a given antigen by repeating immunization at suitable intervals. If immunization is not repeated, IgG antibodies produced show an increase in their average affinity for the antigen. This phenomenon is called affinity maturation.

IgA, the mean adult serum level and serum half-life of which is 1.8 mg ml^{-1} and 6 days, respectively, is unique in its preferential release into external secretions, i.e., colostrum, milk, tears, saliva, sweat, vaginal and intestinal secretions, including bile. Therefore, nonparenteral vaccination with attenuated-live vaccines and microbial infection induce, mainly, the production of IgA, although later than IgM. As already described, some epithelial cells produce a polypeptide called the secretory component (sc), which complexes to the Fc region of IgA antibodies and specifically mediates their transport from interstitial spaces to the epithelial surface.

IgD, the mean adult serum level and serum half-life of which is as low as 0.03 mg ml^{-1} and 2.8 days, respectively, is present as a cell-surface receptor on a majority of circulating B-cells, but its functions are still unknown. IgE, having the least mean adult level of 0.0003 mg ml^{-1} and half-life of 2 days normally, binds tightly via its Fc regions to mast cells in connective tissue and blood basophils. When divalent antigens bridge the specific antigen-binding sites of cell-bound IgE, the mast cells degranulate to release histamine and heparin. By the action of these substances, local vessels are dilated and smooth muscles are contracted to cause acute allergic (type I) syndromes, like asthma, urticaria, and anaphylaxis. However, it was recently elucidated that IgE induces the release of the major basic protein (MBP) contained within the granules of eosinophilic leukocytes and mast cells (refer to Chap. 8) onto the body surface of metazoan parasites (Capron et al. 1980). Because MBP destroys the metazoan tissue, the sole defensive role of IgE resides in protection against metazoan parasitism.

In conclusion, the five classes of humoral globulin are unique in that each molecule is composed of two different portions capable of performing dual functions. One is characterized by its tremendous diversities of the specific epitope or antigen "combining sites" on which Chap. 1 was devoted for description. Another is characterized by the five different biological functions it exerts to realize each different "immune reaction".

Chapter 8 Genesis and Distribution of Cellular Components with Defensive Functions

As the studies of immunology progressed the difficulty in discriminating cellular components with defensive values from nondefensive ones increased. Recently, for example, the defensive role of red cells, i.e., erythrocytes that occupy more than 99% of the total blood volume and thought to be solely important for their oxygen transport, was pointed out (Siegel and Gleichner 1983). Erythrocytes adhere to trypanosomas, a protozoan parasite, in some circumstances also to antigen-antibody-complement complexes to augment the ingestion of the antigen by leukocytes. When fresh human blood containing both erythrocytes and leukocytes was mixed with *Treponema pallidum* and pneumococci and antibody against these bacteria added, 95% of the bacteria disappeared from the blood plasma after 1 h incubation, while practically all of the bacteria remained in the blood plasma when erythrocytes were absent despite the presence of a normal number of leukocytes. Furthermore, the presence of C3b receptors and the enzyme peroxidase on the erythrocyte surfaces might enable the erythrocyte to damage the antigen it adheres to.

Thus, the defensive function of erythrocytes via their cooperation with antibody, complement, and leukocytes, e.g., T-lymphocytes, called the erythrocyte immune adherence phenomenon, should attract more interest for future study. This may supplement the already established defensive role of the leukocyte immune adherence phenomenon that will be discussed later.

However, such a defensive function of erythrocytes was recognized only recently, in contrast to those of leukocytes which have been known for a long time. The ontogenic process of differentiation of blood cells which arise from the common hematopoietic stem cell in the fetal yolk at the embryonic stage is shown in Fig. II.23. At a later stage of gestation, the stem cells migrate to the fetal liver, and the immature mononuclear phagocytes develop. Although divergent, depending on the animal species, hematopoiesis in the liver does not cease until term of the second week after birth. Soon after hematopoiesis begins in the fetal liver, monocytes appear in the circulation; only after that does hematopoiesis start in the bone marrow. Evidence for the differentiation of bone marrow stem cells to form mononuclear phagocyte colonies and granulocytes has been obtained by the in vitro culture of bone marrow cells in the presence of colony-stimulating factor (CSF, Chap. 6; van Furth et al. 1982).

Among the eight diversifications shown in this scheme, the name "leukocytes" encompasses heterologous populations of nucleus-bearing cells, i.e., granulocytes (synonym of polymorphonuclear cells: PMN or neutrophiles),

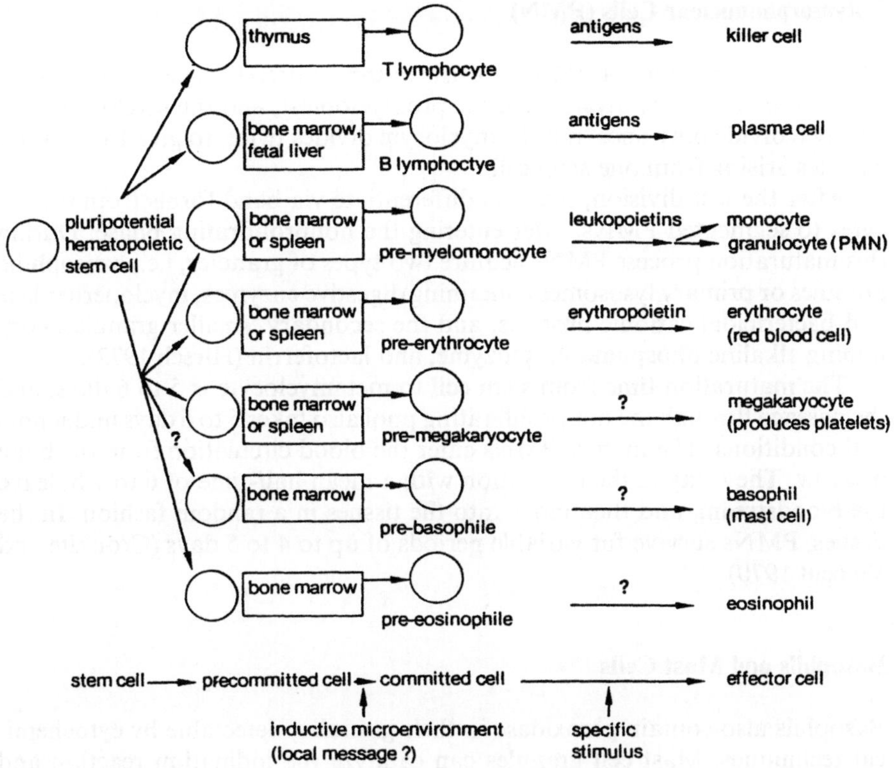

Fig. II.23. Differentiation pathways in the hematolymphoid system. *Branch points* between the pluripotential hematopoietic stem cell and precommitted cells indicate probable precursor cells

the monocyte-macrophage series, T-lymphocytes of various subsets, the B-lymphocyte plasma-cell series, the basophil mast-cell series, and eosinophils. They circulate in the bloodstream after maturation at relatively constant ratios of concentration. But this normal steady state is easily unbalanced as occurs characteristically in response to external stimuli, including those of infecting microorganisms. These patterns of variation are useful for the diagnosis of diseases.

In other words, the specific interaction of the total host leukocyte population with the infecting microorganisms forms uniquely diverse patterns composed of different leukocyte types.

The type of leukocyte involved in these dynamic mechanisms is characterized as follows before their participation in host defenses is discussed, e.g., inflammation.

Polymorphonuclear Cells (PMN)

The proliferating pool of PMNs consists of three different types of cells which arise successively; the myeloblast, the promyeloblast, and the myelocyte. The first two divide only once, but the myelocyte divides twice, to give 16 metamyelocytes arising from one stem cell.

After the last division, the cells differentiate via band forms from myelocytes to segmented PMNs, after entering the nonprolifirating phase. During this maturation process PMNs acquire two types of granules, i.e., azurophilic granules or primary lysosomes containing digestive enzymes, myeloperoxidase and bactericidal cationic proteins, and the secondary, smaller granules, containing alkaline phosphatase, lysozyme, and lactoferrin (Hirsch 1972).

The maturation time from stem cell to metamyelocyte is 5 to 6 days, and the passage through the nonproliferating pool also takes 5 to 6 days under normal conditions. The mature PMNs enter the blood circulation from the bone marrow. They stay in the circulation with a mean half-time of 6 to 7 h, leave the bloodstream, and then move into the tissues in a random fashion. In the tissues, PMNs survive for variable periods of up to 4 to 5 days (Cronkite and Vincent 1970).

Basophils and Mast Cells

Basophils also contain peroxidase in their granules, detectable by cytochemical techniques. Mast cell granules can catalyze the iodination reaction and have bactericidal activity when combined with H_2O_2 and iodide, although these activities are considerably potentiated by the binding of eosinophil peroxidase (EPO, see next section) to the mast-cell granule surface (Klebanoff 1980).

Eosinophils (Acidophils)

Eosinophils stain heavily for peroxidase, which is located within their cytoplasmic granules. The eosinophil peroxidase (EPO) differs in its structure and properties from the neutrophil peroxidase (myeloperoxidase, MPO). Because these two enzymes are coded by different genes, patients with hereditary MPO deficiency lack peroxidase in neutrophils and monocytes, but contain normal amounts in eosinophils (Grignachi et al. 1963). Purified guinea-pig EPO and MPO differ immunologically and in their physicochemical characteristics, subunit structure, and heme prosthetic groups (Desser et al. 1972). EPO and MPO also differ in the rate at which particular electron donors are oxidized and in their response to inhibitors (Migler et al. 1978).

It should be noted with regard to eosinophil-mast cell cooperation that eosinophils are unique because their EPO is toxic to *Schistosoma mansoni,* a parasitic multicellular organism, when combined with H_2O_2 and halide. At a relatively low concentration of EPO, mast cells release histamine without releasing lactic acid dehydrogenase (LDH) by the combined effect of EPO, H_2O_2, and halide. When the EPO concentration was raised, lysis of mast cells occurred with release of both histamine and LDH.

EPO is a strongly basic protein which binds avidly to the surface of mast-cell granules to form a complex with retention of peroxidatic activity. This complex has significantly greater iodinating and bactericidal activity than the free enzyme. For the stimulation of mast-cell secretion, the combination of the mast-cell granule EPO with H_2O_2 and iodide is more effective than EPO only (Klebanoff 1980).

Eosinophil granules also contain a strongly cationic protein (major basic protein, MBP) in high concentration. MBP is toxic to the schistosomula of *Schistosoma mansoni*, and the newborn larvae of *Trichinella spiralis* and is related to the toxic activity of eosinophils in situ.

Mononuclear Cells

Stem cells committed to differentiate into monocytes, pass through only two successive proliferating stages, namely, the monoblast and the promonocyte. Monocytes arise from dividing promonocytes and soon enter the circulation (van Furth 1970). They do not pass through a maturation phase in the bone marrow after their last division, nor is there a large pool of mature monocytes in the bone marrow. The monocytes remain in the circulation longer than PMNs. Their half-life in the bloodstream is variable depending upon the animal species, but with an average of 1 to 3 days. After leaving the blood circulation, they become macrophages within the tissues. Both monocytes and macrophages are nondividing cells under normal conditions (van Furth 1970) and contain primary and secondary lysosomes with acid hydrolases; in addition, the primary lysosomes of the promonocyte and monocyte contain peroxidase (MPO, Nichols and Bainton 1974). Furthermore, Cohn and Scott (1982) showed that the macrophages are secreting cells as noted in Chap. 6 (p. 125). The macrophages in the tissues differ in many respects from the PMNs, e.g., the life span of macrophages is much longer (of the order of months). They differentiate to adapt to the peculiar local circumstances of residence by maturation and modulation, and can form new, secondary lysosomes (Cohn 1970).

Macrophages are called “scavenger cells”, because they move not only to the site of infection with external microorganisms, but also to sites in the body where dead cells and their debris are produced, in order to phagocytose and digest these waste materials. The mechanism by which macrophages discriminate these cells and debris for phagocytosis from those which are not engulfed is unknown.

Lymphocytes and Their Functions

Ontogenically, the sites of B-cell differentiation are multifocal, but fetal liver is the major site of production until take over of the activity by the bone marrow (Owen 1980). At least a proportion of pre-B-cells are in the active cycle to mature during proliferation, and the immunoglobulin biosynthetic differentiations seem to start at this stage, because immunofluorescence assays demonstrate the expression of immunoglobulin-μ chains in the cytoplasma of these proliferating B-cell precursors. They are heterogeneous in size, and it is likely that pre-B-cells synthesize μ-chains only and that these molecules are not inserted into the cell membrane. After becoming B-cells, they are not in an active cell cycle, i.e., many B-cells are short-lived and a constant rate of new cell production is required to ensure that the complete repertoire of various immunological specificities is always available to the individual.

Returning back to the ontogeny, a number of nucleotide base sequences that are exons being separated by introns (Chap. 1), are located along the precursor B-cell chromosomes. The exons that are going to encode V regions of the heavy (H) and the light (L) chains of immunoglobulins are called v genes. Located next to the sequential series of v genes are the D fragment and J fragment on the chromosome DNA-chain-encoding H chain; and the J fragment only on that encoding L chain. During the differentiation of a progenitor cell into a pre-B cell, randomly selected exons of v genes, those of D, J fragments, or J fragment interconnect lineally by deleting introns to form rearranged nucleotide base sequences encoding the V-J and V-D-J joinings. Thus, during the course of new cell production accompanied by the maturation of precursor B lymphocyte to the B-cell, the latter is determined to produce the specific immunoglobulins that can recognize a particular epitope or antigenic determinant with its combining sites locating in the V region.

In parallel with such a nucleotide base sequence rearrangement producing combining site diversities in the V region, five varieties of the nucleotide base sequences, called c genes, i.e., $C\mu$, $C\delta$, $C\gamma$, $C\alpha$, and $C\varepsilon$ are sequentially located on the precursor B-cell chromosome. The V-J and V-D-J joining occurs during the differentiation of progenitor B-cell into a pre-B-cell and always precedes the synthesis of free μ-chain polypeptides in pre-B-cells. After the next stage of maturation of pre-B-cells into the B-lymphocytes, the second form of gene transposition, called class switch, occurs after the pre-B-cells have differentiated into B-lymphocytes. At this stage, the introns separating $C\mu$, $C\delta$, $C\gamma$, $C\alpha$, and $C\varepsilon$ are deleted and the v genes or V-D fragments join to one of five varieties of c genes. In this process the diversities of biological functions are differentiated for the production of IgD, IgM, IgG, IgA, or IgE. Therefore, after the pre-B-cells have differentiated into B-lymphocytes, the latter bear complete IgM molecules on their surfaces.

Thus, newly formed B-cells express Fc receptors to bind with the Ia antigens exposed on T-cells; at least a proportion of B-cells express C3 receptors to interact with complement. In addition, another antigen designated as

Ly 4 is expressed on B-cells. Ly 4 is available for the classification of B-cells into more diverse subsets by serological techniques.

When these mature B-cells belonging to a clone bind to certain antigenic epitopes via their Ig receptors on their cell surfaces, they begin to proliferate and differentiate in two directions; one into expanded clones of plasma cells expressing the unique serological marker Pc-1 antigen on the surface and secreting specific antibodies, the other into clones of memory cells. In contrast to the lifetime of a plasma cell which is measured in days, memory cells produced in a primary response can remain in the lymphocyte population for decades.

That such B-cell maturation is directed by certain microenvironmental influences has been definitively demonstrated in birds. Removal of the bursa of Fabricius, an organ connected with the cloaca, in neonatal birds, does not permit B-cell maturation, as shown by the low concentration of serum antibodies after immunization with antigen. No organ(s) equivalent to the bursa has yet been identified in mammals and many opinions are put forth, e.g., that there exist many such organs scattered along the intestine in mammals.

In contrast, the progenitor lymphocytes which are predetermined to enter a distinct staging house, the thymus (Daires 1980), after arising from such sources as the bone marrow or fetal liver, undergo maturation to differentiate into quite a distinct lineage from B-cells, i.e., T-cells (Miller 1962). Just after the migration of the thymus rudiment in mammalian development, these pre-T-cells enter the epithelium of the thymus being attracted by diffusible substances produced by the gland; in this location they are induced to express some cell-surface molecules characteristic of thymus cells (thymocytes), and other molecules that are common to all T-cells. In mice some differentiation molecules, characterizing the defined maturational stage, have been detected serologically and are designated TL, Thy-1, and Ly-1, Ly-2, and Ly-3. The TL antigen is expressed on almost all thymocytes, but is absent on most, if not all, peripheral T-cells. Thy-1 is a membrane glycoprotein found on thymocytes, peripheral T-cells, and effector T-cells, being absent on B-cells or other hematolymphoid cells. Ly-1, Ly-2, and Ly-3 antigens all are expressed on almost all thymocytes and on nearly 50% of peripheral T-cells; these cells are designated Ly-1, 2, 3. The other 50% of peripheral T-cells carry either the Ly-1 antigen alone, or the Ly-2 and Ly-3 antigens together (designated Ly-1 and Ly-2,3, respectively, Table II.7). No other cells express these antigens. After such differentiation, the small thymocytes leave the thymus and are replaced by new cells every 3 to 5 days, so that in the mouse, for example, there is a daily turnover of 50 to 100 million cells in the thymus.

Mature T-cells thus differentiated migrate out from thymus, with a small proportion remaining there as medullary thymocytes. These migrating T-cells enter the predetermined T-cell areas of lymph nodes and the spleen. These T-cells possess surface receptors for antigens, although it is not known whether they can respond to antigenic stimulation by differentiation into effector cells. After being triggered by specific antigen, these T-cells mature to long-lived, cir-

Table II.7. Classes of effector T-cells

Category	Class of cells	Function	Ly phenotype in mice
T_H	T_H	Stimulation of B-cell differetiation	Ly-1
	T_A	Stimulation of Tc precursor Cell differentiation	Ly-1
	T_{DTH}	Transfer of delayed hypersensitivity	Ly-1
$T_{C.S.}$	T_C	Lysis of specific antigenic target cells	Ly-2,3
	T_S	Suppression of immune response	Ly-2,3

(From Table I.2 of Hood et al. 1978)

culating peripheral T-cells, and then differentiate terminally to effector cells in response to further antigenic stimulation.

In summary, antigen-independent maturation of T-cells in the thymus produces a diverse population of short-lived, antigen-specific T-cells. After being triggered by specific antigen in the periphery, they complete their maturation sequence and gain the property of longevity.

Lymphocytes circulate throughout most of the tissues and organs of the body by two circulatory networks, the vascular and lymphatic systems. Their ratio to all leukocytes is 20–80% in blood, and over 99% in the lymphatic fluids (lymph). But their function to contact and bind with specific antigens is completed effectively only in specialized lymphoid organs that provide the accessory cells and the tissue architecture appropriate for antigen processing and presentation. These lymphoid organs are scattered at many positions that interrupt these vascular systems as was described in Chap. 1. Besides the bone marrow and thymus that are important as the original processing sites of lymphocytes, there are the spleen, Peyer's patches, appendix, adenoids, tonsils, and many lymph nodes for the following functions. The lymphoid system serves (1) to concentrate antigens from all parts of the body into a few lymphoid organs; (2) to circulate the lymphocyte population through these organs so that every antigen is exposed to the organism's repertoire of antigen-specific lymphocytes in a short period of time; and (3) to carry the products of the immune response, antigen-specific effector T-cells and humoral antibodies, to the bloodstream and the tissues (Hood et al. 1978).

Dendritic Cells (DC)

Since around 1980, a cellular component appearing in the white, but not red, pulp of the spleen and lymph nodes of 2- to 4-week-old mice has attracted the attention of immunologists. They are nonproliferating and turn over in situ rapidly with a half-life of 3–5 days, being sensitive to steroids and ionizing irradiation. Upon isolation, they adhere firmly to a glass or plastic wall initially, but then become nonadherent, making enrichment feasible. They are distinguished from the adherent macrophages by their active formation of cell processes from which their name "dendritic cells (DC)" is derived, and by possessing little phagocytic or pinocytic capacity in vitro or in situ (Steinman and Cohn 1973, 1974, 1982; Muramatsu and Inaba 1982). Furthermore, DC are characterized by irregular and motile nuclei, many mitochondria, and few endocytic vacuoles, lysosomes, and ruffles; and by lacking ATPase, acid phosphatase, hemosiderin, and surface markers, such as Fc, C3, Ig (immunoglobulin), Lyt-1, and thy-1. In this context, DC represent a novel differentiation pathway distinct from macrophage-monocyte, granulocyte, and lymphocyte lineages that are illustrated in Fig. II.23 which shows the differentiation pathways in the mammalian hematolymphoid system.

The novelty of DC was furthermore confirmed by the demonstration of epitopes unique to DC. In this experiment, spleen cells were taken out and fused with mouse myeloma cell lines which had been established as tissue cultures. The resulting hybridoma clones produced monoclonal antibodies specific for DC capable of identification of a DC-surface marker antigen.

Because this monoclonal antibody can kill DC in the spleen and lymph nodes with the help of complement, these antibodies were used to remove the DC population from the experimental animals to determine the biological functions of DC.

The so-called mixed lymphocyte reactions (MLR: the mutual activating reaction between lymphocytes from two genetically different mice) were ablated after the removal of DC from the system; the specific sensitization of lymphocyte with antigen processing by the accessory cells was also inhibited after such treatment. On the other hand, purified DC were found to be the primary stimulators of several immune responses. Steinman et al. (1982) could restore these capacities to cause MLR by adding back small numbers of DC (1–2%) to the cultured spleen adherent cells and unfractionated spleen cells from which DC were depleted.

Therefore, DC are the principal cells that are responsible for the accessory cell function of presenting antigenic epitopes to the lymphocytes, although participation of the other cells is not excluded as was described in the previous section. To support this, the surfaces of DC are found to be rich in Ia-antigens, the product of the IA region of the H-2 regulating the immune response (Chap. 2). All trials to convert DC into macrophages or lymphocytes have failed (Steinman et al. 1982).

Platelet

In addition to the erythrocytes and leukocytes derived from the hematopoietic stem cell (Fig. II.23), anuclear particular components, i.e., platelets, are produced by megakaryocytes in bone marrow and circulate throughout the bloodstream. They adhere to the inner surface of blood vessels, when their endothelium is injured, and release various substances, including ADP and thromboxane. The adhering platelets interact with each other by the action of these substances to form platelet-thrombi by aggregation. In addition, platelets release vasoactive amines, above all serotonin, which show vasoconstrictive effects, and play an important role in the inflammation process to be described in Chap. 10.

Chapter 9 Extracellular and Intracellular Destruction of Microorganisms

The strategy of host-defense mechanisms adopts diversified tactics against infections by microorganisms which were previously operationally classified into three categories: those that cause (1) superficial infection; (2) exudative (or partially invasive) infection; and (3) invasive (or penetrating) infection as was described in Part I. The principal mechanisms operating can also be designated correspondingly as superficial defense against the first, extracellular destruction of pathogenic microorganisms against the second, and that encompassing intracellular destruction against the third category of invasive infection. However, this designation of these principal mechanism is not definitive because host-parasite relationships constitute a field as a whole in which all components of host-defense mechanisms can function simultaneously or separately in parallel against a variety of microorganismic virulence factors. From this point of view, the mechanisms responsible for extracellular and intracellular destruction of microorganisms will be discussed before the more general concept of host reaction to the external stimuli, i.e., inflammation.

Reticuloendothelial System (RES)

In addition to PMNs, all tissues especially spleen, lymph nodes, and bone marrow have their "fixed macrophages", and those fixed to liver sinuses or to lung alveoli (septum and space) are so unique in their morphology, surface markers, physiology, etc. that they are called "Kupfer cells" or "alveolar macrophages", respectively.

In spite of their various localizations throughout the body, one can regard all these fixed phagocytes in the context of reaction to the invading microorganisms, as a system of defense mechanisms called the "reticuloendothelial system (RES)".

Phagocytosis

Leukocytes that ingest particles or microorganisms, e.g., neutrophils (PMNs), eosinophils (acidophils), basophils, and monocytes are called phagocytes, although basophils show no phagocytic activity. Phagocytes are distributed throughout the whole body not only in the blood, but also in almost all tissues.

They pass through the endothelial walls of capillary vessels into these tissues, and adapt to the local sites under the influence of their microenvironments.

The major phagocytes among these cells are PMNs and monocytes. When monocytes migrate from the blood vessels into the tissue, they are called macrophages. Eosinophils (acidophils) may be specialized for ingestion and destruction of immune antigen-antibody complexes, but not of bacteria, because they are deficient in lysozyme and phagocytin (Bloom et al. 1968).

In parallel with these phagocytes, mast cells and basophils share a series of related changes. The mast cells of connective tissue and blood basophils have similar properties for their granules. Kitamura et al. (1979) demonstrated that most connective tissue mast cells arise from bone marrow stem cells that circulate in the blood in a precursor state and then slowly differentiate into mast cells in the tissues. These are found widely distributed in the connective tissues and especially abundant along small vessels. Their morphology is variable among mammalian species. They have been found rich in heparin, a protein polysaccharide (mucopolysaccharide) which is an anticoagulant, and histamine. In the rat and mouse, but not in other species, mast cells also contain serotonin (5-hydroxytryptamine), a substance that causes constriction of small vessels and influences blood pressure. In recent years, it has been shown that the activation of rat mast cells by specific divalent antigen-induced dimerization of cell-bound IgE with secondary cross-bridging of IgE-Fc receptors stimulates a cascade of biochemical events which culminates in both the noncytotoxic release of preformed granule-associated mediators and the generation of newly formed lipid mediators. These processes cause not only acute allergic reactions, but also delayed-type hypersensitivity reactions (Askenase 1980).

Endocytosis

Endocytosis is the main process of phagocytosis, accompanied by: (1) metabolic response of phagocytes associated with microbicidal function; (2) release of microbicidal substances; (3) microbial regulation of vacuole formation; (4) extracellular factors affecting phagocyte microbicidal capacity.

Endocytosis includes phagocytosis (i.e., ingestion of particles 1 μm or larger in size), macropinocytosis (i.e., uptake of fluid or particles about 0.1 to 1 μm in size), and micropinocytosis (i.e., uptake of material 0.05 μm or less in diameter) (Sbarra and Karvovsky 1959; Oren et al. 1963).

Phagocytosis can be divided into several phases: attachment, ingestion, fusion of lysosomes with the phagosome, intracellular killing of microorganisms, enzymatic digestion, and disposal of degraded material.

The fusion of phagosome with the lysosomes and other cytoplasmic granules proceeds to complete the formation of the phagosome with a common membrane structure (Fig. II.24).

Immunoglobulins (IgG and IgM) express their opsonizing function to facilitate the attachment of particles to the leukocyte with or without aid of com-

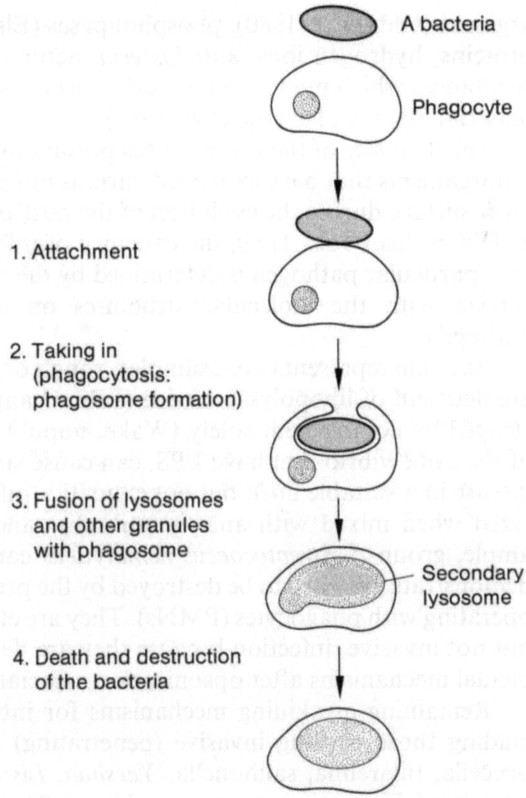

Fig. II.24. Phagocytosis of one bacteria by a phagocyte

plement and other humoral factors. Following attachment of the particle to the cell, pseudopodia or ectoplasm reach out for the particle and fuse around it, thus forming a phagocytic vesicle or phagosome enclosed in the membrane. During and shortly after the engulfment of a particle, the phagocytic vesicle and the granules of the phagocyte fuse, and this is followed by the discharge of their contents into the phagosome. These contents include enzymes like myeloperoxidase, lysozyme, β-galactosidase, acidophosphatase and, in the case of PMNs, lactoferrin.

Mechanisms for the Destruction of Microorganisms

According to the concept of host-parasite relationships, the destruction of infecting microorganisms should result as the consequence of the effective operation of all host-defense mechanisms, including "inflammation", that will be the subject of the next chapter.

Functioning in such microbicidal mechanisms are the following substances: interferon, lysozyme, other lysosomal hydrolases, cationic proteins, arginase (Olds et al. 1980), phospholipases (Elsbach et al. 1979), iron-binding proteins, hydrogen ions, anti-*Listeria* factor (Bast et al. 1974), and perhaps substances which may resemble tumor necrosis factor and which possess anti-malarial activity (Taverne et al. 1981).

The diversity of these substances presumably results from the range of microorganisms that have expressed various molecular structures on their outermost surface during the evolution of the host-parasite relationships defined as a *field* in this book. Thus, the category of infections (refer to Part I) caused by a particular pathogen is determined by the interaction of host-derived substances with the molecular structures on the outermost surface of the pathogen.

As some representative examples, rough organisms of *Vibrio cholerae* that are deficient of lipopolysaccharide (LPS, O antigen) are avirulent and are destroyed by complement solely (Wake, unpublished data); smooth organisms of the same vibrio that have LPS, can cause superficial infection, i.e., cholera disease in a suitable host, but not exudative infection. These vibrios are easily lyzed when mixed with anti-O-antibodies and complement. In another example, group A *Streptococcus hemolyticus* causing an exudative infection is demonstrated in vitro to be destroyed by the presence of anti-M-antibodies cooperating with phagocytes (PMNs). They are effective in causing an exudative, but not invasive, infection because they are destroyed by intracellular microbicidal mechanisms after opsonization associated with specific antibodies.

Remaining are killing mechanisms for intracellular microorganisms, including those causing invasive (penetrating) infections, e.g., mycobacteria, brucella, tularemia, salmonella, *Yersinia, Listeria* organisms and protozoan parasites (viruses are not considered here). These intracellular microorganisms can resist the intracellular killing mechanisms of phagocytes and survive to permit their multiplication and finally destroy these host cells. Only with the help of specific antibody absorbed to the phagocytes (cytophilic antibody) and complement, are some of these microorganisms killed, but other microorganisms can still persist. Additional mechanisms are required to kill these intracellular microorganisms. These require the further enhancement of the microbicidal mechanisms which are the subject of this chapter. However, the mechanisms responsible for these additional effects will be described in an independent chapter (p. 116).

The strict definition of microbial death is important for the distinction of the microbicidal effect from a microbiostatic one. Death of bacteria is the ultimate outcome of a series of events resulting in a bactericidal effect.

For instance, the initial permeability defect in the outer LPS-lipoprotein layer which results in wall damage and release of periplasmic enzymes can be lethal (Feingold et al. 1968). However, the susceptibility of *E. coli* to complement can be diminished when glycerol and acetate are available at this stage for the rapid lipid synthesis by the bacteria, because the damage to the cell membrane can be prevented or repaired (Melching and Vas 1971). Alternatively, the ultimate bacterial death can be protected when Mg^{2+} ion stabilizes the disrupted phospholipid residues in the cell surface and its structural integrity is maintained effectively (Muschel and Larsen 1970). Although the protein synthesis is initially affected, it is the ability to synthesize inductive enzyme, but not constitutive enzyme, that is lost rapidly by the bactericidal reaction (Amano et al. 1955; Amano et al. 1956). All these facts indicate that the lethal event occurs by the effect of bactericidal mechanisms on the cytoplasmic membrane, but not on the cell membrane solely. In addition, the biochemical basis of effective microbicidal mechanisms is more or less divergent depending on the various molecular structures of the target microbial cell surface, and the characteristic cellular metabolism unique to these microorganisms. And these can be modified by the particular host microenvironmental conditions, etc. of extra- or intracellular localizations.

Two pathways of reactions have been recognized as biochemical mechanisms causing microbicidal effects on microorganisms in the infected animal body. The responsible reagents in one of these pathways are called reactive oxygen intermediates (ROI), such as by-products of the respiratory burst reactions of phagocytes with bactericidal activity, e.g., superoxide and hydrogen peroxide (Elsbach et al. 1979; De Chatelet et al. 1977; Nathan et al. 1979a, b; Nathan and Nakagawa 1982; Rosen and Klebanoff 1979; Klebanoff 1980). Another is the protective mechanisms of the host involving long-chain fatty acids (Kanai and Kondo 1979). Both mechanisms have been found to be effective extracellularly as well as intracellularly. Whereas the microorganisms adjacent to the phagocytes can be killed by the release of such active substances from the phagocytic cells (exocytosis), the phagocytosed microorganisms confined within phagolysosomes (endocytosis) can be also killed by these active substances. Again, both phagocytes of PMN and mononuclear type are involved, although the degree of activity is divergent, being higher in the PMNs as a rule and varying among the mononuclear cells depending on their differentiation, e.g., blood monocytes, alveolar macrophages, or Kupfer cells of the liver.

Bactericidal mechanisms relating to the respiratory burst are diagrammatically illustrated in Fig. II.25. The mechanisms occurring in PMNs are characterized by being less affected by immune modifications in comparison to those in mononuclear phagocytes which are remarkably enhanced after direct interaction with specifically activated T_{DTH} cells or the binding of nonspecific lymphokines (Farr et al. 1979). Because the lymphokines are secreted from T_{DTH} cells that are activated by a specific antigenic epitope, the enhanced microbicidal activity of the mononuclear phagocytes constitutes itself a specific T-cell-mediated microbicidal mechanism (p. 116).

The microbicidal systems can be classified into oxygen-dependent and -independent. The former requires oxygen, while the latter operates without oxygen (Table II.8). In fact, patients with a genetic defect in the oxygen-dependent system suffer from chronic infections because their leukocytes are incapable of the phagocytosis-induced respiratory burst and lack microbicidal activity.

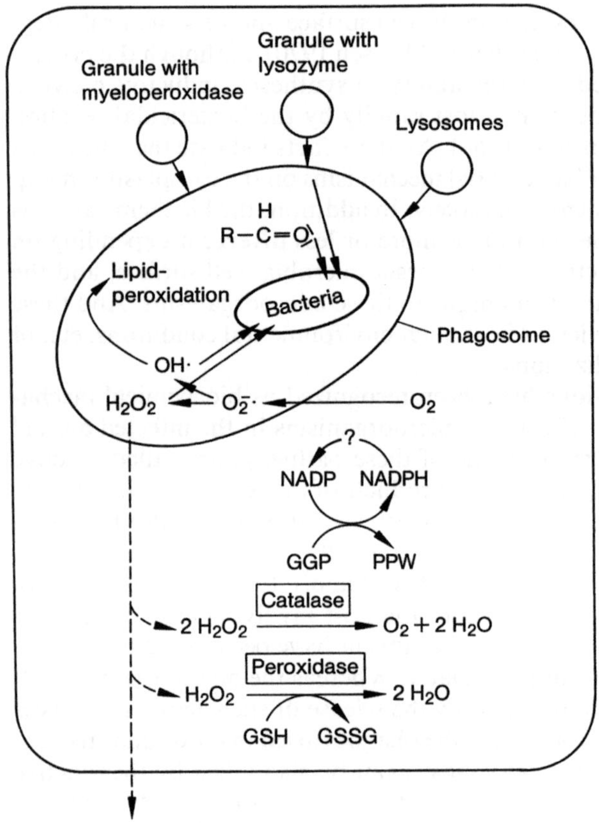

Fig. II.25. Phagocytosis: bactericidal mechanisms of a polymorphonuclear leukocyte. Abbreviations: *PPW*, pentosephosphate pathway; *G6P*, glucose-6-phosphate; *GSH*, reduced (oxidized) glutathion. *Arrows* directed toward the bacteria mean bactericidal activities

Table II.8. Toxic systems of phagocytic leukocytes (Klebanoff 1980)

Oxygen-dependent
Peroxidase-mediated
Peroxidase-independent
Superoxide anion
Hydrogen peroxide
Hydroxyl radical
Singlet oxygen
Oxygen-independent
Acid
Lysozyme
Lactoferrin
Acid and neutral hydrolases

The oxygen-dependent toxic system of phagocytes can be differentiated into those which require the presence of peroxidase (or possibly another heme protein) and those which are operative in its absence. All phagocytic leukocytes respond to perturbation of the plasma membrane with a burst of oxygen consumption. However, this burst also occurs at the time of macro- or micro-pinocytosis, and even in response to a number of soluble agents. The oxygen consumed is initially reduced to its 1 electron reduction product, the superoxide anion (O_2^-), which is both an oxidant and reductant. But, it is more probable that O_2^- exerts its toxicity through the products which it forms, namely, H_2O_2, $OH^\cdot$ and possibly singlet oxygen ($^1O_2$). The divalent reduction of oxygen or the monovalent reduction of the superoxide anion produces H_2O_2 by the activity of such enzymes as glucose oxidase or xanthine oxidase present in the phagocyte. The formation of H_2O_2 from the superoxide anion is direct or an enzyme-catalyzed dismutation. This spontaneous dismutation occurs at lower pH (most readily at pH 4.8). Superoxide dismutase catalyzes the dismutation reaction over a wide pH range (pH 5.0 to 9.5). Since the pH within the phagocytic vacuole and presumably also in the pocket between adherent leukocytes and target cells is acid, spontaneous dismutation would be expected to be rapid.

H_2O_2 is further reduced through catalysis by a trace metal such as iron which is alternatively oxidized by H_2O_2 and reduced by O_2^- with the overall reaction being the Haber-Weiss reaction as follows:

$$\begin{array}{lcl} H_2O_2 + Fe^{2+} & \rightarrow & Fe^{3+} + OH^- + OH^\cdot \\ O_2^- + Fe^{3+} & \rightarrow & Fe^{2+} + O_2 \\ \hline H_2O_2 + O_2^- & \rightarrow & O_2 + OH^- + OH^\cdot \end{array}$$

As a result, the highly reactive hydroxyradical is formed. The validity of this process of hydroxyl-radical formation has been proven by various chemical and physicochemical methods. Another product of the oxygen burst, singlet oxygen, is strongly electrophilic reacting with compounds in areas of high electron density to form characteristic, often oxygenated, products of microorganism cell-surface components resulting in cell damage (Klebanoff et al. 1980).

The presence of an oxygen-independent toxic system in PMNs decreases the pH within phagosomes (and also presumably the pocket formed between adherent phagocytes and particles too large to be ingested). ROI are discharged either into the phagosome or extracellularly. Thus, lysozyme, lactoferrin, acid hydrolases, neutral proteases, and a number of strongly basic proteins attach to the surface components of ingested or extracellular microorganisms.

The interrelationship of the ROI microbicidal systems with the developmental processes of each nonspecific cellular component is as follows:

During the promyelocytic stage of neutrophil development, PMN azurophil granules containing myeloperoxidase (MPO) are formed. Then another

Table II.9. Peroxidase activity of mononuclear phagocytes

Granule peroxidase (myeloperoxidase)
Peroxidase present in the endoplasmic reticulum
Immature phagocytes (myeloperoxidase)
Mature phagocytes
Endocytosed peroxidase
Phagocytosis
Pinocytosis
Catalase

(From Klebanoff 1980) Cytocidal mechanisms of phagocytic cells.

population of secondary granules is formed during the myelocytic stage (Fig. II.25). Among the varieties of ROI microbicidal systems inherent to each of these neutrophil developmental stages, the most potent microbicidal mechanism that destroys bacteria, fungi, mycoplasma, and viruses is the MPO-H_2O_2-halide system belonging to the peroxidase-dependent toxic system of PMN oxygen-dependent toxic systems (Klebanoff et al. 1980).

Of the halides involved, chloride is effective at concentrations present in the PMNs, although iodide and bromide are also effective. Also H_2O_2 (hydrogen-peroxide) is likely to be formed through the pentose-phosphate pathway by the reduction of O_2 with NADPH. MPO combines with H_2O_2 to form an enzyme-substrate complex(es) which oxidizes the halide to form an agent(s) with potent toxic properties for the microorganisms. As a result of the actions of these agents, iodination, bromination, and chlorination of microbial cell components leading to microbial death have been demonstrated.

Besides these microbicidal actions of halides, the antimicrobial role of lactoferrin contained in the neutrophil granules was delineated previously (Chap. 8, p. 140). In addition, however, it was established recently that iron works as a catalyst of the Haber-Weiss reaction in the oxygen-independent toxic system of PMNs to produce OH· molecules which show microbicidal activity (Kaplan and Basford 1979). Furthermore, cationic proteins contained in the PMN granules, are also toxic to microorganisms, by binding to the bacterial outer membrane and causing it to degenerate.

The peroxidases except the endocytosed one of the mononuclear phagocytes are classified into two types as shown in the Table II.9. The hydrogen molecule produced by these peroxidases may contribute to the microbicidal mechanism of monocytes by substituting for the role halide played in the PMN. Another mode of attack is oxidation. The products formed by the peroxidase system are so powerful and oxidant that sulfhydryl oxidation, lipid oxidation, the oxidative cleavage of peptides, and the oxidative deamination and decarboxylation of amino acids with the formation of aldehydes has been reported. The transport of metabolites across the cell membrane appears to be affected by these chemical changes.

Thus, the predominant antimicrobial system during the early postphagocytic period seems to be the MPO-H_2O_2-halide system and it would be anticipated that most organisms would not survive its action.

The genes controlling the MPO-H_2O_2-halide system seem to be the same in both PMNs and monocyte-macrophage cells. The evidence for that is the lack of activity for *Candida,* in this divergent series of leukocytes in patients with hereditary MPO deficiency.

In contrast, the perioxidase of blood monocyte-macrophage cells is not controlled by the same gene as the granule peroxidase of PMNs, because the former is detectable in patients with hereditary MPO deficiency; it is not packed in granules, nor is it released into the phagosome. Another enzyme, catalase, generally catalyzes the degradation of H_2O_2 to oxygen and water, and can oxidize a number of other compounds by H_2O_2 when the H_2O_2 is maintained at low steady-state concentrations.

This characterization of the microbicidal mechanisms of monocyte-macrophage cells shows their critical differences from those of PMNs as stated above. Furthermore, the amounts of ROI detected in the extracellular medium from macrophages are much smaller than from PMNs, although blood monocytes are an exception, and secrete as much ROI as PMNs (Nathan and Nakagawa 1982).

However, the enhancement of ROI-releasing capacity by incubation of macrophages in lymphocyte mediators is an example that a given population of macrophages can differentiate to a state of more active oxidative metabolism, although this differentiation process is fully reversible. Furthermore, the effects of lymphocyte mediators suggests a mechanism through which some of the other agents might work, since some activating agents probably work via lymphocyte products, whereas others appear to be able to affect macrophages directly. Thus, the detailed mechanisms are to be elucidated in addition to those described in the section on cytokines (p. 124).

Among these complex interactions involving macrophages, lymphocytes, and various humoral mediators, such as MIF, MAF, interferons, and CSF (Chaps. 6, 7), secretion of ROI by macrophages is extremely slow when the pathogen itself is used as the augmenting agent until the cells are exposed to membrane-perturbing agents, such as immune complexes, tumor-promoting phorbor diesters, calcium ionophores, lectins or cytochalasins, and many other, but not all phagocytosed particles (Nathan and Nakagawa 1982).

In summary, one aspect of macrophage activation may consist in the acquisition by the macrophages of the capacity to have their respiratory burst triggered by pathogens during their ingestion. On the other hand, however, the biochemical mechanisms inherent to the pathogens are also so divergent as to induce different microbicidal mechanisms in monocyte-macrophage cells so that they deal with pathogens in more complex ways than PMNs.

In contrast with the microbicidal mechanisms of phagocytes involving ROI which are rather effective against Gram-negative bacteria with full length lipopolysaccharide chains, others involving the toxic effects of fatty acids are characteristic in their effectiveness against Gram-positive bacteria, including acid-fast organisms and some special Gram-negative bacteria, such as *Bordetella pertussis, Neisseria meningitis, N. gonorrhoae,* and *Yersinia pestis* (Kanai and Kondo 1979). For example, the bactericidal system detectable in staphylococcal abscesses could be distinguished from the MPO-H_2O_2-halide system by its resistance to heat (100 °C, 30 min), lack of solubility in dilute acid (0.05 N HCl), resistance to strong acid (pH 1), and insensitivity to catalase. It was differentiated from the microbicidal activity of the cationic proteins by its lack of solubility in dilute acid, insensitivity to iron (0.1 mM) or trypsin (5 mg ml^{-1}), and greater activity in solutions of increased ionic strength (Dye and Kapral 1980). Similar antibacterial activities demonstrated with tissue extracts from normal or infected animals, e.g., anti-*Pasteurella pestis* factor (Eisler and Von Metz 1963), antipneumococcal factor (Henley and Nungester 1955), and antimycobacterial activity (Kochan et al. 1972) were attributed to lipid substances which are characterized by being neutralizable by Ca^{2+} or albumin.

Although some experimental observations on the inhibitory effects of fatty acids on microorganisms with regard to the defensive function of the skin have been presented in Chap. 3, further details on the chemical structure of fatty acids and their antibacterial activity can be found in independent reviews (e.g., Kanai and Kondo 1979). In short, the lipophilic and hydrophilic balance of a molecule is an important, if not the sole, factor for the antibacterial activity of long-chain fatty acids. This suggests that at least a part of the antibacterial activity of fatty acids depends upon their detergent activity on the cytoplasmic membrane. In addition, the presence of double bonds in the molecule result in a conformational change so that the surface area is increased to produce a steric effect that destroys the chemical configurations of the membrane.

The cytoplasmic membrane is certainly the target point attacked in the microbicidal activity of long-chain fatty acids. For instance, the "deep rough" mutants of *E. coli* and *Sal. typhimurium,* whose lipopolysaccharides lack most of their saccharide chains, are much more sensitive to fatty acids and hydrophobic antibiotics than the wild-type strain (Nikaido 1976). Hydrophobic substances penetrate by dissolving in the hydrocarbon (a moiety phospholipids) interior to the outer membrane, and the difference in hydrophobic permeability between the wild-type and deep rough mutant is due to radical differences in the organization of the outer membrane, more specifically to the absence of exposed phospholipid bilayer regions in the former (Kamio and Nikaido 1976).

Therefore, the resistance of Gram-negative bacteria to long-chain fatty acids is attributable to the absence of exposed phospholipid bilayer regions on the bacterial surface, and not always due to rejection of fatty acids by the hydrophilic saccharide chains (Nikaido 1976). For example, Gram-negative *Pseudomonas phaseolicola* absorbed toxic fatty acids in the same manner as fatty acid-sensitive Gram-positive *Bacillus megaterium.*

In the case of Gram-negative organisms like *B. pertussis, N. gonorrhoeae,* and *N. meningitis* which possess sufficient length in their lipopolysaccharide chains, fatty acids might be absorbed on the fimbriae through which they are led into the target point of the inner membrane (Kanai and Kondo 1979). In contrast, the hydrophobic cellular surface of acid-fast bacilli has a strong affinity for fatty acids (Kondo et al. 1976 a) and the lipid-rich interior of the cell wall may support the penetration of fatty acids into the cytoplasmic membrane.

Not only the cytoplasmic membrane, but also the cellular organelles are affected by the penetrating fatty acids. Waite et al. (1969) found that acyl bond hydrolysis of exogenous and endogenous phospholipid of mitochondria was stimulated by free fatty acids, among them lauric, myristic, oleic, and linoleic being most effective. After the alteration of the membrane phospholipid layer by these fatty acids, phospholipases can act on individual phospholipids easily. When the impairment of the membrane by the fatty acid is inhibitory to the absorption of nutrients, the effect becomes bacteriostatic. On the other hand, the bactericidal effect is caused by the release of essential intracellular constituents through the damage occurring in the membrane phospholipid layer by the fatty acids. However, the long-chain fatty acids may reversibly react with the membrane protein to alter its structure or to uncouple the electron transport chain. The evidence for such activity of the long-chain fatty acids was obtained from the physicochemical changes of fatty acid-treated protein as revealed by polyacrylamide gel electrophoresis (Shen and Freeze 1972; Fessenden-Raden 1972).

In addition, fatty acids inhibit various enzymatic activities of bacteria, especially of the membrane-associated enzymes for respiration. Two possibilities are considered as the mechanism: the first is the indirect result of the structural change of the membrane and the second is the direct effect on the enzyme protein itself. Fatty acids can act on cytochrome c to produce a partial unfolding of its protein moiety from around the planar heme (Hardesty and Michel 1963); some long-chain fatty acids of the same molecular species as those of the antibacterial fatty acids are highly inhibitory on the activity of the enzyme system for cholesterol synthesis (Kuroda and Endo 1977) or ATPase. Finally, long-chain fatty acids may act as antimetabolites which can be considered in connection with their antibacterial activity (Camien and Dunn 1957).

Such antibacterial and cytotoxic aspects of long-chain fatty acids, are effective in vivo under certain microenvironmental conditions that are free from the neutralizing effects of serum or tissue proteins and this should not be disregarded. Besides, the microenvironmental conditions occurring in the phagosomes, those in the pockets between adherent leukocytes and target microorganisms due to the direct contact of both cell surfaces and in the cell debris within the necrotic abscesses which includes dead leukocytes, should permit those microbicidal mechanisms due to fatty acids to act. Consequently, dead leukocytes at the site of inflammation could be regarded as one of the host-defense mechanisms of a self-sacrificing nature (Kondo and Kanai 1981 a–c).

Furthermore, it should not be overlooked that the activity of long-chain fatty acids on bacteria is of a dual character depending on presence of detoxicants, the concentration of fatty acids in the medium, and length of exposure (Nieman 1954). The fact that the same fatty acids can promote bacterial growth in lower concentrations and inhibit in higher concentration has been well documented (Neal and Strong 1943; Williams and Fieger 1946; Dubos 1947; Williams et al. 1947; J. Lein and P. Lein 1949; Goldschmidt and Koffler 1950).

In conclusion, the two microbicidal mechanisms that relate to the host-cell membrane structure and that are both effective intra- and extracellularly, are effective in eliminating the infecting microorganisms at a definitive stage of the relationship of the host with its invading pathogenic microorganisms. One of them is based on the strong chemical potentiality of some particular atomic state of oxygen, the other on the perturbing action of fatty acids to the cell membrane of microorganisms, which includes phospholipids.

Chapter 10 Inflammation, as a Defense Mechanism

In the healthy state, the mammalian body maintains its physiological homeostasis without disturbance. When a tissue is injured by various causes as a limited locality, the tissue reacts with swelling (tumor) and accompanying pain (dolor), develops a reddish color (rubor), and an elevated temperature (calor) is observed. These four signs characterize the process of defense of the body against a causative event such as mechanical trauma, burns, anaphylactic or delayed hypersensitivity reactions to antigens, electric shock, radiation, local deposition of unusual metabolites, e.g., uric acid, and, particularly in connection with this discussion, adhesion or invasion by parasites. This tissue reaction process is called *inflammation*. One often regards this *inflammatory reaction* itself as a disease and disregards the important function it plays as a defense mechanism. In reality, however, inflammation is a "localized battlefield of the body components so far described against the causative disease agents", and, if the body wins the victory, these agents are neutralized and eliminated. Furthermore, the results affect the whole body to strengthen its resistance; in other words, to make the individual more adaptive to the environment. The acquisition of immunity is only one aspect of this adaptation. However, if the body is defeated, the local site loses its original normal function and is forced to respond with compensatory regeneration of new tissue cells to recover its normal homeostasis. Examples are the reaction to vaccination with BCG and vaccinia virus vaccines by which the host gains immunity to disease, but has residual localized scars.

Reaction of Blood Vessels

The nature of one of the four signs of inflammation, rubor, is the dilatation of blood capillaries caused by the relaxation of the muscles forming the precapillary sphincter. The number of blood capillaries through which the blood flows, is increased within this more or less limited area, resulting in the reddish appearance. Agents causing the relaxation of the precapillary sphincter muscles are called "vasoactive substances" and include histamine, serotonin, kinin, and prostaglandins. The local higher temperature (calor) is the natural consequence of this increase in local flow.

Just after the vasodilatation, the intercellular gaps between the endothelial cells of postcapillary venules are widened to permit blood plasma, containing various plasma proteins, to pass through into the intercellular tissue spaces. Thus, the inflammatory site becomes edematous resulting in the swelling (tumor). This edema adds pressure to activate pain receptors of the nerve end plates causing pain (dolor). The chemical substance responsible for this separation of the endothelial cells is still unknown.

The blood plasma contains various nonspecific and specific humoral components that have been described in the preceding chapters, such as components of complement, iron-chelating proteins, and classes of immunoglobulins. The antibodies may act to cause an opsonization of microorganisms. The ultimate fate of the microorganisms, whether they are destroyed by nonspecific humoral components solely, by the cooperative effects of nonspecific humoral components, e.g., complement and antibodies, by the action of extracellular microbicidal substances, e.g., ROI or fatty acids released from the living or degrading phagocytes, or by the action of intracellular microbicidal mechanisms (Chap. 9), varies depending on the virulence factors inherent to a single type of microorganism as discussed in Part I of this book.

To deal with such divergent virulence factors of microorganisms, the host organisms react with the defense mechanisms mentioned in Part II.

Various kinds of stimulation, e.g., mechanical trauma, irradiation, toxins, proteases, anaphylatoxin produced by complement systems (C3a, C5a) or by the anaphylactic hypersensitivity reaction, trigger the degranulation of mast cells to release histamine from the granule contents, containing the heparin-protein-histamine complexes which bind with the histamine receptors of the target cell surface (e.g., H-receptor of smooth muscles) to cause anaphylactic reactions due to vasoconstriction and increase in the permeability of capillary walls.

On the other hand, serotonin is released at the site of an injured vessel during the inflammatory reaction by the adhesion and aggregation of blood platelets. Many animals, except man, have mast cells that contain serotonin in cellular granules, and thus serve as a source of serotonin release. Serotonin increases the permeability of capillary walls and causes vasoconstriction.

When the capillary walls become permeable for blood plasma, the Hageman factor of subendothelial collagen is activated to form plasmin from plasminogen to cause blood coagulation. Thus, a series of cascade reactions occurs starting from plasmin and leading to the synthesis of kinines that have vasoactive as well as leukocyte-chemotactic activities via kallikrein and bradykinin with the participation of various enzymes contained in plasma.

In addition to these degranulation products of mast cells, it was found that a variety of molecular species which are the split products of IgG are produced to stimulate the blood vessel permeability in the inflamed area as the result of interactions among leukocytes, particularly of monocyte-macrophage series and tissue cells via cytokines.

These IgG split products show diverse activities, e.g., induction of chemotaxis for PMNs, monocytes-macrophages, or lymphocytes depending on the sort of the inflammation, e.g., Arthus-type inflammation, burns, human rheumatoid arthritis, and the delayed-type hypersensitivity-inflammatory reactions involving T-cell-mediated immunity (Hayashi et al. 1969, 1974; Hayashi 1980; Yoshinaga et al. 1971). The different nature of these inflammatory reactions is determined by these diverse bioactive substances. Similarly, the complement-derived C5, C6, C7 complex, thermolabile plasmin-degraded C3 product, split product of C3 convertase, and degradation products of C5 can show chemotactic activities for PMNs or macrophages.

Besides, various polypeptides and proteins originating from degradation products of parenchymal tissues, lysosomal enzymes, extracellular substrates, collagenous proteolytic polypeptides of host origin, bacterial endotoxins, and proteins and even bacteria themselves may play vasoreactive and/or chemotactic roles in the inflammatory reaction.

Dynamics of Leukocytes in Inflammation

With regard to the discussion of the dynamics of leukocytes in inflammation, three fundamental types of microcirculation structures can be recognized with respect to endothelial cells and the basement membrane: (1) Both endothelial cells and the basement membrane are continuous; the interendothelial connection is a tight junction and vessel permeation is most difficult. Examples of this are found in subcutaneous tissues, striated muscles, the central nervous system, reproductive organs, smooth muscles of digestive organs, and fatty tissues. (2) Both endothelial cells and the basement membrane are continuous, but there are small passages through the endothelium. Examples are found in the glomeruli of kidney and endocrine organs. Permeation occurs easily here. (3) Both endothelial cells and the basement membrane do not form continuous layers. Examples are seen in the sinusoidal vessels of liver, spleen, and bone marrow. Permeation take place easily here (Bennet et al. 1956).

The inflammatory reaction occurring in various sites of the body thus inevitably shows modified modes of expression as influenced by these various characteristics of the local microcirculation structures. Therefore, the following description of the dynamics of leukocytes in inflammation is no more than a basic mechanism underlying the general feature of the reaction that is modified depending on the peculiarities inherent to each local microcirculation type.

To delineate the common features among these various types, phagocytic cells accumulate at the site of the lesion during an inflammatory reaction: usually PMNs appear first, followed by an increased number of mononuclear phagocytes. This increase in the number of phagocytic cells in the tissues can be brought about by an increased influx of cells from the circulation, a local cellular proliferation, or a combination of the two.

Since PMNs are end cells which cannot divide, these cells must come from the circulation. Acute inflammatory reactions are often accompanied by an increase in the number of circulating PMNs, cells which can be mobilized from the bone marrow and are on their way to the site of the lesion. Although the initial supply can be drawn from the storage and maturation pools of PMNs in the bone marrow – as indicated by the appearance of immature cells, such as band forms and metamyelocytes, in the circulation – a sustained demand requires increased production of PMNs.

The mechanisms of this increase can be an accelerated rate of stem cell proliferation or an increase in the number of proliferating cell precursors of mature PMNs.

Inflammatory reactions to some kinds of stimuli are accompanied by monocytosis. In animal experiments, the change in monocyte production during such reactions have been studied. Initially, the increased production of monocytes in the bone marrow is caused by a decrease in the cell-cycle time of the promonocytes. The influx of monocytes from the bone marrow into the circulation also increases, as does their efflux from the peripheral blood into tissues; two-thirds of the increment of monocytes leaving the circulation arrive at the inflammatory site.

Besides the acute inflammatory reaction, mononuclear phagocytes participate in two other forms of inflammation, i.e., chronic inflammation and chronic granulomatous inflammation. In the former, the mononuclear phagocytes found at the site of the lesion are usually recruited from the circulation, but under certain conditions, as in processes accompanied by cell-mediated immune reactions, local proliferation of macrophages can occur. The chronic granulomatous form of inflammation can be classified as high-turnover or low-turnover granulomas. The high-turnover granulomas show a high level of recruitment of monocytes and proliferation of macrophages, whereas the low-turnover granulomas are characterized by a low rate of recruitment, longevity of macrophages, and infrequent divisions of these cells.

Several humoral factors have been described as regulating PMN release from and production in the bone marrow. Recently, a specific humoral factor regulating monocytosis during an acute inflammation has also been established as was described in Chap. 7.

The dynamics of lymphocyte recruitment to the inflammatory site are characterized by the nature of the agent inducing the reaction. This is particularly remarkable in the case of infectious inflammation. For example, most of the leukocytes that infiltrate the inflammatory site in the lung induced by influenza viruses are lymphocytes. Therefore, remarkable individuality is seen in the leukocyte dynamics of inflammatory lesions depending upon the causative microorganisms; this aids in the histopathological diagnosis.

Intracellular parasitism is often accompanied by the induction of granulomatous inflammation and the frequent appearance of "giant cells" caused by cell fusion. Thus, the recruited monocyte-macrophages, including these giant cells and the development of surrounding fibroblastic tissues, lay siege to the causative agents which are mixed with living and dead phagocytes resulting in the formation of encapsulated necrotic foci (refer to Chap. 9) thus producing a characteristic local reaction for these parasites.

Result of Inflammation

When the blood supply is inhibited at the inflammatory site, the tissues become anoxic and cells die to form areas of necrosis. But, new proliferating capillaries extend into the necrotic tissues after the inflammatory process is completed and the blood supply is thus restored.

The dead tissues are replaced by collagen and mucopolysaccharides derived from the proliferating fibroblast cells and these form scars. Thus, the restoration of apparently normal tissue structure with some scarring usually occurs after inflammation if the causative agents are completely removed. However, as a result, the host-immune system, including the reticuloendothelial system (RES), attains a higher level of activation, the grade of which is controlled by the genetic mechanisms to be described in the next chapter.

Chapter 11 Genetic Regulation of Defense Mechanisms: Cell-Mediated Immunity in Particular

All the functions of nonspecific humoral and cellular components as well as the specific elements so far mentioned in Part II are regulated under polygenic control and each function has an independent genetic regulation. For example, Biozzi et al. (1980) succeeded in establishing both homozygous high and low responder lines of mice by selective breeding for the relevant immunological functions, such as antibody response, macrophage activity, and T-cell response to phytohemagglutinin. High antibody responder mice are more resistant than low responders against the type of infections in which antibody plays a protective role, such as malaria, pneumococcal infections, or trypanosomiasis. Lines of mice with high macrophage activity are more resistant than low activity lines against the types of infection in which macrophage function constitutes the principal protective mechanism (e.g., *Salmonella, Brucella,* and *Leishmania*).

Both innate resistance or susceptibility and the efficacy of prophylactic immunization have been found to be under major genetic control, in a wide range of different microbial and parasitic infections. While the genes involved are predominantly non-H-2-linked (Chap. 2) in mice, one example of the converse is the involvement of the MHC-linked regulation of susceptibility to infection with a nematode (*Trichinella spiralis*). The data supporting this have been presented by Wassom et al. (1980) in which the inbred-mouse strain B10 · BR(H-2^k) has been shown to be highly susceptible to infection and the strains B10 · S(H-2^s) and B10 · M(H-2^f) have been shown to be highly resistant (see later paragraphs with regard to these designations). The strains B10 · RIII(H-2^r) and B10 · (H-2^b) showed intermediate resistance. Furthermore, the involvement of a gene in the I-A or I-B subregion of the H-2 complex was demonstrated by several congenic recombinant strains. However, other genes outside the H-2 region are also clearly involved (Wassom et al. 1980).

Among the genes located on non-MHC-loci, some are known to be linked to the allotype, to the coat color, or to the X-chromosome. These must be defined by individual, genetic-analytic experiments. Thus, experiments showed the presence of immune responsive genes (Ir genes), histocompatibility antigen loci, and loci controlling T- and B-lymphocyte surface antigens together on the X-chromosome of mice. These associations are homologous to those of the MHC complex on the sixth chromosome in man.

However, the mechanisms whereby genes controlling the innate resistance and efficacy of prophylactic immunization exert their effects are as yet little known. The target-cell species which may be controlled directly by these genes are listed below.

The higher incidence of a series of diseases in humans of some certain HLA types (Chap. 2) has been recognized and the term "relative risk" indicating how many times more frequently a disease develops in individuals with a certain HLA-type antigen as compared to the frequency of the disease in individuals not possessing this antigen, has been used (Svejgaard et al. 1980, refer also to Part III). One of the mechanisms proposed to explain the association of HLA and disease is that of mimicry between HLA-gene products and bacterial antigens. For example, much work has been done on the association of HLA with various streptococcus and staphylococcus protein infections to illustrate such mimicry phenomena. With regard to the former, streptococcal M proteins inhibit some anti-HLA antisera, but not all of the same specificity. Lipopolysaccharides from many Gram-negative bacteria behave in the same way. As another example, a cross-reactivity between HLA-B27 lymphocytes and *Klebsiella aerogenes* was demonstrated. Extracts of cultured *Klebsiella* K43 are able to transform B27 normal lymphocytes into reactive lymphocytes. In addition, there are some indications that during epidemics in the past, a part of the population – the most susceptible – was eliminated. For example, the HLA-B27 antigen has almost disappeared from a Dutch colony established in Surinam which was attacked by yellow fever and typhoid fever epidemics (De Vries et al. 1977).

As the result of the establishment of inbred-mouse strains and by using the unique H-2 designations as the criterion (p. 88), the concept of "haplotype" can be understood. In these inbred mice, all gene alleles on the H-2 region of the 17th chromosome are designated by the affix, for example a, b, d, or k, unique to each haplotype, for example, $H\text{-}2^a$, $H\text{-}2^b$, $H\text{-}2^d$, or $H\text{-}2^k$. This designation is used to mark each gene allele, including the H-2 gene maps of hybrid mice, the genetic characteristics of which are known on the basis of breeding programs used for their production. Thus, for example, the F1-hybrid of female $H\text{-}2^a$ and male $H\text{-}2^b$ is designated $H\text{-}2^{a/b}$ along the whole allele map of the H-2 region. This method of designation enables one to express the genetic events, e.g., recombination, crossing-over, translocation, etc. on the basis of results obtained in experimental genetic analyses.

For further analytical studies of the genetics in mice, a new concept of *congenic inbred* mice was introduced. According to this concept, certain isolated H-2 alleles of a known series of H-2 markers are put one by one within the H-2 region of the 17th chromosome of an inbred-mouse strain, e.g., the C57BL/10Sn strain, then the so-called B10 series of congenic mice is obtained (Klein 1975). [For example, a chromosomal crossing-over occurred when B10 · A mice with the H-2 map of ($K^kI\text{-}A^kI\text{-}B^kI\text{-}C^dS^dD^d$) was produced by breeding B10 · A mice with the H-2 map of ($K^kI\text{-}A^kI\text{-}B^kI\text{-}C^kS^kD^k$) with B10 · D2 with the H-2 map of ($K^dI\text{-}A^dI\text{-}B^dI\text{-}C^dS^dD^d$), where the affix on the shoulder of the allele designates the haplotype marker.]

The use of various inbred-mouse strains, each of which has unique haplotype markers, as well as the use of a series of congenic mice, enabled the determination of high or low responder mice with regard to a given antigenic stimulation. Thus, high or low responders as well as intermediate responders to simple synthetic haptens or antigens and complex antigens, e.g., sheep erythrocytes have been identified to open up the field of immunogenetic analysis (Benacerraf and McDevitt et al. 1972; McDevitt et al. 1972). Furthermore, these studies have been employed to analyze the cell-to-cell cooperation in the various aspects of the immune system using in vivo cell-culture techniques, because each cell taken from an immunogenetically known animals is marked with a known haplotype marker.

This chapter introduces such contributions from the immunogenetic analysis of T-cell-mediated defense mechanisms.

T-Cell-Mediated Intracellular Destruction of *Listeria* Monocytogenes

With the sole exception of the mouse yersiniosis model that will constitute Part III of this book, mouse listeriosis has been the only experimental model so far studied in connection with T-cell-mediated destructive mechanisms for intracellular parasitic bacteria.

The facultative intracellular parasitism of *Listeria monocytogenes* has been described in Part I of this book. Accordingly, *Listeria* infection sensitizes specific T-cells which in vivo activate macrophages to a nonspecifically increased bactericidal capacity (Mackaness 1962, 1964, 1969; Lane and Unanue 1972; Blanden and Langman 1973; North 1973).

Application of immunogenetic knowledge on the inbred-mouse haplotypes for the study of the interaction between the specifically activated T-cells and macrophages has produced a new breakthrough in cellular immunology. The technique of adoptive transfer of spleen cells was developed in these studies (Zinkernagel 1974). Two inbred-mouse strains of the haplotype $H\text{-}2^d$(BALB/C) and $H\text{-}2^k$(CBA/H) served as the donor as well as recipient animals in various reciprocal combinations (Fig. II.26). Six days after intravenous immunization of the donor with a sublethal dose (¼ LD_{50}, i.e., ca. 5×10^3) of *Listeria* organisms, dispersed suspensions of spleen cells were transplanted into the recipient mice according to the experimental design (Fig. II.26). Both T-cells and non-T-cells were used.

Two hours previous to such adoptive transfer, a lethal dose of (ca. 6×10^3) *Listeria* organisms was injected i.v. for the determination of the protection conferred by these adoptively transferred cells. By sacrificing all mouse groups 24 h after the challenge infection, the number of viable *Listeria* organisms in the spleens was enumerated for analysis. When the presence or absence of immunoglobulin (Ig) on the cell surface was used as the criterion to distinguish non-T from T-cells, the non-T-cell fraction conferred a much lower degree of protection (0.2–0.8 $\log_{10}$) which was significant only in some combinations ir-

Group	Listeria immune donor	Transferred spleen cell population (cell numbers)	Recipients	Log_{10} Listeria per recipient spleen
1	--	--	BALB/c	5.41 ± 0.20
2	--	--	CBA/H	5.67 ± 0.15
3	BALB/c	Ig-negative (2.1×10^7)	BALB/c	3.43 ± 0.11
4		"	CBA/H	5.89 ± 0.20
5		Ig-positive (4.0×10^7)	BALB/c	4.60 ± 0.04
6		"	CBA/H	5.22 ± 0.22
7		Ig-negative (2.1×10^7) + Ig-positive (4.0×10^7)	BALB/c	3.85 ± 0.34
8		Unseparated	BALB/c	3.65 ± 0.10
9		"	CBA/H	4.93 ± 0.13
10	CBA/H	Ig-negative (2.0×10^7)	BALB/c	5.59 ± 0.11
11		"	CBA/H	3.72 ± 0.08
12		Ig-positive (3.2×10^7)	BALB/c	4.88 ± 0.10
13		"	CBA/H	5.61 ± 0.19
14		Unseparated (6.0×10^7)	BALB/c	4.87 ± 0.03
15			CBA/H	4.02 ± 0.12

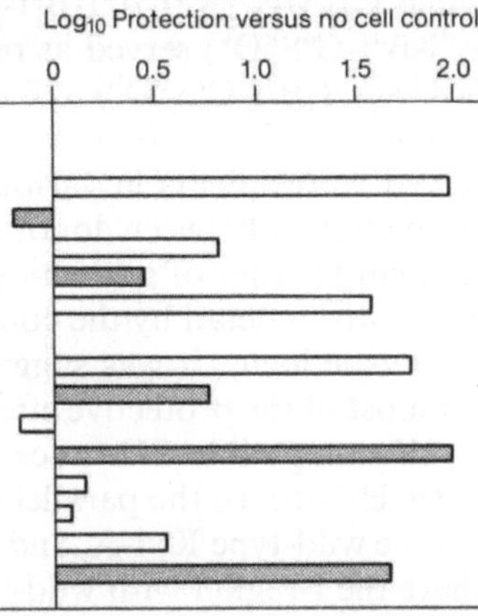

Fig. II.26. H-2 restriction of the anti-*Listeria* activity of surface Ig-negative lymphocytes assayed in an adoptive transfer system. Donor mice were infected with approx. 5×10^3 *Listeria* 6 days previous to cell harvest. Ig-negative spleen cells were separated from Ig-positive ones by the method of Parish et al. The number of transferred separated spleen cells corresponded approximately to the relative numbers before separation. Recipient mice were injected with about 6×10^3 *Listeria* 2 h previous to cell transfer. Viable bacteria counts were determined 24 h later. Means $\pm$SEM of groups of three to five mice were compared statistically: group 3 vs 7 or 8: not significant (NS); 3 vs 1, 5, 10, 12, 14: $P<0.001$; 4 vs 6: NS; 4 vs 9: $P<0.01$; 10 vs 12, 14: $P<0.01$; 1 vs 12: NS; 1 vs 14: $P<0.05$; 11 vs 2, 4, 6, 9, 13: $P<0.01$; 11 vs 15: NS. (From Zinkernagel et al. 1977)

respective of the H-2 compatibility between the donor and recipient; T-cells transferred about 2 $\log_{10}$ of protection measured as the difference in viable bacterial counts per spleen in experimental groups as compared with control groups receiving no cells. At the doses tested, the T-cells were restricted to operate optimally only in the H-2 compatible recipients (Fig. II.26).

In such experiments, T-cells could be recruited in relatively great numbers into the infectious lesions and thus cause non-H-2-restricted anti-*Listeria* protection. This possibility was denied, using each BALB/C and CBA/H control group which received adoptive transfers of lymphocytes stimulated not with bacterial antigens, but allogenically (i.e., with the interaction between the cells of different haplotypes). In these experiments shown in Fig. II.26, the protection versus control values calculated on the basis of $\log_{10}$ viable *Listeria* numbers in the CBA/H-CBA/H combination was 1.09, while in BALB/C-BALB/C combination it was 2.39. Therefore, it is apparent that the relative specific activity of spleen cells to confer protection is inherent to the given inbred-mouse strain and strain differences exist.

The next step of investigation using the congenic mouse series represents the routine way of genetic mapping of the H-2 regions involved in defining T-cell specificity as follows:

Among B10 series, B10·BR($K^kI\text{-}A^kI\text{-}B^kI\text{-}C^kS^kD^k$), B10·A($K^kI\text{-}A^kI\text{-}B^kI\text{-}C^dS^dD^d$), and B10·D2($K^dI\text{-}A^dI\text{-}B^dI\text{-}C^dS^dD^d$) were used as spleen donors, while CBA/H ($K^kI\text{-}A^kI\text{-}B^kI\text{-}C^kS^kD^k$), B10·D2, B10·A, and C3H·OL($K^dI\text{-}A^dI\text{-}B^dI\text{-}C^dS^kD^k$) served as recipients. In the other two experiments, C57BL/6($K^bI\text{-}A^bI\text{-}B^bI\text{-}C^bS^bD^b$), CBA/H, and A·TL($K^sI\text{-}A^kI\text{-}B^kI\text{-}C^kS^kD^d$) were used as the donors and B10·A, C57BL/6, CBA/H, and SJL($K^sI\text{-}A^sI\text{-}B^sI\text{-}C^sS^sD^s$) served as recipients in various combinations. The results indicated that H-2 compatibility between donors and recipients at the K-end results in much more efficient transfer of protection than if the D-end is shared. After summarizing all results in detail by the comparison of compatibility or noncompatibility at each gene locus, it was suggested that the I-A subregion may be responsible for most of the protective effect, because A·TL-immune spleen cells protected H-2K compatible SJL mice markedly less than I plus S-region compatible CBA/H mice. In the parallel experiments, immune T-cells from mice possessing the wild-type K, I-A, and I-B regions protected H-$2K^b$ mutant mice which share the I-region with wild-type mice, as well as they protected wild-type recipients. These results suggest that compatibility as the I-region is sufficient for adoptive transfer of immunological protection against *Listeria*.

Returning to the previous report of Zinkernagel (1974), the adoptive transfer of *Listeria* immune spleen cells, taken from heterozygotic F1-hybrids between two distinct H-2 haplotypic mice into respective parental H-2 haplotype recipients, was performed in addition to the confirmation of the presence of H-2 restriction. This is also evidence that the self-marker recognized is probably coded by the I-A subregion of H-2. Therefore, at least two separable T-cell specificities seem to exist in the donor F1 heterozygous mice – one each for *Listeria* antigen in association with the two respective parental H-2 haplotypes. In this way, the H-2-associated specificities of effector T-cells generated by the infections with intracellular bacteria differ with respect to the H-2 subregions involved; the effector activity of *Listeria*-immune spleen cells is specific for an I-region coded structure, which is analogous to the situation for helper T-cells.

When taking the molecular structure of the polymorphic Ia antigen into consideration, T-cells specific for bacterial antigenic determinants and sharing the common Ia with macrophages, can recognize the signals from both Ia molecules and the antigenic determinant (Fig. II.26). Whether T-cells recognize Ia molecules modified by antigen (altered self-model) or both determinants concomitantly (dual recognition model) is too sophisticated a problem to discuss here.

Intracellular bacteria never lose their anatomical and functional identity even when parasitized host cells are lyzed, and the result is the release of intact bacteria. Thus, growth and spreading of *Listeria* infection is controlled only when phagocytic cells have acquired the capacity to digest the bacteria efficiently. T-cells which recognize antigen in association with I-coded structures may send "activating" signals to the interacting phagocytic cell through this I-structure (Ia molecule). Alternatively, this Ia-associated T-cell recognition may trigger the T-cell itself to release lymphokines, which, in turn, could activate the macrophage. In conclusion, T-cells are triggered specifically to re-

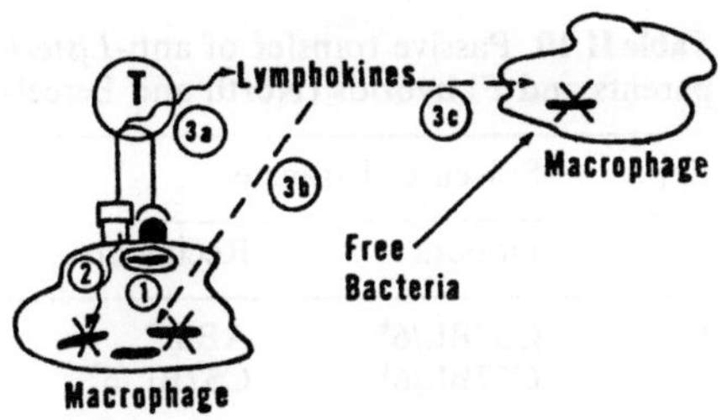

Fig. II.27. Models of T-cell-mediated macrophage activation. Although only the dual recognition model for T-cell recognition is depicted, the model is also applicable to the single recognition model, in which the T-cell is equipped with one clonally expressed receptor for the complex of Ia plus *Listeria* antigen or *Listeria* antigen modified by Ia determinants. In both, antigenic specificity is given by the *Listeria* (*1*) specific part of recognition and the activation signal is delivered through the I-region-coded self-structure (*2*). This could result in direct macrophage activation resulting in digestion of bacteria. Alternatively, the recognition of *Listeria* antigen plus I-coded self-marker could result in lymphokine production (*3a*). Such factors could then trigger the activation (*3b*) of surrounding (*3c*) macrophages

lease lymphokine. However, the macrophages are nonspecifically recruited and activated to result in the increase of bactericidal capacity. In addition, it is also possible that the triggered T-cells may activate macrophages so as to express the relevant foreign and self-surface antigens by direct contact (Fig. II.27). The mechanisms by which Ia-negative macrophages readily become Ia-positive were delineated in Chap. 6.

Apart from the contributions of the in vivo passive-transfer experimental systems for the elucidation of the mechanisms of cell-to-cell interaction in the immune system, the limitation of this experimental system in drawing a general conclusion that the passive transfer of T-cell-mediated immunity to the bacterial pathogen, *Listeria monocytogenes* is H-2 restricted, has been pointed out recently. For example, North and Bereche (1982) intravenously immunized an inbred strain of mice C57BL/6 ($H\text{-}2^b$) and three F1-hybrid mouse strains, i.e., AB6($H\text{-}2^{a/b}$), B6D2($H\text{-}2^{b/d}$), CB6($H\text{-}2^{b/d}$), all containing the C57BL/6 genome and transferred 10^8 donor spleen cells to the same variety of recipient mouse groups which had received an intravenous challenging infection with a lethal *Listeria* inoculum ($1\text{–}2 \times 10^5$) several hours previously. When the level of adoptive immunity expressed in the recipients was determined by subtracting the 48 h $\log_{10}$ growth of the challenge organism in the spleens of recipients of immune spleen cells from its 48 h $\log_{10}$ growth in the spleens of recipients of normal cells, severe restriction occurred. However, such a result had been unexpected because the parent and the F1-hybrid cells share common IA-encoded determinants enabling their interaction (Table II.10). In view of these results obtained with parent mice and their F1-hybrids, these authors performed similar reciprocal passive-transfer experiments between different inbred-mouse strains sharing the same H-2 haplotype. In contrast to the results obtained by Zinkernagel (1974), passive transfer of immunity between differ-

Table II.10. Passive transfer of anti-*Listeria* immunity between parents and Fl hybrids (North and Bereche 1982)

Exp.	Spleen cell transfer		Protection-index in the spleen
	Donors	Recipients	
1	C57BL/6[b]	AB6[ab]	1.3
	C57BL/6[b]	C57BL/6[b]	3.3
2	C57BL/6[b]	B6D2[bd]	1.6
	C57BL/6[b]	CB6[bd]	1.2
	C57BL/6[b]	C57BL/6[b]	3.6
3	AB6[ab]	C57BL/6	2.6
	AB6[ab]	AB6[ab]	4.2
4	CB6[bd]	C57BL/6[b]	1.0
	CB6[bd]	CB6[bd]	3.3
5	B6D2[bd]	C57BL/6[b]	1.8
	B6D2[bd]	B6D2[bd]	3.5

Table II.11. Passive transfer of protection between strains sharing the same H-2 (The standard bacterial challenge was $1–2 \times 10^5$ *L. monocytogenes* i.v., except in experiments C and D, where it was $5–7 \times 10^4$)

Exp.[a]	Spleen cell transfer		Protection index	
	Donors	Recipients	Spleen	Liver
A 1	C3HSW[b]	C3HSW[b]	2.9	2.4
2	C57BL/6[b]	C3HSW[b]	2.6	1.3
3	C57BL/6[b]	C57BL/6[b]	3.9	1.9
4	C3HSW[b]	C57BL/6[b]	1.6	0.2
B 1	BALB/c[d]	BALB/c[d]	3.3	1.7
2	C57BLKs[d]	BALB/c[d]	1.0	0.7
3	C57BLKs[d]	C57BLKs[d]	1.9	1.3
4	BALB/c[d]	C57BLKs[d]	1.3	1.5
C 1	C57BL/6–H2K[k]	C57BL/6–H2K[k]	1.8	0.8
2	AKR[d]	C57BL/6–H2K[k]	1.8	0.4
3	AKR[d]	AKR[d]	2.4	1.4
4	C57BL/6–H2K[k]	AKR[d]	2.2	1.3
D 1	DBA/2[d]	DBA/2[d]	1.3	2.3
2	BALB/c[d]	DBA/2[d]	2.9	1.7
3	BALB/c[d]	BALB/c[d]	3.3	1.7
4	DBA/2[d]	BALB/c[d]	2.1	0.9
E 1	C3H/He[k]	C3H/He[k]	2.9	1.3
2	AKR/J[k]	C3H/He[k]	3.3	1.2
3	AKR/J[k]	AKR/J[k]	2.0	1.4
4	C3H/He[k]	AKR/J[k]	2.6	1.3

Table II.12. H-2 restriction of adoptive protection against *L. monocytogenes* (Kaufmann and Hahn 1982)

Recipient mice	H-2 complex					Log_{10} protection in spleen[a]
	K	I-A	I-E	I-C	D	
C57BL/6	b	b	b	b	b	1.34
B10.A (4R)	k	k	b	b	b	0.17
B10.A (5R)	b	b	k	d	d	1.04
B10.MBR	b	k	k	k	q	0.18

[a] Recipient mice of different haplotype were infected with 9×10^4 viable *L. monocytogenes* 2 h before intravenous transfer of 3×10^6 T-cells. Means of four experiments; SD < 10%.

ent strains with the same H-2 haplotype resulted in severe restriction, no restriction, or even an enhanced expression of anti-*Listeria* immunity, depending on the mouse-strain combination, the direction of transfer, and whether protection was measured in the liver and spleen (Table II.11). Taking into consideration that the three mouse-strain combinations, i.e., C3HSW(H-2^b) and C57BL/6(H-2^b); C57BLK$_s$(H-2^d) and BALB/C(H-2^d); and DBA/2(H-2^d) and BALB/C(H-2^d) differ with respect to their M1 and Ig-1 loci that are located on the mouse chromosomes outside the H-2 region, it is obvious that the restriction is not necessarily by H-2 incompatibility. Based on these experimental data, North and Bereche were (1982) alarmed that the exaggerated evaluation of the in vivo passive-transfer experimental system should occur, although these authors' interpretation of these results is not that H-2 restriction does not occur, but that the restriction caused by non-H-2 differences is so severe that caution should be used in interpreting the results of in vivo passive-transfer experiments. Therefore, more definitive in vitro experimental systems have been awaited eagerly.

The in vitro technique for growing the specifically activated T-cell lines in a microtiter microculture plate containing special tissue-culture medium with accessory cells (presumably macrophages), TCGF, or interleukin 2 (Chap. 7) and the specific antigen, has replaced the in vivo spleen cell-transfer experimental system as a less biased improved technique and it circumvented the disadvantages of the in vivo spleen cell-transfer systems (Kaufmann and Hahn 1982).

Peritoneal-exudate T-cells from mice immunized with live *Listeria monocytogenes* were thus cloned in double-layer soft agar containing heat-killed *L. monocytogenes* (lower layer) and syngeneic accessory cells (upper layer).

With regard to six such T-cell lines and several sublines derived from them, in vitro proliferation, interleukin secretion, and the function to cooperate with B-cells for the production of anti-sheep-erythrocyte antibodies were found to be antigen-dependent and restricted by the H-2 IA locus of the mouse MHC region (Table II.12). When these T-cell lines were adoptively transferred intra-

Table II.13. Systemic (intravenous) transfer of protection against *L. monocytogenes* and DTH to listerial antigens by *Listeria*-specific T-cell lines (Kaufmann and Hahn 1982)

T-cell line	Cell numbers injected	Log_{10} protection in spleen [a]	DTH (0.1 mm) [b]
9– 6	2×10^4	0.13	1.0
9– 6	2×10^6	1.06	6.4
–	None	–	0.8
9–16	5×10^4	0.24	2.0
9–16	5×10^6	1.87	ND [c]

[a] Syngeneic recipient mice were infected with 8×10^4 viable *L. monocytogenes* 2h before intravenous cell transfer. Means of four experiments; SD < 15%.
[b] Syngeneic recipient mice were challenged with listerial antigen immediately before intravenous cell transfer. Means of four experiments; SD < 15%.
[c] Not done.

venously into syngeneic recipient mice, delayed-type hypersensitivity was induced by listerial antigen and the recipient mice were protected against a lethal challenge dose (8×10^4) of *L. monocytogenes* (Table II.13). Thus, not only the H-2 restriction, but also the different biological functions of acquired antibacterial immunity can be mediated by a single T-cell population (Kaufmann and Hahn 1982).

A similar study was performed by Kearns and De Freitas (1983) and they compared the specific activities of *L. monocytogenes* crude whole-cell antigen and the *Listeria* intracellular product (LIP) which is a filtrate of a centrifuged supernatant fraction of disrupted *L. monocytogenes* bacterial cells. Whereas as few as 10^6 LIP-stimulated specific T-cells conferred significant protection against a lethal challenge of *L. monocytogenes*, but did not against that with *Salmonella typhimurium*, cultures induced against crude whole-cell antigen showed little or no protective function (refer to p. 72). This result suggests that the specific antigenic epitope that can induce the cell-mediated immunity is contained in the intracellular *Listeria* antigen designated by Kearnes and de Freitas (1983) as LIP.

Chapter 12 Summary

All examples of host-parasitism so far mentioned in Part II showed how infecting microorganisms encounter the body, humoral and cellular factors of defense mechanisms in a sequential time course. Due to these multifaceted defense mechanisms, only a relatively small number of microbial species equipped with a full set of virulence factors (Part I) succeed in producing lethal infectious disease in the host animals who do not have any genetic or temporary deficiency in some part of their defensive mechanisms. Here, "genetic deficiencies" are caused not only by mutations in the nucleotide base sequences of the chromosomal DNA resulting in the hereditary disease, e.g., sickle cell anemia, agammaglobulinemia, but also by the abnormalities in the number or structure of the chromosome, e.g., Downs or Turner's syndrome. In contrast, "temporary deficiencies" means the incapability to synthesize enough of the constituent factors to maintain a healthy state of homeostasis. The representative cause of such temporary deficiencies are malnutrition, severe fatigue or exhaustion from overwork or mental stress, or combinations with other diseases, including those of an infectious nature. Therefore, temporary deficiencies can be overcome by the removal of causes in their reversible stages.

A large number of microbial species equipped with partial sets of virulence factors succeed in causing infectious diseases by circumventing some facets of host-defense mechanisms, but are prevented from proceeding to a lethal outcome by other effective facets and, thus, the host recovers and develops immunity. Lastly, even microbial species with no virulence factors can cause infections due to the genetic or temporary deficiencies of at least some facets of host-defense mechanisms. These microbial species can kill the patients whose facets of defense mechanisms are impaired, e.g., in the terminal stage of cancer with "opportunistic infections".

Another important concept for the correct understanding of host-parasite relationships is that the expressions of virulence activities by the parasites and that of defensive activities by the host organisms proceed time-competitively. Because of the temporal nature of nongenetic deficiencies and the various lengths of time required for the appearance of individual facets of defense mechanisms (e.g., interferon versus specific immune reactions), the defensive potentialities of the host organisms change in time course. On the other hand, the biochemical expressions of some virulence factors of disease-causing microorganisms require divergent lengths of time and variable physicochemical microenvironmental conditions which must be provided by the host organisms. Therefore, host-parasite relationships must not be considered as two biological processes which are occurring independently in parallel, but must be regarded as an integrated *field* in which the dynamic changes in each component, i.e., host and parasites, are going on. However, each component is accessible for study by analyses performed in animal model infections or in vitro experiments. Thus, Part I and Part II have been presented to provide background for the understanding of the host-parasite relationship with regard to the particular pathogen *Yersinia* in Part III.

In this context, it is important to define the concept of "intracellular bacteria" as follows.

Some microorganisms are neutralized by neither nonspecific, nor specific humoral defensive factors or cellular defensive factors, i.e., (PMNs and macrophages). Some representative bacteria of this class are the causative agents of leprosy, murine leprosy, tuberculosis, brucellosis, tularemia, listeriosis, glanders, melioidosis, and salmonellosis. Plague also should be included as will be described in Part III. The first four can multiply exclusively within parasitized cells to be called obligate "intracellular bacteria": the remaining are "facultative intracellular bacteria".

Finally, the speculations of Zinkernagel (1979) on the evolutionary significance of MHC polymorphism are principally in accord with one of the fundamental thoughts of our book that host-parasite relationships have evolved as the result of intimate mutual interactions between the host and microorganisms. According to his statement, "MHC polymorphism, together with gene duplication, maximizes overall responsiveness in the species population and in the individual, minimizing the chance that the population (or any individual) is a nonresponder to, for example, a highly pathogenic virus. The extent of polymorphism (and duplication) of MHC products and the size of the T-cell receptor repertoire in a given population therefore must be linked and, probably, have coevolved. The two following sets of experimental observations illustrate these arguments and are crucial for our explanation of MHC-associated diseases (p. 165, this book). One set of observations concerns the role of T-cell-mediated immunity and the MHC products in handling intracellular parasites, and the second elucidates Ir gene functions in this context."

Although this statement elected "pathogenic virus" for the example of intracellular microorganisms, it is clear from the discussions of Chap. 2 that the extent of polymorphism (and duplication) of MHC products and the size of the T-cell receptor repertoire in a given population must have coevolved with the intracellular "pathogenic bacteria" cited above, too.

The discussions presented in Part I elucidated the importance of "virulence determinants" which are located on bacterial chromosome and plasmids. The mechanisms by which the "virulence factors" are expressed to cause diseases by neutralizing some facet(s) of the host-defense mechanisms were also discussed. Furthermore, we know that these "virulence determinants" are certain nucleotide base sequences which can be transposed or deleted within the same bacteria and have been found to be distributed among different species of bacteria. It is conceivable that the infectious diseases have selected those bacteria equipped with virulence factors by allowing their multiplication within the host body. Thus, we can propose the evolution of bacterial pathogenicity according to the Darwinian theory of biological evolution during the historical time course of interactions between bacteria and host organisms.

When we integrate the summary of Part I and that of Part II, it is possible to conclude that the intracellular parasites are one extreme of such microorganismic evolution. On the other hand, the extent of polymorphism (and duplication) of MHC products and the size of the T-cell receptor repertoire in a given population are another extreme of such animal evolution. Along each process of both (i.e., microorganismic and animal) evolutions, host-parasite interaction have played, a great many roles in the diversifications in the microorganismic virulence factors and host-defense mechanisms.

Part III
The *Yersinia*-Host Relationship

The choice of the *Yersinia*-host relationship was not arbitrary, although there are many host-parasite relationships in which the interactions of the functions of each component system, host and parasite, could have been chosen. Any particular biological model is incapable of unraveling all the details of the general picture of host-parasite relationships, because both components of parasites and hosts have been evolutionally diversified into species equipped with unique characteristics, therefore, there are many complex and diverse systems. However, the advances in genetics suggest that the selection of an existing biological model, i.e., flowering sweet peas by Gregor Mendel, *Drosophila* (fruit fly) by T.H. Morgan, *Neurospora crassa* (bread mold) by G.W. Beadle and E.L. Tatum, and *Escherichia coli* by Joshua Lederberg, is an effective methodology which results in real progress in our understanding of biological systems. All of the aforementioned models presented a system in which dissection and manipulation of factors was possible.

The nature of the virulence of pathogenic microorganisms has been progressively elucidated at the molecular level as described in Part I. However, few scientists are aware that the *Yersinia*-mouse model is applicable to the study of intracellular parasite-host relationships.

Accordingly, several characteristics of these microorganisms belonging to genus *Yersinia* will be illustrated with regard to their applicability to the study of host-parasite relationships in order to correct the misconception that this bacterial genus, including the historically famous *Yersinia pestis* (now officially designated *Yersinia pseudotuberculosis subsp. pestis* according to the recent agreement of the International Committee of Systematic Bacteriology), is an inappropriate subject for such studies which is one of the major objectives of this book. In addition, however, the importance of public health through advances in this model system will be discussed because of the difficulty in the complete eradication of plague and the frequent occurrence of diseases resulting from infections with other *Yersinia* species.

In this context, the following characterizations deserve introduction: (1) *Y. pestis* is a facultative anaerobic bacterium, belonging to the family of *Enterobacteriaceae,* which is easy to cultivate on conventional media and is recognizable by specific antisera. (2) Natural contamination of a laboratory with the *Y. pestis* organism without evidently traceable routes is not likely, so we can draw accurate conclusions from our experimental results without bias. (3) The mouse is a good host for experimental plague infection. Although experimental mouse plague is artificially produced under strict conditions; both wild and commensal varieties of mice can serve as carriers of *Y. pestis*. In the monograph by Pollitzer (Pollitzer 1954) there are 22 descriptions of *Mus musculus* involvement in natural plague foci in various areas of the world, as well as mention of involvement of other species of rodents. For example, plague-positive animals found at Dakar, Senegal, West Africa in 1935 and 1944 belonged to the following species:

Year	1935	1944	Total
R. ratus alexandricus	10	21	31
R. rattus rattus	1	33	34
R. norvegicus	1	7	8
M. musculus	2	3	5
Total	14	64	78

These animals were detected as part of natural epizootics. (4) The virulence factors for *Y. pestis* F1 VW and Cal, P, P1, and Pu (see later sections) have been defined and each facet of the pathogenesis (virulence expression) of the plague bacillus can be explained by the action of one or more virulence factors. Not only has the localization of gene(s), controlling virulence factors been elucidated on the bacterial chromosome or plasmid, but also in vitro assay methods for each factor have been established making the determination of the conditions for the expression of each factor feasible. (5) *Y. pestis* is highly virulent for the mouse ($LD_{50} < 10$). (6) The usual cause of death is septicemia. This makes the experimental procedures very simple, because the causative pathogens can be diagnosed easily by simple staining of the bacilli in a tissue-smear slide preparation, although definitive confirmation requires serological methods and/or isolation of a pure culture. (7) It has already been proven that fully virulent *Y. pestis,* introduced after growth at 25 °–27 °C, enter monocytes and multiply within the cell (Cavanaugh and Randall 1959). Although once considered as an extracellular parasite, *Y. pestis* is now firmly established as a facultative intracellular infection (Brubaker 1972). Therefore, we can study the host-parasite relationships at the intracellular level, i.e., the interactions between the parasitic microorganisms and phagocytic cells, using this model. (8) However, their intracellular persistence is only manageable for short periods in laboratory experiments (30 days, Wake et al. 1978), and therefore in immunological studies the investigator need not be concerned with cryptic carriage of living *Y. pestis* organisms by the mice at the time of reinfection which complicates the study. In line with the above characterizations of the *Y. pestis*-mouse model it is reasonable to select it as a biological system suited for the study of the bacteria-host relationships among the many candidates mentioned in Parts I and II.

In addition, although regarding *Y. pestis* as most important principally, the other members of genus *Yersinia; Y. pseudotuberculosis* and *Y. enterocolitica* also possess some of the same characteristics (virulence factors) when freshly isolated from diseased mammals. Data demonstrating many of these characteristics have been published recently as will be described in the following chapters and, therefore, it is likely that the experiments using *Y. pseudotuberculosis*- and *Y. enterocolitica*-mouse models will supplement the knowledge derived from the *Y. pestis* model. For this reason, Part III was entitled: "The *Yersinia*-host relationship".

In accordance with the point of view mentioned in the summary of Part I, the subject, "plasmid controlling bacterial virulence", is becoming popular in the world of medical and veterinary science, but here also, the underestimation of the *Yersinia*-mouse model deserves pointing out. For an example of such a serious mistake, a chapter entitled "plasmid-mediated factors associated with virulence of bacteria for animals" (Elwell and Shipley 1980) includes no review mentioning the *Yersinia* model. So why was this model left out? One reason seems to be the recent decrease in the number of investigators of the organism which may, in part, be due to the fear of biohazards. But for experimental microbiologists with adequate experience, the danger of laboratory infection with *Y. pestis* is less than that in studies of other bacterial diseases, such as typhoid, tuberculosis, tularemia, brucellosis, and less than that for many rickettsial and viral diseases like tsutsugamushi, small pox, etc. Thus, experimental studies of *Yersinia* organisms are not particularly dangerous, except for experiments that involve the airborne or fleaborne infection routes unless adequate protective equipment is available. However, these kinds of experimental projects are not required for many fundamental studies and can be avoided.

Chapter 1 Human Disease Caused by *Yersinia*

An overview of the human diseases caused by *Yersinia* organisms which offer appropriate experimental models for the study of host-parasite relationships in general will be presented in the following chapters.

These organisms constitute the genus XIV of the family *Enterobacteriacae,* Gram-negative facultatively anaerobic rods (Bercovier and Mollaret 1984), which are 0.5–1.0 by 1–2 μm and are characterized by unique bipolar staining in all cases and polymorphism.

Particularly, *Y. pestis* is the causative agent of bubonic plague in rats and man. The latter decimated the European population during the Middle Ages, the surviving population was altered both socially (Langer 1964) and genetically (Vogel et al. 1960). Plague is normally transmitted to man from infected rodents by a flea bite, but upon development of secondary plague pneumonia the organisms may be further disseminated by aerosol as primary pneumonic plague without aid of the vector. *Y. pestis* can grow within the gut of the flea or in the mammalian host, but does not lead a saprophytic existence in nature, i.e., *Y. pestis* is an obligate parasite of these animals (Pollitzer 1954; Meyer 1958). In spite of its unique properties, the International Committee on Systematic Bacteriology has decided that this pathogen does not constitute an independent species, but should be called *Yersinia pseudotuberculosis subsp. pestis.* There is a good deal of controversy over this decision and the designation *Y. pestis* is used in this book for clarity and due to historical precedence.

Y. pseudotuberculosis proper is closely related to *Y. pestis* as evidenced by similar sensitivity to bacteriophages (Stocker 1955; Smith and Burrows 1962) and possession of at least 13 common antigens (Lawton et al. 1960; Burrows and Bacon 1960). This organism is also a pathogen and has been isolated from a variety of birds and mammals (Meyer 1958), which serve as natural reservoirs (Mair 1973, 1975). Rats, in general, however, are resistant to infection with *Y. pseudotuberculosis,* but sensitive to *Y. pestis*. Guinea pigs are subject to lethal infections when challenged intraperitoneally with either organism, but only *Y. pestis* is infective for guinea pigs following percutaneous challenge (Beesley et al. 1967). Nevertheless, *Y. pseudotuberculosis* causes severe disease in many rodents as well as occasional fatal septicemic infections (Knapp 1963). In man, pseudotuberculosis follows a milder course involving the mesenteric lymph nodes and mimicking appendicitis (Mashoff 1953; Knapp 1963), or causing enteritis, or erythema nodosum (Mair 1975; Mollaret 1968). Recently, the first conclusive study was published reporting that a community outbreak

Table III.1 Some distinctive properties between typical *Y. pestis* (oceanic) and *Y. pseudotuberculosis* (Brubaker et al. 1965)

Property	*Y. pestis*	*Y. pseudotuberculosis*
Glucose 6-phosphate dehydrogenase	0	+
Phosphofructokinase	+	0
Fermentation of glycerol	0	+
Fermentation of rhamnose	0	+
Fermentation of melibiose	0	+
Murine toxin	+	0
Capsular antigen	+	0
Fibrinolytic factor	+	0
Coagulase	+	0
Urease	0	+
Motility at 20°C	0	+

of *Y. pseudotuberculosis* (serovar VA, refer to the next paragraph) infection involving 535 school pupils, 5 teachers and 1 food attendant occurred in Japan (Inoue et al. 1984). This indicates its peroral route of infection versus the percutaneous one of *Y. pestis.*

Y. pseudotuberculosis has been divided into ten serovars, IA, IB, IIA, IIB, III, IVA, IVB, VA, VB, and VI (Tsubokura et al. 1971), depending upon the linkage and type of 3.6-dideoxyhexose in the O antigen (Davies 1961). The distribution of these serovars has certain geographical characteristics. In European countries, strains of serovar IA predominate in humans and many animals (Mair 1968). In contrast, no serovar IA organisms have been isolated in Japan and serovars IVB, VB, and VI, which are detected in Japan (Tsubokura et al. 1971), have not been reported in other countries (Hubert 1972; Langford 1972; Mair 1968; Mollaret 1968; Wetzler and Hubbert 1968). Recently, serovar IIA was also isolated in Japan.

Actually, the discrimination of *Y. pseudotuberculosis* from *Y. pestis* is of practical importance because, nowadays, the epizootics and human outbreaks of plague are confined within several definitive areas of the world since the last wave of pandemics. The isolation of *Y. pestis,* but not of *Y. pseudotuberculosis,* outside these areas would precede the arrival of the next pandemic to which all efforts for prevention would be needed. The most important difference is that *Y. pseudotuberculosis* causes rather localized infections in their pathogenic natures, whereas *Y. pestis* produces generalized involvement referred to here as acute invasive (penetrating) infection (refer to Part I). Table III.1 shows some distinctive properties of typical *Y. pestis* and *Y. pseudotuberculosis.* However, such distinction is not always clear, because there are some exceptional *Y. pestis* and *Y. pseudotuberculosis* strains that have properties typical of the other organism. Accordingly, Martinevski (1967) proposed the foundation of a new species designated *Yersinia pestoides* to include such intermediate *Yersinia* organisms before the international agreement to include *Y. pestis* within a single species *Y. pseudotuberculosis*. In fact, the possibility of interconversion between *Y. pseudotuberculosis* and *Y. pestis* has been long debated. However, a hypothesis may be assumed that some transposons, insertion units (ISs), plasmids, or phages are involved in such interconversion phenomena as referred to Part I of this book.

With regard to another separate *Yersinia* species, the first human cases of *Yersinia enterocolitica* infection were reported in 1939 (Schlifstein and Coleman 1939). Human and veterinary cases of *Y. enterocolitica* infections were sporadic, their disease forms somewhat diverse, e.g., acute mesenteric lymphadenitis, terminal ileitis mimicking appendicitis (pseudoappendicitis), and septicemia. In recent years, two community outbreaks of *Y. enterocolitica* enteritis occurred in Japan involving 182 children and 1 teacher in a primary school among 390 persons and another involving 554 of 1042 primary school children (Asakawa et al. 1973). At about the same time, another outbreak involving 198 of 1086 junior high school pupils occurred in another prefecture (Zen-Yoji et al. 1973). The source of infection could not be demonstrated, but was attributed to contaminated food or water (Asakawa et al. 1973; Zen-Yoji 1975). Zen-Yoji concluded that contaminated pork was probably the main source of the outbreak he reported.

Large outbreaks of *Y. enterocolitica* also have been reported from Canada (Anonymous 1976) and New York State (Anonymous 1977). Wauters (1981) classified *Y. enterocolitica* into 57 serovars of which only 4 are known to be pathogenic. Geographically three of the pathogenic *Y. enterocolitica* (03, 05B, and 09) have been isolated in Europe, Canada, and Japan and the fourth (08) is the predominant serovar isolated in the USA.

In the summary of the above characterizations of all pathogenic *Yersinia* species, one can recognize the clearer difference between *Y. pestis*, on the one hand, and *Y. pseudotuberculosis* plus *Y. enterocolitica*, on the other, than *Y. pestis* plus *Y. pseudotuberculosis*, on the one hand, and *Y. enterocolitica*, on the other. Such differences are also recognized because *Y. pestis* is always found to be host-associated, while the latter two species are sometimes found to be saprophytic (Schütze 1929; Brubaker 1972). Remarkable are also the unique patterns of geographical distributions of the *Y. pseudotuberculosis* and *Y. enterocolitica* serovars (p. 82). In the case of *Y. pestis*, however, such geographical distributions are rather related to their biochemical characteristics (biovars), i.e., (1) the oceanic type which does not ferment glycerin, but produces nitrous acid found in India, Burma, South China, North America, South Africa, and rural areas of South America; (2) the continental type which ferments glycerin and produces nitrous acid found in Central Asia and Central Africa; and (3) the intermediate type which ferments glycerin, but does not produce nitrous acid found around the Caspian Sea in southeast Russia.

With regards to the interrelationships among these *Yersinia* organisms, Alonso et al. (1980) proposed that *Y. enterocolitica* which is widely spread among mammalian populations in various parts of the world, might act as a silent, but efficient, barrier against the transfer of *Y. pestis* from the natural epizootic foci. A similar hypothesis that *Y. pseudotuberculosis* also might act as a barrier was presented orally by Bercovier at the XIIth International Congress of Microbiology (held in Boston 1982) with slides showing the geographical and historical (i.e., in the long-measured time scale) distribution of the recorded infections caused by the three *Yersinia* organisms. He suggested that plague was more frequent when and where the isolation of pathogenic *Y. pseudotuberculosis* and/or *Y. enterocolitica* were less frequent.

For the explanation of such phenomena, the recent finding that all pathogenic *Yersinia* species are harboring commonly either one of a plasmid family of 40 to 48 Mdal molecular size (Portnoy et al. 1981; Portnoy and Falkow 1981), deserves attention. The virulence factor coded by these plasmids is responsible for the induction of protective cell-mediated immunity against *Y. pestis* infection (Wake and Sutoh 1983; Wake et al. 1983). Presumably the wild rodents infected with *Y. pseudotuberculosis* and/or *Y. enterocolitica* could become immune to *Y. pestis* infection through these mechanisms and play the role of a natural barrier against the latter pathogen.

Chapter 2 Virulence Factors of *Yersinia pestis*

The concept of virulence factors delineated in Part I play a crucial role in the understanding of *Yersinia* and host interactions. In fact, Burrows (1962) first proposed to study the genetics of virulence in bacteria generally, and to employ suitable sensitive hosts, e.g., mice and guinea pigs. Accordingly, the following quotations from his article (Burrows 1962) express correctly the approach he adopted for the establishment of the concept, "virulence determinant":

"Just as in formal genetic studies the existence of inherited determinant can be recognized only following its mutation to loss or alteration of activity, so a virulence determinant is available for comparison. The fact that two strains differ in virulence is easily established; precisely what is the difference between them leading to this dissimilarity is less easily disclosed. Before classical genetic studies can be initiated, however, it is necessary that some observable character distinguishing the two strains in vitro be identified and proved to be responsible for, or invariably associated with, their difference in virulence."

As the particular bacteria which are feasible to study genetically, he defined a *Y. pestis* strain having an average lethal dose of less than ten organisms for mice and for guinea pigs injected intraperitoneally (i.p.) or subcutaneously (s.c.) as a "typical" fully virulent strain to compare with a "typical" avirulent strain having an average lethal dose exceeding 10^6 organisms. He continued: "If a virulent representative strain might be shown to possess a character not shown by an avirulent strain, this character being provisionally assumed therefore to determine their difference in virulence. Supporting evidence would be obtained by the demonstration that all virulent strains possessed this character, while many avirulent strains did not. If, then, derivatives of the virulent strain selected on the basis of their inability to show this character were universally avirulent, without being detectably altered in other ways, and if, on their regaining this ability, virulence was fully recovered, one would have confidence in the assumption that one had recognized a virulence determinant, or at least a property associated with virulence. The decision between these alternatives would require investigation, including the isolation and purification of the material responsible for the observed difference between the virulent and avirulent representatives and demonstration that such material had properties which reasonably could act in promoting virulence. If it were antigenic, specific antiserum would be expected to remove its biological activity and thereby afford some protection against virulent infection, provided that material was accessible to antibody action in vivo.

Having identified a character as a virulence determinant, or associated therewith, one would then compare in detail strains which differed in virulence despite the fact that both possessed this particular determinant. In this stepwise manner, one could hope to formulate the minimum properties required to permit the pathogen to show high virulence for the selected hosts. In confirmation of one's conclusions, it would be desirable to select, by presumed single-step mutations from a single fully virulent parent, a series of derivative strains lacking the identified determinants singly and in all permutations, and to show that the virulence of such derivatives conformed to prediction. This single step selection would ensure the greatest possible genetic identity of the parent and its avirulent derivatives and help to eliminate possible errors arising from cryptic differences between the virulent and avirulent representatives originally investigated."

In line with this approach stated above, he and his associates published a series of experimental studies which led to the establishment of three virulence determinants of *Y. pestis;* VW (two antigens produced at 37 °C under specific culture or in vivo conditions), P (capability to adsorb hemin on a defined medium), and Pu (capability to biosynthesize purine endogenously).

However, prior to these extensive studies, another approach had found other factors, i.e., Kurauchi et al. (1937) focused on the "envelope substance" present around *Y. pestis* cells and regarded this substance as the protective antigen useful for human vaccination. Meyer (1953) and his associates studied its isolation, characterization, and use for protective vaccination against plague and called it Fraction I (F1). F1 is the envelope substance and is also a virulence determinant as well as the VW, P, and Pu described by Burrows (1962). Another virulence factor designated P1 (pesticin 1 production: pesticin is a bacteriocin which inhibits the growth of serovar 1 *Y. pseudotuberculosis*), was later added to the list (Ben-Gurion and Hertman 1958).

Prior to Brubaker (1972), Surgalla et al. (1970) formulated standard methods to detect the expression of F1, VW, P1, P, and the production of fibrinolysin by *Yersinia* organisms, employing in vitro assay methods. The fibrinolytic activity of *Y. pestis* designated F_f or F, was first described by Madison (1936) and was particularly evident when tested with rat or guinea pig fibrin. Jawetz and Meyer (1944) reported that both avirulent and virulent *Y. pestis* organisms are able to coagulate plasma from rabbits, but not from mice, guinea pigs, squirrels, horses, sheep, or man. This characteristic was designated C. Domaradski et al. (1963) had found that 116 of 173 strains of *Y. pestis* coagulated rabbit plasma with subsequent lysis, the remainder were negative for both activities, indicating the F_f and C are linked in *Y. pestis*. Sixty-eight strains of *Y. pseudotuberculosis* were all negative. Beesley et al. (1963) noted that the PI- strains isolated by Brubaker and Surgalla (1961) also were deficient in F_f and that 12 of 46 strains of *Y. pestis* lacked F_f, C, and PI; the remainder exhibited all three activities. The phenotypes of six of these mutants and some other atypical strains are shown in Table III.2. The only known properties that are always associated with the PI determinant are F_f and C

Table III.2. Phenotypes of some PI+ and PI− strains of *Y. pestis* (From Brubaker et al. 1965)

Property	Strains										
	A12	JAVA	TRU	B610	A4/PR2	Harbin	A1122[a]	M23	Q14[b]	B14[c]	Siam
Pi production	−	−	−	−	−	−	+	+	+	+	+
Capsular antigen	+	+	0	+	+	+	0	+	+	+	+
Murine toxin	+	0	0	0	+	+	+	+	+	0	+
Rhamnose fermentation	0	0	0	+	−	−	0	0	0	0	+
Glycerol fermentation	0	0	0	0	0	+	0	0	0	0	+
Fibrinolytic factor	0	0	0	0	0	0	+	+	+	+	+
Coagulase	0	0	0	0	0	0	+	+	+	+	+
Glucose-6-phosphate dehydrogenase	0	0	0	0	0	0	0	0	0	0	0
Urease	0	0	0	0	0	0	0	0	0	0	0
Motility at 20°C	0	0	0	0	0	0	0	0	0	0	0

[a] Prototroph of strain A12.
[b] Strain 14 obtained from Dr. T.S. Quan.
[c] Strain 14 obtained from Dr. T. Burrows.

[strains lacking only one or two of the three activities had not been encountered as of 1965 (Brubaker et al. 1965)].

The constant linkage of PI production (PI) with fibrinolysin (F_f) and coagulase production (C) in all of the three studies resulted in the adoption of the operational designation of PI to include associated expression of fibrinolysin and coagulase as well as PI sensu stricto. The role of pesticin I (PI) itself as a virulence factor of *Y. pestis* will be discussed later.

Returning to Burrow's first proposal of the concept of virulence factors, it is remarkable that he included the toxic activity (T) inherent to *Y. pestis* cells, or released from them into the culture medium, as one of the virulence factors. Fraction II that was isolated by Baker et al. (1952) from the extract of acetone-killed *Y. pestis* organisms had attracted much attention because micrograms of this substance showed intraperitoneal (i.p.) 50% lethal doses (LD_{50}) in mice and rats. When injected with this "murine toxin" or infected with *Y. pestis*, the sensitive animals were killed without showing extensive visible abscesses or necrotic foci. In contrast, however, there are a number of rodents that are insensitive to Fraction II, i.e., murine toxin, that show high susceptibility to *Y. pestis* infection (LD_{50}, less than ten organisms). Animals which are resistant to the murine toxin succumb to septicemia revealing visible abscesses and marked necrotic foci on autopsy. The pathological findings of human plague infection are similar to those found in such murine-toxin-resistant animals, e.g., guinea pigs (Kasuga 1977) and clinical experience has demonstrated that repeated administrations of antimurine-toxin serum to the plague patients are not protective. On the other hand, almost all avirulent and virulent *Y. pestis* strains produce similar amounts of murine toxin. Accordingly, the importance of the murine toxin had to be reinvestigated in more detail. Furthermore, the host-species differences in sensitivity to murine toxin must be always referred to when the investigators use mice as the experimental animal model to represent all mammalian hosts.

Concluding from the above-delineated literature dealing with the virulence factors of *Yersinia* organisms, Brubaker (1972) discriminated the five *established virulence factors of Yersinia pestis*, i.e., F1, VW, PI, P, and Pu from the other *accessory* ones. The former is a detectable phenotype that can be expressed under certain environmental conditions in all fully virulent *Y. pestis* organisms defined as such by Burrows (1962, p. 85). The deficiency of the single phenotype solely in these organisms was demonstrated to accompany the decrease of lethal activity to one or both sensitive animals, i.e., mice and guinea pigs. Furthermore, he proposed to distinguish "*virulence factor* as phenotype" from "*virulence determinant* as genotype" and to designate the latter with three small letters in italic form as had been employed in the discipline of genetics. Thus, the virulence determinant for the virulence phenotype F1 will be designated *fra;* for VW, *vwa;* for PI, *pst;* for P, *pgm;* and for Pu, *pur* in this book, although some degenerated uses of the designations, e.g., *Fra, Vwa, Pst, Pgm,* and *Pur* without discrimination of phenotypes from genotype, are frequently used in most publications and, therefore, will appear in this book. The following discussions of each *established virulence factor* and *determinant* of *Y. pestis* following this rule of designation will be useful for the understanding of not only host-parasite relationships of *Y. pestis,* but also of that of other *Yersinia* organisms as well. Also included here are discussions of *accessory virulence factors* that deviate from the aforementioned strict definition of established virulence factors, e.g., toxin, etc.

The following sections are the particularities of these virulence factors and determinants.

F1

This virulence factor is encoded by a plasmid *fra.*

This phenotype, i.e., Fraction I, is the only important characteristic of practical diagnostic importance. One of the remarkable autopsy findings in animals dying of *Y. pestis* septicemia is the thin viscous layer covering the liver and spleen surface which, when touched with a bacteriological loop, reveals a continuous threadlike viscous material. That the latter was produced by *Y. pestis,* is easily recognized by a simple staining of an impression smear (Wake 1978, Fig. III.1).

The colorless (white) envelope surrounding the typical bipolar *Y. pestis* cell, seen following this procedure, is responsible for the viscous material, although additional characterization, e.g., fluorescent microscopy using anti-F1 specific serum, of course, is required for definitive identification.

Y. pestis does not produce F1 in a recognizable amount at temperatures lower than 26 °C, in spite of their optimal range for growth of 27 °C to 28 °C (Fukui et al. 1960). The in vitro optimal temperature for F1 production is 37 °C, in parallel with the body temperature of most susceptible animals. However, Chen et al. (1975) demonstrated that other in vivo conditions enhance the production of F1. The author studied preparations made in situ on agar surfaces using a novel method of fixation followed by ultra-thin sectioning for electron microscopy. After a 24 h incubation at 37 °C on blood agar containing 3% sheep blood or heart-infusion agar containing 3 mM $CaCl_2$, 20 mM $MgCl_2$, and 0.2% xylose, cells of a virulent strain, T_1, were surrounded by an extensive, extracellular fibrillar matrix. In contrast, a genetically fra^+, but avirulent *Y. pestis* strain [designated EV76 (Paris) R] produced this matrix in recognizable quantities only after 48 h or 72 h incubation; and another fra^+ avirulent strain (designated EV406) even after 72 h of growth produced little if any matrix material.

However, the last strain produced as much or more matrix material as the virulent T_1 strain after one passage through the guinea pig and regrowth for 72 h, under the same conditions (Fig. III.2). Thus, the production of this extracellular matrix material was stimulated by in vivo environmental factors the effects of which persist for a short period even after return to the in vitro environment. Under the latter conditions, the production was limited by the

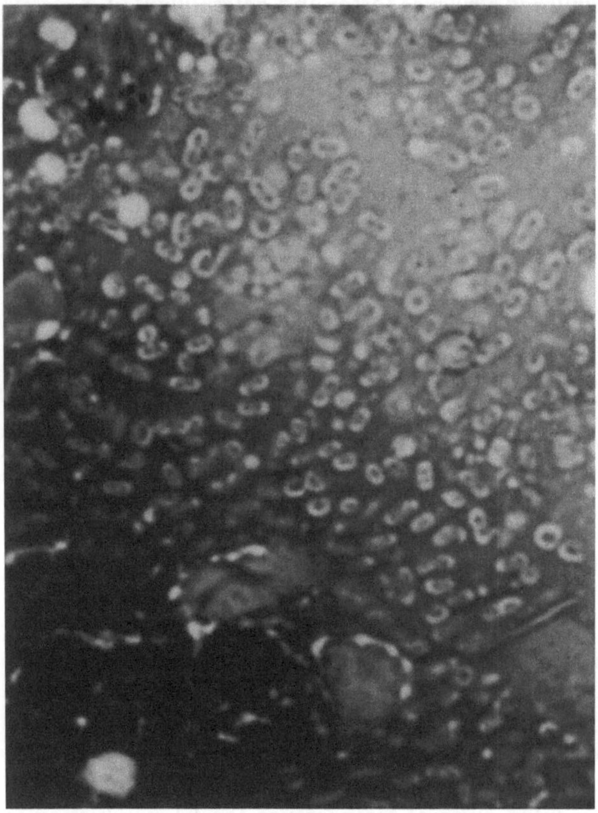

Fig. III.1. A simple method for revealing the envelope substance of in vivo *Y. pestis*. A spleen cut surface of a plague-killed mouse was stamped on the surface of a slide glass. Without fixation, the glass was immersed in amido-black solution for 15 min. The stained sample was decolorized by repeated immersing of the slide into 15% acetic acid until an appropriate staining was obtained before washing with tap water and drying. The bacteria are killed during this procedure, but the used solutions and the container of wash-water must be sterilized before discarding

quantities of Ca^{2+} and Mg^{2+} ions; increased concentrations of these ions showed a remarkable enhancing effect (Chen et al. 1975), which suggests that the in vivo change is not the result of selection. This experiment also suggests some positive correlations between the virulence of *Y. pestis* and the speed of biosynthesis of this matrix by the earlier appearance of this matrix in virulent (T_1) rather than in avirulent *Y. pestis* organisms. Thus, this matrix material corresponds to the F1 capsule.

The F1 capsule is composed of a population of heterogeneous proteins sharing a single immunogenic specificity, i.e., antigenic epitope (determinant), consisting of a series of molecular aggegrates (Habig et al. 1971). The antigen consists of two components with isoelectric points (pI) of 4.6 and 4.8. One component (pI 4.6) is a protein bound to a small carbohydrate moiety identified as an oligomeric galactan; the other component (pI 4.8) is a simple protein. These antigenically identical components exist as aggregates of molecular weights larger than 300,000 in buffered solution.

The aggregates dissociate into a variety of smaller molecular weight forms depending on the nature of the treatment for dissociation and each aggregate

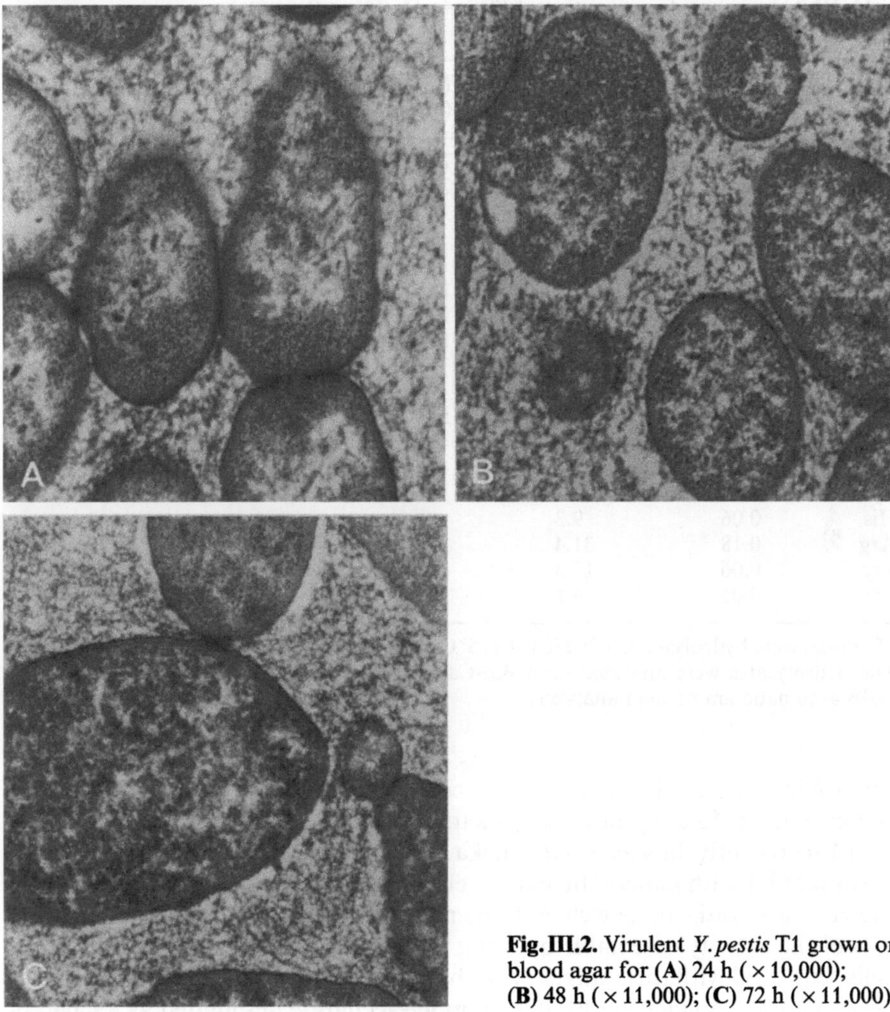

Fig. III.2. Virulent *Y. pestis* T1 grown on blood agar for (**A**) 24 h (× 10,000); (**B**) 48 h (× 11,000); (**C**) 72 h (× 11,000)

can be further dissociated into a single antigenic subunit containing protein and glycoprotein species with molecular weights in the range from 15,000 to 17,000, respectively. Subunits obtained by a dissociation treatment with 0.1% mercaptoethanol in 0.25% sodium dodecyl sulfate (SDS) at 95 °C for 5 min, will readily reaggregate into a variety of larger molecular weight forms on the removal of the SDS. Moderate differences in the isoelectric patterns were recognized between the F1 isolated from an avirulent strain A1122 and that from a fully virulent strain 195/P, in which F1A (pI 4.6) comprises about 40% of F1, representative amino acid compositions of both 195/P and A1122 strains are shown in Table III.3. F1 is 94% protein and based on gas-liquid chroma-

Table III.3. Amino acid composition of Fl protein[a]. (From Bennett and Tornabene 1974)

Amino acid	Fl (μmol mg^{-1})	Fl (μg/mg^{-1})	Nearest integral No. (residues/molecule)
Asp	1.28	170.4	21
Thr	1.31	156.0	22
Ser	0.61	64.1	10
Glu	0.53	78.0	9
Pro	0.22	25.3	4
Gly	0.92	69.1	15
Ala	0.71	63.3	12
Val	0.68	79.7	11
Met	0.10	14.9	2
Ile	0.43	56.4	7
Leu	0.65	85.3	11
Tyr	0.24	43.5	4
Phe	0.37	61.1	6
Lys	0.50	73.1	8
His	0.06	9.3	1
Arg	0.18	31.4	3
Try	0.06	12.3	1
Cys	0.05	6.1	1

[a] Samples were hydrolyzed in 6 N HCl at 115° C for 16, 24, and 42 h. The hydrolysates were analyzed on a Bio-Cal modified Beckman 120B automatic amino acid analyzer.

tographic analysis, the carbohydrate component which is nearly lost to F1B, is a galactan consisting of eight galactose residues.

More recently, however, Głośnicka and Gruszkiewicz (1980) found that digestion of F1 with pancreatin caused elimination of the protein component of the envelope antigen, as well as some nucleotide contamination, without any loss of the activity detectable by using the agglutination reaction of sheep red cells sensitized with this antigen with the specific antiserum. The pancreatic digest contained a macromolecular lipopolysaccharide designated as a galactolipid by the authors. This galactolipid contained galactose, fucose, phosphatidylserine, and phosphatidyl ethanolamine, but no glycolipid. The galactolipid showed high specificity in the passive hemagglutination reaction and in *Y. pestis* phage-receptor activity.

These findings may explain the stability of the antigenicity of F1 to heating in a boiling water bath for 30 min, extremes of pH (from 1 to 14), mild oxidation with performic acid, reduction with mercaptoethanol in 8 M urea, and high concentrations of denaturants, such as 10% SDS, 10 M urea, and 8 M guanidine hydrochloride. The antigenic stability to boiling is used to extract specific F1 antigen (from the spleens or livers of animals killed by plague) for serological diagnosis using capillary or tube precipitation tests to supplement the passive hemagglutination test (Chen 1952).

Y. pestis F1+ organisms can be easily identified by both direct and indirect fluorescent microscopic techniques and colonies of *Y. pestis* can be identified on antiserum-agar plates by a F1-antibody precipitin ring that develops around F1+ *Y. pestis* colonies after exposure to chloroform vapors (Surgalla et al. 1970).

With regard to the functional role of F1 in the virulence expression of *Y. pestis,* the bacterial attachment onto the mammalian cell surfaces is not inhibited by F1, because less F1 is recovered from the centrifuge-precipitated spleen cell pellets than from the supernatant saline when *Y. pestis* organisms grown at 37 °C, but not at 27 °C were i.v. injected into mice prior to the removal of the spleen from the mouse body (Wake, unpublished data). Accordingly, the bacteria were often trapped in the remaining macrophage membranes when a macrophage monolayer inoculated with *Y. pestis* was lysed by suitable methods (Charnetzky, private communication). After this attachment, however, uptake of *Y. pestis* by professional phagocytes is inhibited (Janssen et al. 1963).

Thus, F1 was once thought to be critical in the pathogenesis of plague, but despite the lack of F1, the i.p. LD_{50} of F1−VW+P+P1+Pu+ *Y. pestis* is less than ten organisms for mice, and the i.p. LD_{50} of about 10^4 organisms of this strain for guinea pigs, indicates that F1 is not essential.

Turning to the *fra* determinant for F1, Tsukano et al. (1982) was the first who observed the plasmid controlling the F1 phenotype by electron microscopy. The plasmid molecules were isolated from ten diverse derivative avirulent strains of a fully virulent *Y. pestis* strain Yreka and a control strain designated YD2 (N3) with a drug-resistance R-plasmid of 33 Mdal. Each of the derivatives was confirmed to lack some established virulence factors by in vitro assays.

After isolating DNA with a phenol extraction, the dialyzed DNA solutions were subjected to buoyant ultracentrifugation within a cesium chloride (CsCl) solution containing ethidium bromide. The resultant bands were removed from ethidium bromide by treatment with isopropanol saturated with CsCl solution, and dialyzed against 50 mM Tris (hydroxymethyl) aminoethane-HCl to prepare the samples for electron microscopy. The molecular weight of each detected plasmid was calculated using SV-40 DNA added to the sample as the standard. The calculated value was the average length of 50 molecules of each plasmid. YD1, a P− mutant derived from Yreka was found to have an average of 0.01 45 Mdal-plasmid copies per cell, few copies of the 1 Mdal-, but many copies of the 7 Mdal-, 13 Mdal-, and 23 Mdal-plasmids at a ratio of 3:6:1, respectively. YD3-a, a VW− mutant derived from YD1 and eight other VW− strains all lacked 45 Mdal plasmid.

Mutation from a F1+ strain into F1− strain is accompanied by the loss of the 13 Mdal plasmid. Furthermore, the decrease in the copy number of 13 Mdal plasmids paralleled the decreased production of F1 substance by that strain (Wake 1984). It was also noticed that Yreka cells segregate into F1− organisms at a high frequency following growth in the presence of chemicals

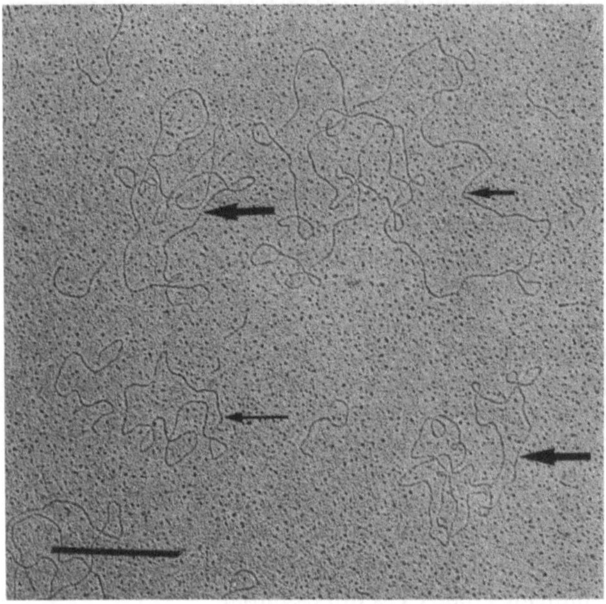

Fig. III.3. Electron micrographs of 13 Mdal plasmids (*large arrows*) found in the phenol-extracted lysate of *Y. pestis* strain YD1. The scale indicates 4.7 μ. One each of the 23 Mdal (*smaller arrow*) and the 7 Mdal (*fine arrow*) plasmid are also recognized

that inhibit the replication of plasmid DNA. These segregants lost 13 Mdal plasmids as determined by measurements on electron micrographs (Fig. III.3). Spontaneous F1 − mutants of Yreka organisms were also devoid of the 13 Mdal plasmid. However, these F1 − mutants could revert to the F1 + phenotype upon reappearance of the 13 Mdal plasmid following treatment with SDS. Therefore, it is highly probable that F1 production is controlled by a plasmid of 13 Mdal molecular size, although expression may be influenced by certain intracellular situation(s), e.g., integration into the chromosomal DNA or by the function of some distinct regulatory gene(s).

For definitive confirmation, the transfer of F1 + to a F1 − VW − P1 − P − *Y. pestis* organism from a F1 + VW + P1 + P − *Y. pestis* organism by conjugation was performed as follows.

A conjugative R-plasmid RP4 ($Tc^rKm^rAp^r$) was transferred from *E. coli* to a F1 + VW + P1 + P − *Y. pestis* mutant of Yreka origin for the use as a donor strain. For use as a recipient strain, a F1 − VW − P1 − P − *Y. pestis* strain was isolated from Yreka.

Finally, these donor and recipient *Y. pestis* strains were mixed and cultivated for their conjugation and the colonies selected on the streptomycin-containing agar plates were tested for agglutination with anti-F1 specific antiserum. Because some isolates were agglutinated and the donor strain could not grow on this agar, it was concluded that the plasmid controlling F1 production was mobilized by RP4 and transferred to the recipient *Y. pestis* organisms which are streptomycin-resistant.

VW and Cal

These virulence factors are encoded by a plasmid *vwa=cal,* composed of at least two regions of DNA nucleotide base sequences; one encoding the V protein; another vesting the host-bacteria with such physiological character that their growths at 37 °C, but not at 27 °C, are inhibited under calcium-deficient environments as will be mentioned later.

The designation VW was derived from the description of Burrows and Bacon (1956 b) that the production of another antigen W accompanies that of the V-antigen, which symbolizes "virulence antigen". In fact, the loss of this factor results in a great increase in the s.c. and i.p. LD_{50} for both mice (i.p. $LD_{50} > 10^7$) and guinea pigs (i.p. $LD_{50} > 10^8$).

However, VW is unique to not only *Y. pestis,* but also all virulent strains of *Yersinia* (Gemski et al. 1980 a, b; Carter et al. 1980) vesting some of them, e.g., serovar 0:8 *Y. enterocolitica,* with the virulence of 10 to 100 organisms for i.v. or i.p. LD_{50}.

Burrows and Bacon (1956 b) first discovered VW when they compared the interaction of a fully virulent *Y. pestis* MP6 and an avirulent strain TS with the intraperitoneal exudate PMNs of mice taken at intervals after i.p. injection with 5×10^8 organisms of either strain using Giemsa-staining of smear preparations. The number of intracellular TS microorganisms increased until 2 h and only slightly decreased, thereafter, until 5 h. On the other hand, the MP6 organisms increased in intracellular numbers within 1 h, then decreased. After 5 h, most MP6 organisms were extracellular suggesting the acquisition of antiphagocytic activity. When these MP6 organisms from the peritoneal cavity were reinjected into mice, no phagocytosis by PMNs occurred. Because TS as well as MP6 are F1+ strains, this antiphagocytic activity cannot be attributed to the function of F1 and, thus, the appearance of new antigen(s) that enhance resistance to phagocytosis is indicated. Burrows and Bacon (1956 a) determined in vitro conditions for obtaining the phagocytosis-resistant phase of the strain MP6. These were rotatory cultivation at 37 °C with a particular broth containing a carbon source and 25–400 µg cystine. Cells of the TS strain remained phagocytosis-sensitive under the same conditions in spite of the production of F1. Thus, the acquisition of the antiphagocytic character was due to the expression of a characteristic unique to the virulent strain. Burrows and Bacon (1956 b) succeeded in detecting this characteristic by demonstrating one or two precipitation line(s) on an agar-gel diffusion test between the suspensions of various virulent strains of *Y. pestis,* including an attenuated P− *Y. pestis* strain EV, using a rabbit serum prepared by immunizing with the phagocytosis-resistant phases of MP6. The serum had been absorbed with the acetone-dried powder of TS cells, grown under the same conditions as for the growth of phagocytosis-resistant MP6. The first recognized antigen, a single precipitin line, was named V, a second line seen when V-producing bacteria were cultivated for longer periods was designated W; the genes controlling these antigens were assumed to be linked and were designated *vwa* by Brubaker (1972). Thus, the finding of VW had been based on the belief that the biological function of VW resides in its antiphagocytic activity, i.e., the resistance to uptake by professional phagocytes. According to the view prevalent at that time, this implied that *Y. pestis* should be considered as an extracellular parasite unable to multiply intracellularly.

However, a different approach was taken by Higuchi and Smith (1961) who devised an agar medium for the selection of virulent *Y. pestis* organisms. They found if the agar medium contains more than 20–40 mM Mg ions and less than 2.5 mM Ca^{2+} ions, fully virulent *Y. pestis* organisms form separated colonies at 26 °C, but not at 37 °C, the body temperature of mammals. If the Ca^{2+} content exceeded 2.5 mM, fully virulent *Y. pestis* organisms also grew at 37 °C. This unique physiological characteristic was named "calcium dependency" (at 37 °C) and the differential agar medium used was called "magnesium oxalate agar". Figure III.4 shows the efficiency of this method.

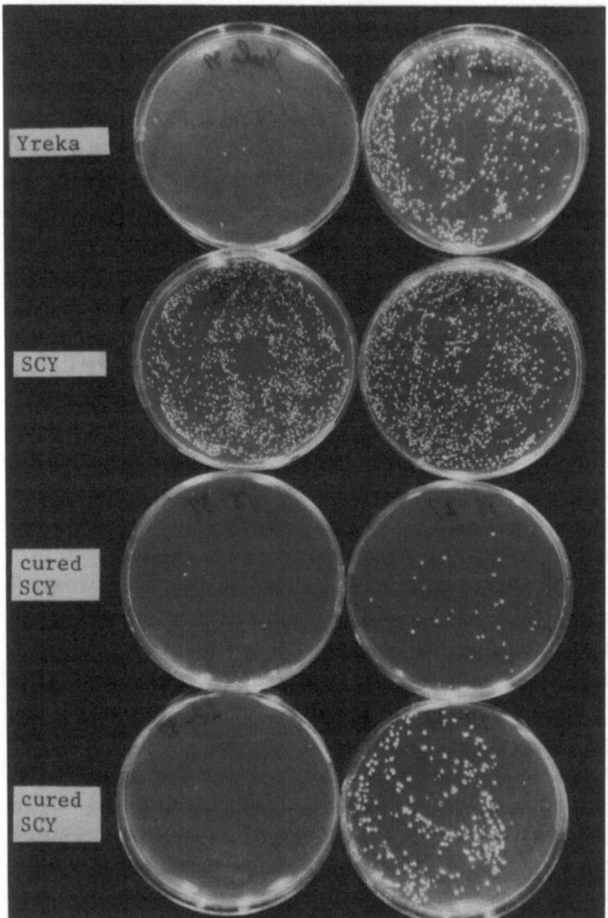

Fig. III.4. Calcium-dependency tests of *Y. pestis* strains using magnesium oxalate agar plates. The plates of the *left column* were incubated at 37 °C, while those of the *right column*, at 27 °C, for 48 h. *Upper to lower rows:* A fully virulent strain Yreka, an avirulent strain SCY, and two avirulent calcium-dependent derivatives

In the earlier period of research on virulence these two markers of plague virulence, VW and calcium dependency, were always associated. Therefore, many papers reported calcium-dependent *Y. pestis* strains as VW+.

However, Brubaker and Surgalla (1962) later described *Y. pestis* strains exhibiting all of the known established virulence factors of *Y. pestis*, but which were avirulent for mice and guinea pigs ($LD_{50} > 10^8$ cells). These strains produced VW antigens, but formed rough colonies of normal size at 37 °C on Ca^{2+}-deficient medium, as opposed to the typical avirulent strains which formed smooth colonies on this medium and failed to produce VW antigens. Growth of these strains in a Ca^{2+}-deficient broth medium was slow when compared to that of typical avirulent strains, but was more rapid than that of virulent strains. The new strains were obtained (1) at 26 °C by selecting for resistance to streptomycin (SM) or (2) at 37 °C by isolation from rough colonies found on a Ca^{2+}-deficient agar medium plated with either a virulent SM-sensitive or -resistant culture. Usually, isolates obtained at 26 °C exhibited a marked reduction in the ability to form colonies upon primary plating at 37 °C. Of 108 SM-resistant isolates obtained from six virulent strains, 20% were of the new type. In addition, five isolates were of reduced virulence (mouse $LD_{50} = 10^4$ cells), but were Ca^{2+}-dependent and possessed all of the known virulence properties. Some SM-resistant strains failed to give rise to avirulent colonies when plated on a Ca^{2+}-deficient agar medium containing SM. In other words, the major phenotypic change associated with avirulence was the loss of the Ca^{2+} requirement.

Since these isolates produced VW antigens, they were referred to as VW+ avirulent mutants to distinguish them from the typical VW−, Ca^{2+}-independent strains. Thus, this might indicate that Cal rather than VW is the important characteristic relative to virulence, although VW−Ca^{2+}-dependent strains have never been encountered.

Thus, the phenotype VW is dissociable from calcium dependency; and these phenotypes appear to be controlled by distinct genes. Accordingly, Brubaker (1979) designated the newly found VW+ avirulent *Y. pestis* strains as F1+VW+Cal−P1+P+Pu+.

Precisely, however, the inducibility of VW+ avirulent mutant was variable among virulent strains of different origin (Brubaker and Surgalla 1962). In addition, some of their mutants were characterized by only partially reduced virulence of 10^3–10^4 LD_{50}; the others required streptomycin for growth, but were fully virulent. Therefore, the interrelationships among virulence, calcium dependency, and SM resistance must be more complex than the all-or-none law.

Furthermore, it is important to learn the environmental conditions that affect the *population shift* to avirulence in order to define the physiological role of these linked, but dissociable, factors, since confusion sometimes results, when population changes are not distinguishable from the changes occurring in an individual microorganism. Bicarbonate (Delwiche et al. 1959) or pyrimidines (Baugh et al. 1964) can replace calcium in preventing the *population shift* to avirulence. Surgalla et al. (1968) found that virulent and avirulent microorganisms would respond in opposite directions to some compounds, i.e., faster growth of one and slower multiplication of the other. Each compound had an optimal concentration range for the "activity".

After a number of compounds had been found to maintain consistently virulent populations at certain optimal concentrations, they were all tested at the same time at these concentrations. All of these "active" compounds, i.e., dinitrophenol, thiocyanate, iodide, salicylate, oleate, biliverdin, and deoxycholate, alter cell surfaces or membranes and cause either uncoupling of oxidative phosphorylation or promotion of ATPase activity and permit *vwa*$^-$ avirulent segregants to overgrow *vwa*$^+$ organisms which are sensitive to this alteration.

One prerequisite for the elucidation of physiological and functional relationships between VW and Cal was to characterize VW antigens chemically. For that purpose *vwa*$^+$ organisms must be cultured in large mass while preventing a *population shift* to *vwa*$^-$ organisms. Taking these technical requirements into consideration, Lawton et al. (1963) were originally successful in purifying these antigens.

1. An F1-antigen negative strain *Y. pestis* M23 *fra*$^-$*vwa*$^+$*pst*$^+$*pgm*$^+$*pur*$^+$ was selected for VW-antigen production to avoid contamination with F1 antigen. Incubation of virulent *Y. pestis* organisms at 37 °C with aeration, which results in the selection of *vwa*$^-$ avirulent segregants, but which promotes VW-antigen production, was used (Fukui et al. 1957).
2. A 16 h incubation period was used for the production of VW antigens to minimize the number of *vwa*$^-$ organisms present.
3. The quantitative agar-gel double-diffusion test of Ouchterlony was used to evaluate each step of the purification procedure, i.e., the fractionation of proteins with ammonium sulfate and repeated DEAE-cellulose chromatography of the dialysate. Specific anti-VW serum was prepared by immunizing animals with *vwa*$^+$ *Y. pestis* organisms and extensively adsorbing nonspecific antibodies by the use of an acetone-dried powder of *vwa*$^-$ organisms derived from the *vwa*$^+$ immunizing bacteria.

The V, which was purified 100-fold, was found to be a protein with the molecular weight of 90,000 and W, which was purified 1000-fold, was found to be a lipoprotein appreciably more acidic than V with the molecular weight of 145,000, and 38% of the dry weight of W was lipid. Both antigens were stable at 60 °C, but not at 80 °C. W antigen, but not V, was lost upon extensive dialysis against distilled water, or evaporation in vaccum. The titer of both antigens was reduced by prolonged storage at 5 °C or by lyophilization, but not by storage at −20 °C.

Specific antiserum to V or W was obtained when each of these antigen preparations were injected intramuscularly into guinea pigs or rabbits. A mouse passive immunization test using rabbit antisera produced against V or W followed by challenge with 100 LD_{50} of a fully virulent *Y. pestis*, showed that V antibody, but not W antibody, can protect mice against plague infection (Table III.4).

Table III.4. Passive protection of mice against plague with rabbit antisera containing V or W antibody. (From Lawton et al. 1963)

Rabbit serum[a]	Dead/total Challenged[b]			
	Expt. 1	Expt. 2		Expt. 3
Normal	10/10	10/10		10/10
Anti-V[c]	0/10			0/ 5
Anti-W[d]		10/10	5/5	5/ 5

[a] Five-tenths milliliter rabbit serum was injected i.p. concomitantly with the challenge dose. Three days later, another 0.5 ml serum was injected via the same route.
[b] Mice were challenged i.p. with approx. 1000 organisms of *Yersinia pestis* strain Alexander (100 times the LD_{50} dose).
[c] Two different antisera were tested.
[d] Three different antisera were tested.

Important is the fact that VW-antigen productions are no more than an aspect of the general alterations occurring in a *vwa*$^+$ organism in response to particular environmental conditions characterized above. The alterations of all-around aspects simultaneously occurring in the bacterial body are summarized as follows.

The morphological and biochemical studies disclosed that an individual *vwa*$^+$ *Y. pestis* organism, but not a *vwa*$^-$ one, cultivated in a Ca^{2+}-deficient medium shows marked changes in cellular structures and biochemical capabilities immediately following a shift from 26 °C to 37 °C (Hall et al. 1974; Brubaker 1972; Charnetzky and Brubaker 1982). Those changes include: (1) cessation of cell division after less than two generations resulting in bacteriostasis (Figs. III.5 and III.6); (2) alteration of bacterial nucleotides from compact to filamentous axial forms (Fig. III.7); (3) and increase in cell size (Fig. III.8); (4) V- and W-antigen biosynthesis and release; and (5) decrease in ribonucleotide pools, adenylate energy charge, and the rates of synthesis of DNA, RNA, and protein.

The first item, i.e., bacteriostasis, is potentiated by Mg^{2+} (Zahorchak et al. 1979). With 2.5 mM Mg^{2+} and no added Ca^{2+}, growth restriction is relaxed at elevated pHs at which maximal-cell yields occur in the presence of 20 mM exogenous ATP, but the nucleotide is neither transported into the organism nor hydrolyzed extracellularly. This ATP-dependent growth is inhibited by Ca^{2+} or 20 mM Mg^{2+}. Intracellular V antigen was detected 1 h after a temperature shift in Ca^{2+}-deficient medium containing 20 mM Mg^{2+}, a time corresponding to the earliest known event associated with shutoff of the cell division, the reduction in the rate of stable RNA synthesis. Adenylate energy charge occurs later. Maximal yield of V was obtained 2 h later when cell division ceased; the titer of the antigen remained constant thereafter. The recovery of V in cells grown with ATP was significantly reduced, especially at elevated pH. These results would be expected if exogenous nucleotides promote growth

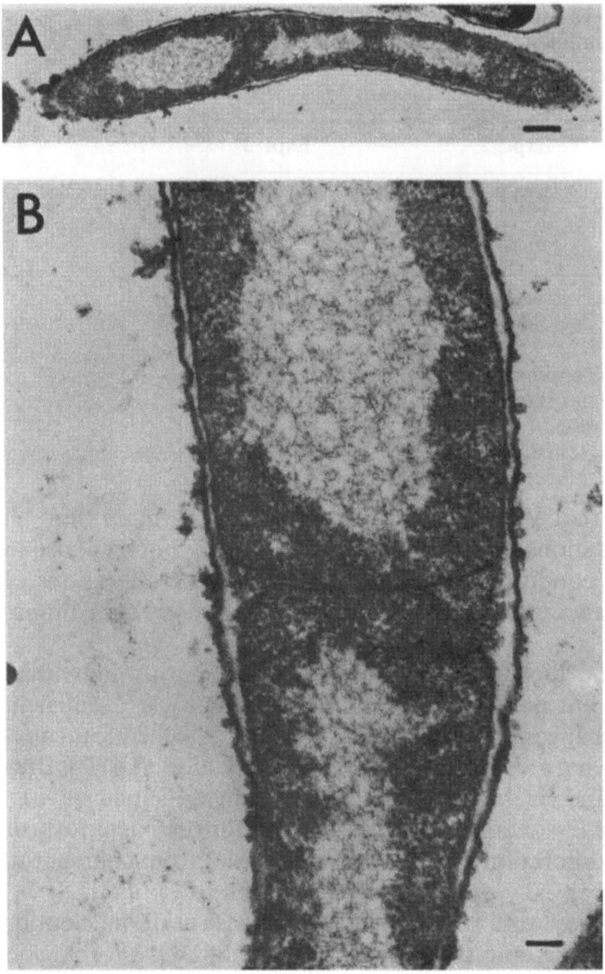

Fig. III.5. A typical division pattern in cell of *Y. pestis* strain EV76 cultivated for 12 h at 37 °C in Ca^{2+}-deficient synthetic medium (**A**) oblique section (*bar*, 0.1 μm). **B** Higher magnification of area of atypical separation (*bar*, 0.001 μm). (Hall et al. 1974)

by sequestering sufficient Mg^{2+} to prevent restriction of cell division mediated by V-antigen production.

This switch over to V-protein synthesis is realized by the following changes in the rate of different RNA molecular species: the messenger RNA (mRNA) increased from 57% to 95% of the RNA synthesized, while ribosomal RNA (rRNA) decreased from 37% to 4.7%, transfer RNA (tRNA) decreased from 5% to 0.7% of the total following a temperature shift of a vwa^+ *Y. pestis* from 26 °C to 37 °C in a Ca^{2+}-deficient medium. The earliest known alteration in normal metabolism associated with this transition then, is a termination of net RNA synthesis (Charnetzky and Brubaker 1982).

To summarize these biochemical alterations, the temperature shift from 26 °C to 37 °C in a Ca^{2+}-deficient medium blocks the synthesis of stable RNA,

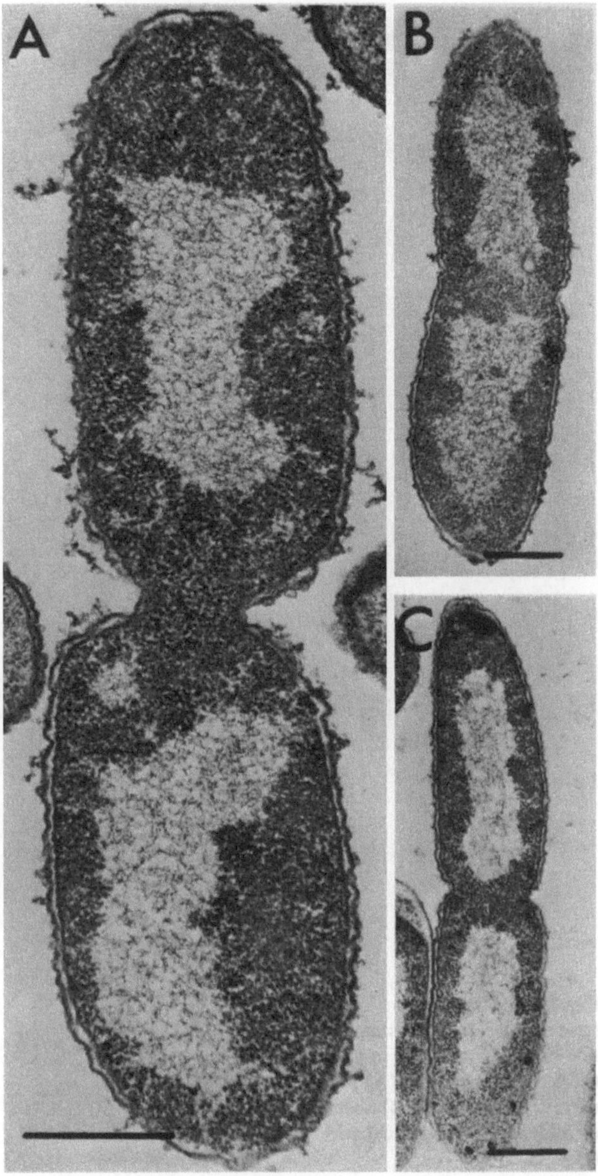

Fig. III.6. Dividing cells of *Y. pestis* strain EV cultivated at 37 °C in synthetic medium; **A** wild-type cell grown with added Ca^{2+} (2.5 mM); **B** Ca^{2+}-independent mutant grown with added Ca^{2+}; and **C** Ca^{2+}-independent mutant grown without Ca^{2+}; *bars*, 0.1 µm. (Hall et al. 1974)

in a manner not mediated by guanosine tetraphosphate or guanosine pentaphosphate, restricts the growth of the vwa^+ *Y. pestis* organism, and results in subsequent decreases in the DNA and protein synthesis rates. Further evidence for the bacteriostasis, but not for death, is that bacterial multiplication can be restored within 40 min by the addition of Ca^{2+} at 37 °C or a decrease to 26 °C. Along with the metabolic alterations, the vwa^+ gene (or genes) is de-

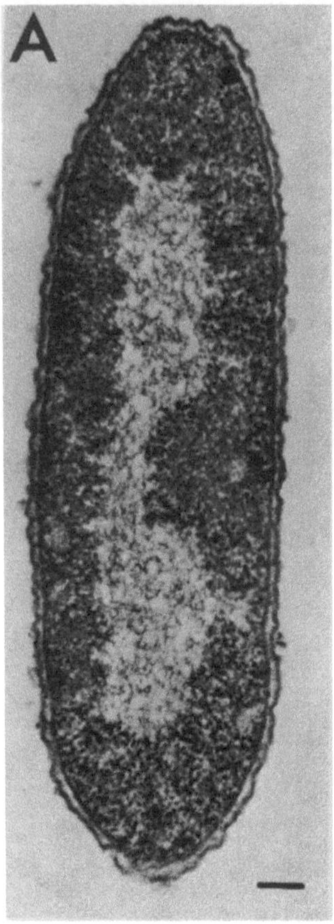

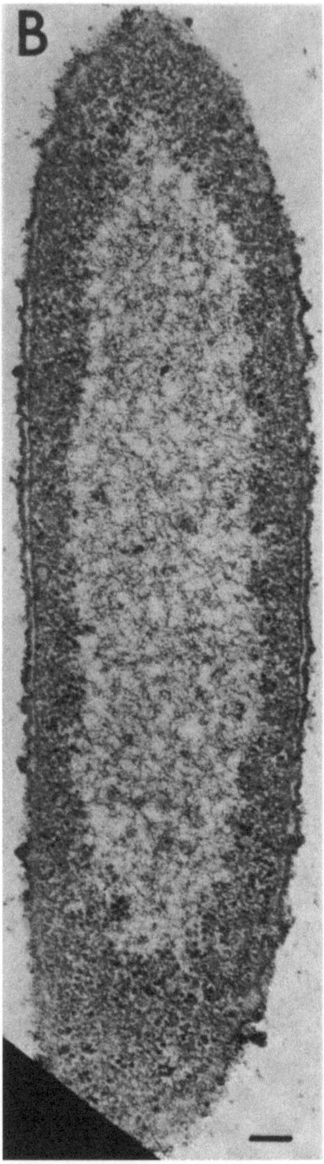

Fig. III.7. Thin sections of wild-type cells of *Y. pestis* strain EV76 grown at 37 °C in synthetic medium in the (**A**) presence (2.5 mM) and (**B**) absence of added Ca^{2+}; *bars*, 0.1 μM. (Hall et al. 1974)

repressed and the biosynthesis of VW antigens initiated. However, this process does not always accompany the derepressions of the other genes of the same replicon. Therefore, the VW+ avirulent *Y. pestis* organisms seem to synthesize VW independently of these physiological alterations. This may be a result of the deleted or depressed *cal* gene, or such mutants may have an altered control

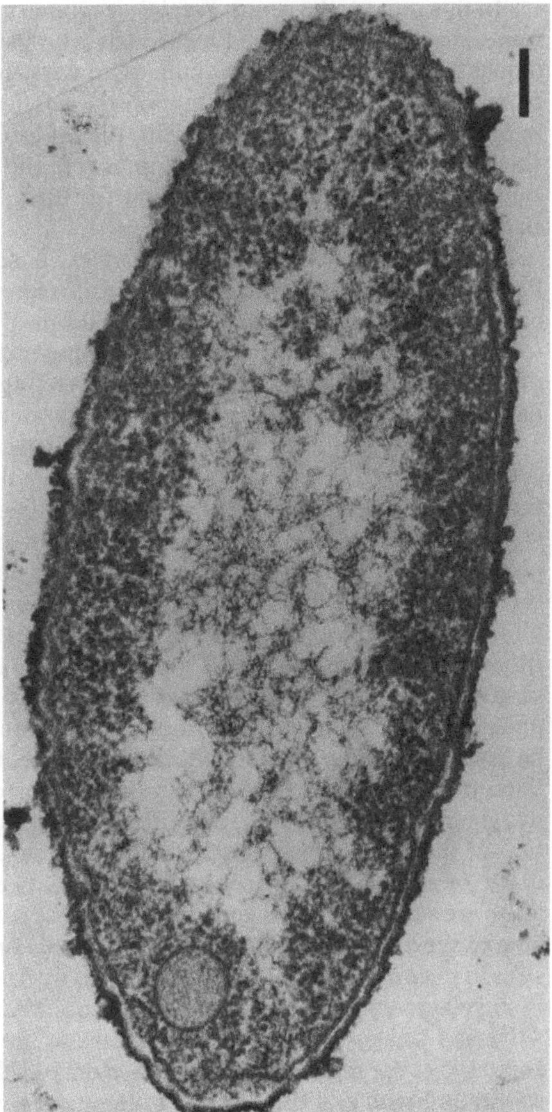

Fig. III.8. Cells of *Y. pestis* strain EV76 after 16 h of cultivation in Ca^{2+}-deficient medium; *bar*, 0.1 μm. (Hall et al. 1974)

region for VW production, i.e., *vwa*^{c} rather than *vwa*$^+$. The avirulence of these organisms presumably results from dysfunction of the *cal* gene, which otherwise would regulate cellular metabolism in response to temperature and the presence or absence of Ca^{2+}, thus regulating the bacterial multiplication essential for the expression of virulence.

In this regard, the genus *Yersinia* is unique among other bacteria of *Enterobacteriaceae,* although the biochemists who used in vitro techniques and fundamental knowledge of metabolic pathways, started their studies of *Yersinia* from the hypothesis of close taxonomic relationships of *Yersinia* with *Escherichia coli* (Brubaker 1984). As with the other *Enterobacteriaceae* in common, Ca^{2+} is transported from the cytoplasm to the extracellular environment, because it is such an essential nutrient for the genus *Yersinia* (Charnetzky and Brubaker 1982) as seen above.

In line with the discussion of *Yersinia*-host relationships, i.e., the title of Part III, the question remains why the aforementioned sequence of events occurring in the *vwa*$^+$ *Y. pestis* cell as the result of temperature shifts from 26 °C to 37 °C under a Ca^{2+}-deficient environment can serve as an indicator of one of the virulence factors involved in protecting *Yersinia* organisms from the bactericidal defense mechanisms of the mammalian host.

By the use of guinea pigs, virulent *Y. pestis* organisms, both of the phagocytosis-sensitive and -resistant types were found to be rapidly removed from the vascular system of guinea pigs by fixed macrophages of liver and spleen (Janssen et al. 1958). Accordingly, the ability to resist ingestion by circulating phagocytes could have limited effects on early survival in the blood, although the progeny of initially phagocytosis-resistant cells subsequently appear in blood and organs about 10 h earlier than those of initially phagocytosis-sensitive cells. These experiments did not address the possibility that phagocytosis-resistant cells can, nonetheless survive and multiply within fixed macrophages or that such cells can multiply at a rate that exceeds their capacity to be destroyed by the reticuloendothelial system. Brubaker (1972) supported the latter possibility by reviewing the works of Burrows and Bacon (1956 b), Cavanaugh and Randall (1959), and others and suggested that virulence is a function of intracellular survival rather than resistance to phagocytosis. In summary, he stated, "it has been established that cells cultivated at room temperature are susceptible to ingestion by free phagocytes but that the organisms are able to survive intracellularly. Growth *in vivo* or at 37 °C prior to infection results in phenotypic resistance to ingestion by free phagocytes and this property is correlated with expression of F1 but not V and W. Both phagocytosis-sensitive and phagocytosis-resistant organisms are rapidly cleared by fixed macrophages of the reticuloendothelial system but the progeny of the latter appear sooner in blood and tissues." This statement is supported by our data (Wake, unpublished) for recovery of fully virulent *Y. pestis* organisms from the blood of normal, killed vaccine and EV-live-vaccine immunized rabbits (Fig. III.9). The concentration, decreased and then increased, subsequently killing the normal rabbit. If the rabbit was immunized with killed vaccine, the main effective component of which was F1, the blood concentration decreased more rapidly suggesting enhanced ingestion of *Y. pestis* organisms by the fixed macrophages of liver and spleen after opsonization by anti-F1 antibodies. But the concentration of *Y. pestis* organisms again rose in the blood and the animal died. On the other hand, this recovery in concentration of organisms in the blood was

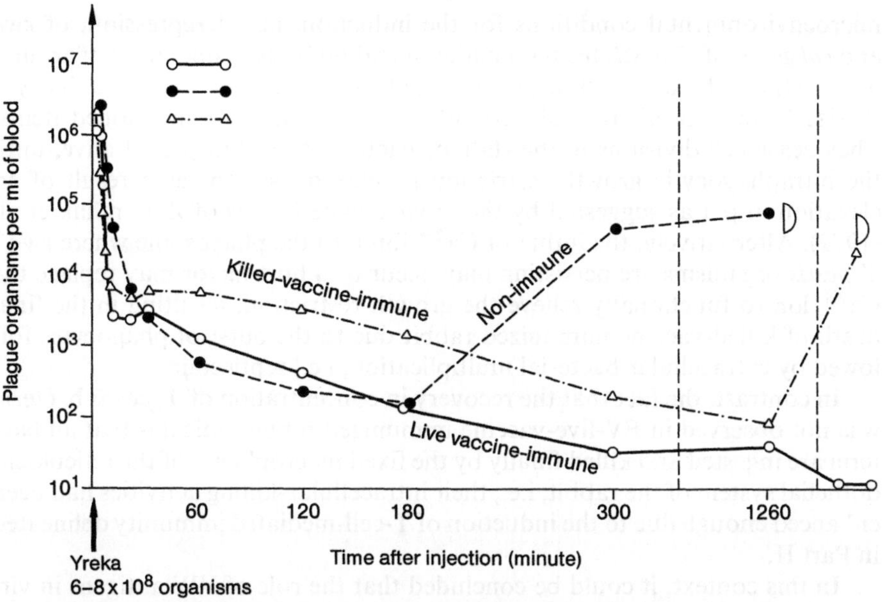

Fig. III.9. Blood bacterial concentration after i.v. injection of virulent *Y. pestis* strain Yreka organisms into nonimmunized, killed-vaccine-immunized, and live-attenuated, vaccine-immunized rabbits. Three rabbits, a normal one, one that had received two s.c. injections of 1 ml formalin-killed antiplague vaccine (prepared by the National Institute of Health, Tokyo, Japan) at one week intervals, and another that had received one injection with 10^8 organisms of *Y. pestis* EV 51, were fixed on an apparatus 3 weeks after the first vaccination. A sterilized polyethylene catheter connected to a special syringe was surgically inserted into the jungular vein of each rabbit and fixed under Nembutal anesthesia. Viable bacteria in blood sample were determined by diluting and plating on three blood-agar plates at the indicated time intervals after i.v. injection of 10^8 organisms of *Y. pestis* Yreka

not observed in the rabbit immunized with live *fra*$^+$*vwa*$^+$*pst*$^+$*pgm*$^-$*pur*$^+$ *Y. pestis* EV51.

The interpretation of these experimental results can be provided by all the above explored in vivo behaviors of *vwa*$^+$ organisms as seen in the following.

Presumably, all facets of defense mechanisms delineated in Part II failed to eliminate *vwa*$^+$ *Y. pestis* organisms finally in the bodies of the normal rabbit and the killed-vaccine immunized rabbit, although the fixed macrophages of their reticuloendothelial system were effective in trapping some bacteria in the bloodstream of a normal rabbit and all the bacteria in that of a killed-vaccine immunized one. This difference resulted from the opsonic action of anti-F1 humoral antibodies. The bacteria that escaped from the ingestion by the macrophages of normal rabbit, multiply extracellularly to kill the animal, because its plasma and lymph are rich in the Ca^{2+} ion. In contrast, the mammalian intracellular fluids are characteristic of a deficiency in the ionized form of Ca^{2+} and an abundance of Mg^{2+} (20–40 mM, Kugelmass 1959) providing optimal microenvironmental conditions for the induction, i.e., derepression, of *vwa* and *cal* genes of *Y. pestis* bacteria under the rabbit body temperature of around 37 °C. Under these conditions, the ingested bacteria are not killed, but biosynthesize V-antigen, altering all-around aspects of the aforementioned items. They cease cell divisions in the state of bacteriostasis, but persist alive, until the intraphagocytic growth restriction is relaxed, perhaps as a result of an elevation in pH as suggested by the in vitro experiments of Zahorchak et al. (1979). Alternatively, the influx of Ca^{2+} ion into the phagosome where *vwa*$^{+}$ *Y. pestis* organisms are persisting may occur or a host factor may replace the Ca^{2+} ion to functionally relieve the growth restriction, resulting in the final death of killed-vaccine immunized rabbit due to the burst of phagocytes followed by extracellular bacterial multiplication, i.e., septicemia.

In contrast, the fact that the recovery in concentration of *Y. pestis* bacteria was not observed in EV-live-vaccine immunized rabbit indicates that all bacteria are ingested and killed finally by the fixed macrophages of the reticuloendothelial system of the rabbit, i.e., their intracellular-killing activities had been enhanced enough due to the induction of T-cell-mediated immunity delineated in Part II.

In this context, it could be concluded that the role of VW antigens in virulence resides in their enhancement of resistance to the intracellular-bactericidal activity of macrophages, rather than in their enhancement of resistance to uptake by professional phagocytes. Consequently, the VW phenotype associated with Cal is one of those virulence factors that confer upon *Y. pestis* the ability to invade or penetrate the host animal body (Pollitzer 1954). Presumably the monocytes or macrophages which harbor the viable but, perhaps nondividing *Y. pestis* organisms, migrate via the blood and lymph vessel network throughout the body, bacterial multiplication is restored, and the host cell is killed at any site resulting in septicemia as was the case of fixed macrophages.

On the other hand, however, VW and Cal alone are not sufficient to allow this dissemination as shown by Burrows et al. (1960) using strains of *Y. pseudotuberculosis,* and by Zink et al. (1980) and Gemski et al. (1980 a) using a tissue-invasive *Y. enterocolitica.* Actually, these two *Yersinia* species rarely cause invasive (penetrating or disseminating) infections (refer to Part I) when compared with the acute septicemic infections commonly seen by *Y. pestis,* particularly following introduction via the subcutaneous route (Part III, Chap. 1).

In spite of this difference in pathogenicity between *Y. pestis* and the other two *Yersinia* organisms, i.e., *Y. pseudotuberculosis* and *Y. enterocolitica,* the VW phenotype associated with Cal distinguishes the genus *Yersinia* as intracellular parasites in contrast to other Gram-negative pathogens. In other words, these virulence factors are unique to the genus *Yersinia* (Zahorchak and Brubaker 1982).

In order to clarify the structural basis for this function, Darveau et al. (1980), determined the outer membrane protein composition of *Y. pestis* organisms at different growth stages and at incubation temperatures of 26 °C and 37 °C. By the use of one-dimensional sodium dodecyl sulfate-polyacrylamide gel electrophoresis of heated preparations of isopycnic sucrose density centrifugation isolated outer membranes, they detected five proteins in the outer membrane of 37 °C-grown cells not evident in 26 °C-grown cells. Also, envelopes from 26 °C-grown cells yielded two peptidoglycan-associated proteins, but those grown at 37 °C contained an additional protein which was not readily apparent in either the bound or free form at 26 °C. Thus, the change in outer membrane protein composition in response to incubation temperature may relate to known nutritional and antigenic changes which occur under the same conditions. These authors discussed the possibility that the peptidoglycan-associated proteins might be porins, i.e., form channels for the passage of hydrophobic compounds through the outer membrane. But a porin role for only one of these peptidoglycan-associated proteins of *Y. pestis* has been demonstrated (Darveau et al. 1983).

To determine the location of V-antigen biosynthesis, Straley and Brubaker (1981) prepared spheroplasts from *vwa*$^+$ and *vwa*$^-$ *Y. pestis, Y. pseudotuberculosis,* and *Y. enterocolitica* organisms cultivated at 37 °C in synthetic medium containing 2.5 mM Mg^{2+}, but no added Ca^{2+}. The spheroplasts were disrupted by sonication and the membranes were collected by centrifugation. The supernatant from the latter step was saved and designated the soluble-protein fraction. SDS-polyacrylamide gel electrophoresis showed that the only noticeable qualitative difference between fractions from *vwa*$^+$ and *vwa*$^-$ organisms was a protein of apparent molecular weight 38,000 present in the soluble protein fractions of *vwa*$^+$, but not *vwa*$^-$ organisms. When the separated membranes and soluble protein fractions of *vwa*$^+$ and *vwa*$^-$ organisms were analyzed by two-dimensional electrophoresis, the resulting protein patterns revealed numerous qualitative and quantitative differences between *vwa*$^+$ and *vwa*$^-$ cultures, the most striking qualitative difference being in the soluble-protein fraction. This molecular species was the major component of the protein precipitated by anti-V antiserum. In one-dimensional electrophoresis, the major band from the labeled immunoprecipitate migrated with the same apparent molecular weight (38,000) as the band described earlier in *vwa*$^+$-soluble proteins. There was no *vwa*$^+$-specific spot in a location similar to that for V in two-dimensional gels of either inner or outer membrane fractions from *Y. pestis*. Thus, Straley and Brubaker (1981) concluded that V antigen is a cytoplasmic protein, or, if it is membrane-associated, it is substantially altered from its cytoplasmic form or is released essentially quantitatively into the soluble-protein fraction during membrane isolation. With regard to the discrepancy in molecular weight of the V antigen between Lawton et al. (1963) who estimated it to be 90,000 and Straley and Brubaker (1981) whose estimation was 38,000, the possible existence of a dimer form or association with another protein in the bacterial cell was assumed.

In parallel with these biochemical studies dealing with the three *Yersinia* species in comparison, the serological relatedness among V and W antigens produced by *vwa*$^+$ *Y. pestis, Y. pseudotuberculosis,* and *Y. enterocolitica* was in-

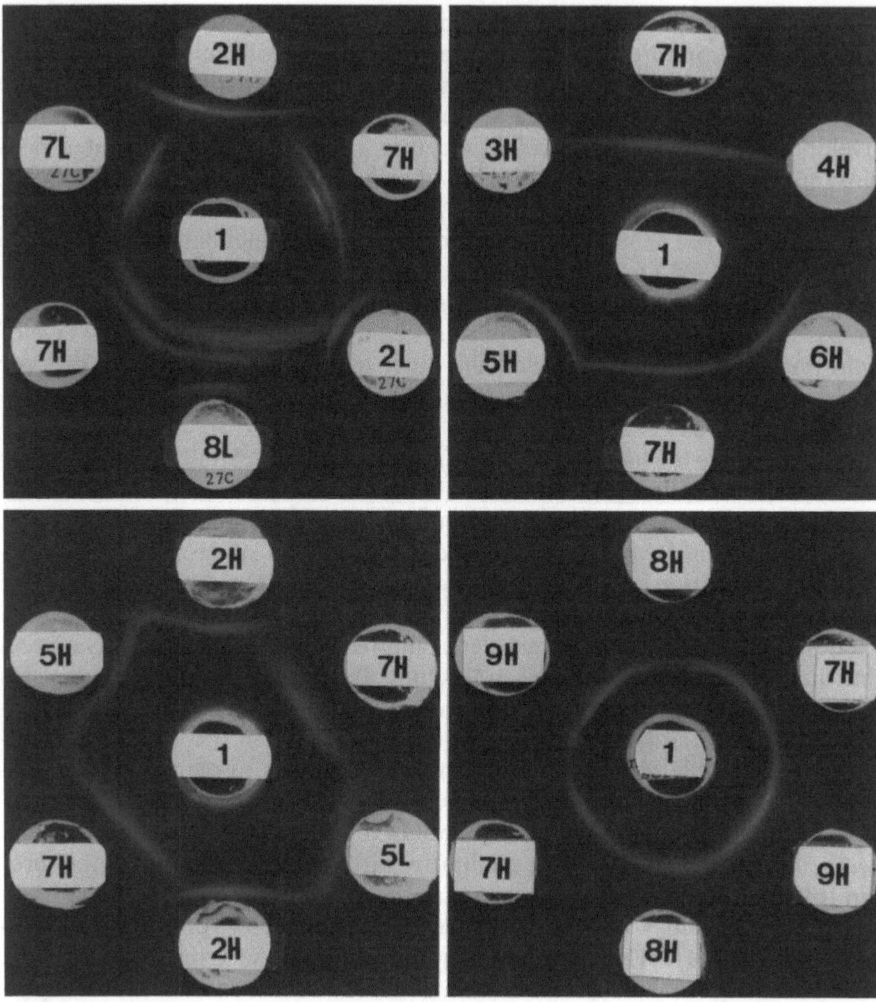

Fig. III.10. Agar-gel immunodiffusion tests between anti-*Y. enterocolitica* WA2-1 (calcium-dependent) rabbit serum (*central wells*) and bacillary suspensions of *Y. pestis*, *Y. pseudotuberculosis*, and *Y. enterocolitica* strains. (*1*): Wells containing anti-*Y. enterocolitica* WA2-1 (calcium-dependent) rabbit serum. (*2H*): Wells containing bacillary-saline suspension of *Y. enterocolitica* WA2-1 grown at 37 °C. (*2L*): Wells containing bacillary-saline suspension of *Y. enterocolitica* WA2-1 grown at 27 °C. (*3H*): Well containing bacillary suspension of *Y. pseudotuberculosis* 1779 grown at 37 °C. (*4H*): Wells containing bacillary suspension of *Y. pseudotuberculosis* Saru 1-1 grown at 37 °C. (*5H*): Wells containing bacillary suspension of *Y. pseudotuberculosis* 12 grown at 37 °C. (*5L*): Well containing bacillary suspension of *Y. pseudotuberculosis* grown at 27 °C. (*6H*): Well containing bacillary suspension of *Y. pseudotuberculosis* R2 grown at 37 °C. (*7H*): Wells containing bacillary suspension of *Y. pestis* EV grown at 37 °C. (*7L*): Well containing bacillary suspension of *Y. pestis* EV grown at 27 °C. (*8H*): Well containing bacillary suspension of *Y. pestis* Yreka grown at 37 °C. (*8L*): Well containing bacillary suspension of *Y. pestis* Yreka grown at 27 °C (*9H*): Well containing bacillary suspension of *Y. pestis* MII40 grown at 37 °C

vestigated using the agar-gel diffusion test of Ouchterlony and anti-*Y. enterocolitica* VW serum prepared by extensive i.v. immunization of rabbits with *vwa*$^+$ *Y. enterocolitica* WA2-1 then absorbed or not absorbed with the *vwa*$^-$ isogenic *Y. enterocolitica* WA5-1 (Wake et al. manuscript in preparation). Figure III.10 is a set of photographs showing the results. The precipitin lines between this anti-*Y. enterocolitica* VW serum and suspensions of *Y. enterocolitica* or *Y. pseudotuberculosis* fused forming arc(s), while those between this serum and suspension of *Y. pestis* formed spurs against the precipitin lines between this serum and the other two *Yersinia* organisms.

Therefore, the virulence determinants controlling the VW phenotype with Cal, i.e., *vwa*$^+$ and *cal*$^+$ genes, should be an important subject for review from the point of view of molecular biology and the microevolution of bacterial pathogenicity, as exemplified by some other pathogens described in Part I. Furthermore, many problems introduced in this section at the phenotypic level of VW and Cal can be reexamined by considering the genetic mechanisms controlling these phenotypes.

Tsukano and Wake (1982) using procedures described under *fra* in this chapter, determined the localization of *vwa*$^+$ and *cal*$^+$ genes of *Y. pestis* strain Yreka on the plasmid of 45 Mdal size. Figure III.11 shows an electron micrograph of this covalently closed-circular molecular form.

However, the functional aspect of this plasmid was more extensively elucidated when not *Y. pestis,* but less virulent *Yersinia,* i.e., *Y. enterocolitica* was studied due to its feasibilities of safer treatments. In this species and *Y. pseu-*

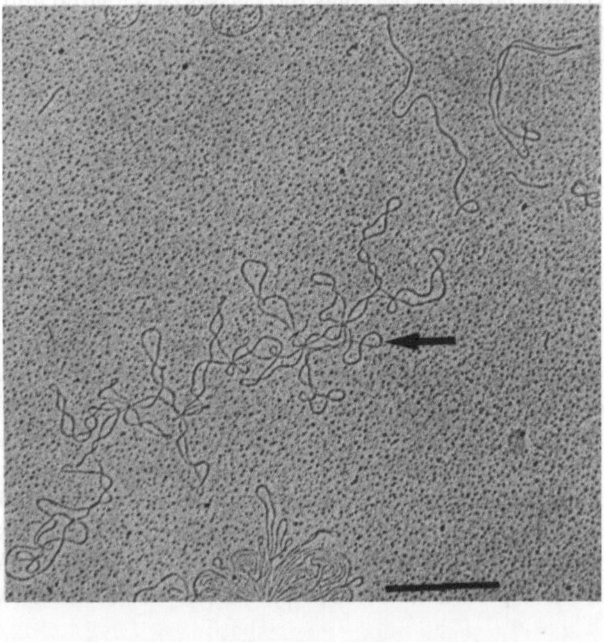

Fig. III.11. An electron micrograph of a covalently closed-circular form of a 45 Mdal plasmid controlling VW and Cal phenotypes (*arrow*) found in the phenol-extracted lysate of *Y. pestis* strain YD1; *bar*, 3 μ

dotuberculosis, the V and W are serologically related with those of *Y. pestis* as aforementioned, but the virulences are expressed in association with tissue invasiveness to guinea pig corneal mucous membranes (Serény test) and cultured mammal cell layers in addition to their detectability by Ca^{2+} dependency on magnesium oxalate agar as follows:

Zink et al. (1980) reported that the plasmids were detected by agarose-gel electrophoresis; their molecular weights were calculated using R-plasmids S-a, RP4, and R6-5 as the standards. The results suggested that tissue invasiveness is related to 41 Mdal plasmids, because all of the invasive strains, but none of the noninvasive strains harbored this plasmid. An isolate from an invasive strain which lost 41 Mdal after treatment with ethidium bromide was no longer tissue invasive, and did not regain the ability to invade tissue after repeated subculturing. The 41 Mdal plasmid can also be tested upon growth of the host strain at 35 °C, resultant isolates are noninvasive. These experiments confirm a relationship between tissue invasiveness and the 41 Mdal plasmid since loss of the plasmid is always accompanied by loss of tissue invasiveness, regardless of the curing procedure. Many tissue-invasive strains were found to harbor a 35 Mdal plasmid in addition to the 41 Mdal plasmid, but any relationship of 35 Mdal plasmid to pathogenicity remains to be determined. Gemski et al. (1980a) found that three *Y. enterocolitica* (08) strains which he tested had similar invasive and toxinogenic properties and were lethal to mice challenged either orally or intraperitoneally (i.p. $LD_{50} = 10^2$ organisms), while three noninvasive strains derived from them failed to cause conjunctivitis in guinea pigs or fatal infections in mice (i.p. $LD_{50} > 10^8$ organisms). Electrophoretic as well as electron microscopic studies revealed that all three invasive strains harbored two plasmid classes of 42 Mdal and 36 Mdal. A derivative of these strains, which lost the 42 Mdal plasmid, but retained the 36 Mdal plasmid, was noninvasive and all three avirulent, but still enteropathogenic derivatives lacked the 42 Mdal plasmids. Thus, the results of Gemski et al. (1980a) and those of Zink et al. (1980) both show that a plasmid is involved in the control of invasiveness of *Y. enterocolitica.* Gemski et al. (1980a) assayed these six strains for Ca^{2+} dependency using the procedures developed for *Y. pestis* (Chap. 1), but added blood-agar plates to exclude the possibility of temperature-sensitive (Ts) mutants. It was evident that the growth on magnesium-oxalate agar of the three 42 Mdal plasmid-bearing invasive strains was inhibited at 37 °C, but not 26 °C. Noninvasive derivatives which lacked the 42 Mdal plasmid were not inhibited at either temperature. Gemski et al. (1980b) reported essentially similar results for a *Y. pseudotuberculosis* strain III using an electrophoretic technique for plasmid isolation and the production of keratoconjunctivitis in guinea pigs to assay for tissue invasiveness. He designated the 42 Mdal plasmid of *Y. pseudotuberculosis* strain III as similar to *Y. enterocolitica* in that it contains a plasmid associated with pathogenicity and Ca^{2+} dependence. On the other hand, a dissimilarity between these two plasmid species, pWR41 isolated from *Y. enterocolitica* WA and pWR43 isolated from *Y. pseudotuberculosis* strain III, was disclosed when their restriction enzyme fragmentation patterns were compared using the procedure of Roberts (1976). To exclude the possibility that the pWR42 harbored in *Y. enterocolitica* Y7P was contaminated with another species, such as the cryptic 36 Mdal plasmid, Gemski et al. (1980b) selected *Y. enterocolitica* WA harboring pWR41, but not the 36 Mdal plasmid. Treatment of DNA from both plasmids with the restriction enzyme *Bam* HI yielded some fragments that are the same size, but in general the fragment-banding patterns were distinct for the two plasmids. *Eco* R1 restriction-enzyme digestion of these plasmids also yielded different electrophoretic patterns. Clearly, the plasmids from *Y. pseudotuberculosis* and *Y. enterocolitica* have different specific cleavage sites for restriction endonuclease and are, therefore, nonidentical at the molecular level.

In spite of this divergence of the plasmids at the molecular level, the agar-gel diffusion tests using rabbit antiserum against VW+ *Y. pestis* which had been exhaustively adsorbed with VW− *Y. pestis* revealed a mutually-fused precipitin line between Ca^{2+}-dependent *Y. pestis* EV76, *Y. pseudotuberculosis* PB1, and *Y. enterocolitica* WA, but no precipitin lines with avirulent Ca^{2+}-independent derivative isolates of these species (Carter et al. 1980). This discordance between the results showing divergent nucleotide sequence patterns in the genotype and those suggesting homologous amino acid sequence patterns among various species of *Yersinia* indicate that the nucleotide sequences encoding the VW phenotype in each species are similar, but comprise only a part of the plasmids found in the different species. However, similar tests using rabbit antiserum against VW+ *Y. enterocolitica* WA2-1 (08) adsorbed with the isogenic VW− organisms revealed precipitin lines with spur formation between Ca^{2+}-dependent *Y. pestis* Yreka, *Y. pseudotuberculosis* 222/+, and *Y. enterocolitica* Te1001 (03), but no precipitin line with Ca^{2+}-independent organisms of these species (Fig. III.12, Aoki and Wake, manuscript in preparation). These results show that antigens produced by Ca^{2+}-dependent cells are not entirely homologous in *Y. pestis* VW, *Y. pseudotuberculosis* VW, and *Y. enterocolitica* VW of 03 and 08 serovars.

Independent of these studies, Portnoy et al. (1981) and Portnoy and Falkow (1981) examined 90 *Y. enterocolitica* isolates of a wide variety of serovars of human, animal, and environmental origin from diverse geographical locations. Those isolates associated with human disease all contained plasmids of molecular weights ranging from 40 to 48 Mdal. The similarity suggested the possibility that a family of related plasmids was associated with the pathogenicity of these strains. Their relatedness was determined using DNA homology with plasmid DNA from an arbitrarily elected isolate as probe. The distribution of DNA homology thus measured ranged from 58% to 87%, reflecting a diversity of nucleotide sequences. He also cleaved each isolated plasmid DNA with the restriction endonucleases *Hind* III, *Bam* HI, and *Eco* R1 and separated the restriction fragments by agarose-gel electrophoresis. The individual DNA fragments were transferred to a nitrocellulose filter and hybridized with heat-denatured [^{32}P]-labeled probe DNA, according to the blotting method of Southern (1975). Homologous sequences were distributed

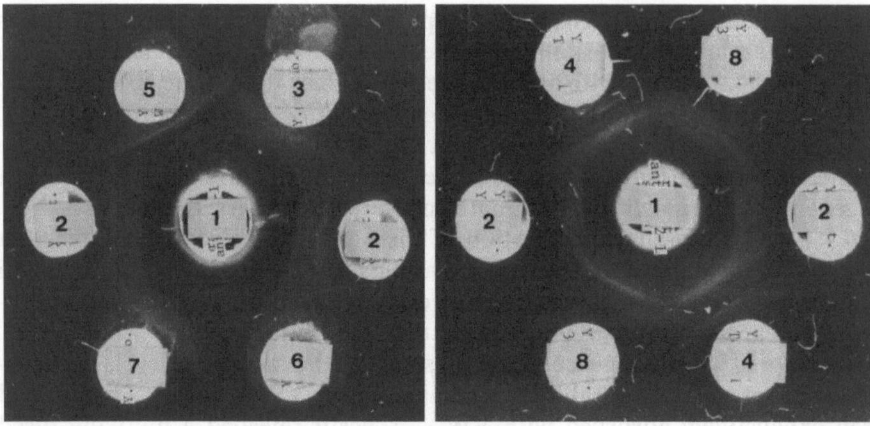

Fig. III.12. Agar-gel immunodiffusion tests between anti-*Y. enterocolitica* WA2-1 (calcium-dependent) rabbit serum (*central wells*) and bacillary suspensions of *Y. pestis*, *Y. pseudotuberculosis*, and *Y. enterocolitica* strains. (*1*): Wells containing anti-*Y. enterocolitica* WA2-1 (calcium-dependent) rabbit serum. (*2*): Wells containing bacillary saline suspension of *Y. pestis* Yreka. (*3*): Well containing bacillary saline suspension of *Y. pseudotuberculosis* 222/+. (*4*): Wells containing bacillary saline suspension of *Y. enterocolitica* Te1001. (*5*): Well containing bacillary saline suspension of *Y. enterocolitica* WA2-1. (*6*): Well containing bacillary suspension of *Y. enterocolitica* WA5-1 (calcium-independent). (*7*): Well containing bacillary saline suspension of *Y. pseudotuberculosis* 377. (*8*): Wells containing bacillary suspension of an avirulent *Y. pestis* strain 376e9. All bacterial strains were grown at 37 °C

throughout most of the plasmid molecules and did not merely represent a shared subset of sequences. This indicated the diversity of this plasmid family, presumably resulting from microevolution of the genetic material.

In order to confirm the validity of this assumption, the fragment patterns of endonuclease-cleaved plasmid DNAs isolated from these *Y. enterocolitica* strains were compared by agarose-gel electrophoresis. Many fragments of identical size were shared by multiple plasmids, and strains isolated from adjacent geographical areas showed the greatest conservation in restriction patterns.

This suggests that analysis of the plasmid associated with bacterial pathogenicity at the molecular level may elucidate the evolutional diversification, i.e., niche segregation of some bacterial infections at the global level, a new aspect of the general theory of host-parasite relationships at the epidemiological level.

Schiemann and Devenish (1980) also associated this family of plasmids with the pathogenesis of *Y. enterocolitica* using the Mongolian gerbil as the animal model. He found that all strains isolated from cases of human disease harbored a 40 to 48 Mdal plasmid and caused lethal infections in Mongolian gerbils at doses of 10^3 organisms. In contrast, Portnoy et al. (1981) found one strain harboring a 46 Mdal plasmid and a strain lacking a 40 to 48 Mdal plasmid, but otherwise isogenic to the virulent one, did not kill gerbils even when 10^6 organisms were injected.

The presence of one avirulent strain that harbors a 46 Mdal plasmid is perhaps relevant to the finding of avirulent VW+ strains described in Chap. 2. These examples suggest nonfunctional states of this plasmid family (40 to 48 Mdal molecular size) relative to virulence. In this regard, Portnoy et al. (1981) also examined the relatedness of this plasmid family with tissue invasiveness by the use of a human epithelial-tissue cell-line model. All strains, whether they harbored or lacked a member of this plasmid family associated with this cell monolayer and were found intracellularly. Cells with a member of this plasmid family induced the detachment of the cell monolayer from the surface of the culture flasks, where cells lacking these plasmids did not. Therefore, the plasmid is not essential for the invasion of host cells by *Y. enterocolitica,* but apparently influences a factor responsible for monolayer detachment.

The possibility that toxic substances released from plasmid-bearing *Y. enterocolitica* are involved in cell detachment was excluded, because neither cell-free supernatant culture fluids nor filter-sterilized sonic extracts of these bacteria caused detachment of the host-cell monolayers. Then, it is conceivable that these toxic substances are produced under the regulation of a chromosomal gene(s). Because the process of the host-cell monolayer detachment may be involved in tissue invasiveness, the latter is influenced by a genetic determinant(s) localized on chromosomes. In conclusion, tissue invasiveness as expressed in the keratoconjunctivitis of the guinea pig model may be the cooperative effects of the products of plasmid and chromosomal genes.

To compare these differences at the molecular level, he examined membrane proteins of plasmid-bearing and plasmid-lacking strains grown at 37 °C and 25 °C. Fractions extracted from total envelopes as well as enriched outer membranes were examined by SDS-PAGE (polyacrylamide-gel electrophoresis). A minimum of three additional polypeptides were apparent in the outer-membrane profiles derived from plasmid-bearing cells grown at 37 °C compared with the same strains grown at 25 °C, consistent with that found in *Y. pestis* (Darveau et al. 1983). When these strains were cured of the 40 to 48 Mdal plasmid, the ability to express these additional polypeptides was lost, regardless of the growth temperature. From these results, Portnoy et al. (1981) concluded that strains of *Y. enterocolitica* harboring the 40 to 48 Mdal plasmid are drastically altered in their outer-membrane protein composition relative to cells lacking this plasmid when grown at 37 °C and that this may play a role in their pathogenesis. Another conclusion was that the addition of 2.5 mM calcium to the growth medium completely repressed the expression of all the plasmid-associated polypeptides.

Thus, a family of related plasmids controlling Ca^{2+} dependency is found in the major human pathogens of the genus *Yersinia.* Plasmid DNA isolated by the same procedure from an avirulent F1+VW+Cal+P1+P−Pu+ *Y. pestis* EV76 (Portnoy et al. 1981) revealed three species of plasmids of 6, 61, and 47 Mdal. The 6 Mdal plasmid has been known to be an associated P1 phenotype according to Ben-Gurion and Shafferman (1981). In a succeeding paper, Portnoy and Falkow (1981) sought to determine the extent of interspecies relatedness of the plasmids belonging to the 40 to 48 Mdal family by their shared DNA-sequence homology. A derivative *Y. pestis* EV76, cured of the 47 Mdal plasmid EV766 was also used. As the result of solution-hybridization assays, he concluded that when DNA is deleted, the *Y. pestis* organisms become Ca^{2+}-independent.

The reports by Portnoy et al. (1981) and Portnoy and Falkow (1981) caused abandonment of the dogmatic view that virulence plasmids are stable and do not suffer from insertion or deletion modifications. Thus, the loss of phenotypes does not always mean the loss of the corresponding plasmid. Alternatively, a limited length of the DNA segment is inserted into or deleted from the region of plasmid DNA base sequences that are regulating phenotype expression thus resulting in the loss of function. In other words, not only the loss of the whole-plasmid molecule, but also the submolecular modification of the base sequence at the functionally critical site, can cause the apparent loss of phenotype.

In this regard, the knowledge introduced in Part I (Chap. 3, p. 2) provides a wider scope of the involvements of transposon and IS that promote the diversification of an ancestral plasmid into a plasmid family in *Yersinia*. In fact, the extensive comparative analysis of *Y. enterocolitica vwa = cal* plasmids revealed that the existence of not the same, but rather more or less heterogeneously-sized, plasmids constituting a family throughout the species, is the result of evolution of this molecular species via diversification occurring in nature by the participation of transposon or insertion units (IS). The validity of this hypothesis will be experimentally supported later when the interaction between virulence- and drug-resistance (R)-plasmids is discussed (p. 228).

A concrete example of the participation of IS was given by Portnoy and Falkow (1981) when they detected the same DNA segment, i.e., a "conserved region" of 2.2-kb size in both *Y. pestis* EV and *Y. enterocolitica* 8081 using restriction endonuclease analysis.

To characterize the origin of the 2.2-kb DNA insertion found in some Ca^{2+}-independent strains a 7.5-kb fragment was isolated from an agarose gel of *Bam* HI-digested plasmid DNA from a Ca^{2+}-independent strain of EV76 (EV7627) and labeled with [^{32}P] in vitro. This fragment contained the entire *Bam* fragment 5 (Fig. III.13) and the 2.2-kb insertion sequence. *Bam*-5 was also isolated and labeled with [^{32}P]. Therefore, one could identify homologous sequences at the insertion site, because the fragment that can hybridize with [^{32}P]-labeled 7.5-kb, but not the [^{32}P]-labeled *Bam*-5 contains the 2.2-kb insertion sequence. Based on this principle, sequence homology with the insertion sequence was found at approximately ten sites in the bacterial chromosome and on the 6 Mdal plasmid, the 61 Mdal plasmid, and the *Bam*-2 fragment of the 47 Mdal plasmid (Fig. III.13). All of the other Ca^{2+}-independent mutants caused by insertions showed homology with the 2.2-kb DNA sequence. The

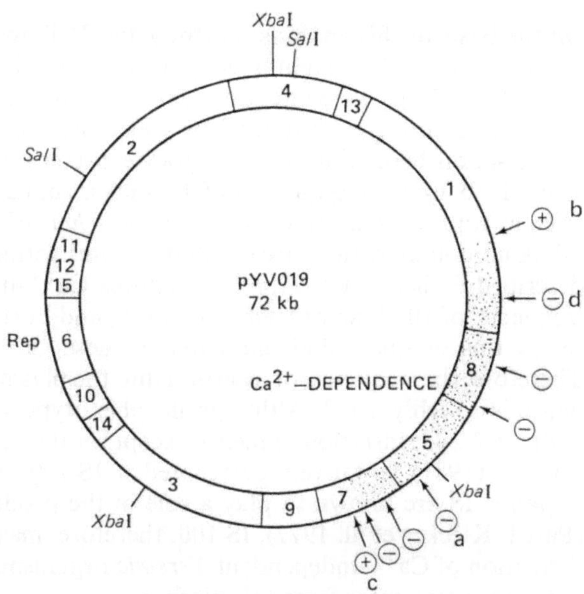

Fig. III.13. *Bam* HI restriction endonuclease map of plasmid pYV019 in strains of *Y. pestis*. The *numbers* refer to *Bam* HI restriction fragments (6). The *stippled area* refers to the border of the Ca^{2+}-dependence locus, which ends somewhere between insertion sites *b* and *d*. Symbols: (−) = Tn5 insertions that abolish Ca^{2+} dependence; (+) = Tn5 insertions that do not alter Ca^{2+} dependence; *a* = site of Tn5 insertion in strain 195-P3 and of IS100 insertion in strain 195-P2; *b* = site of Tn5 insertion in strain 195-P5; *c* = site of Tn5 insertion in strain 195-P5; and *d* = site of Tn5 insertion in strain 195-P6. Strain 195-P2, -P3, -P4, -P5 and -P6 are substrains of *Y. pestis* strain 195-P

2.2-kb sequence was hybridized with *Bam* HI-digested plasmid DNA from the deletion mutants, and no homology existed between the remaining plasmid sequences and the 2.2-kb sequence. The 2.2-kb sequence was also hybridized with cellular DNA from Ca^{2+}-dependent *Y. enterocolitica* 8081 and no sequence homology was detected.

Apart from the insertion mutant aforementioned, the majority of the Ca^{2+}-independent deletion mutants of the plasmid of EV76 (pYV76) had lost 95% of the plasmid, including all *Bam* HI-restriction sites, but retained a miniplasmid. One of these, a 2.5 Mdal (2.3 kb) miniplasmid was isolated from an agarose gel, labeled with [^{32}P], and hybridized to the *Bam* HI-restriction digest of the parental plasmid. The miniplasmid shared homology with the *Bam*-6 (4.9 kb) and *Bam*-13 (1.4 kb) fragments. Portnoy et al. (1981 b) concluded that this miniplasmid was derived from sequences internal to *Bam*-6, because it lacked *Bam* HI sites. This miniplasmid should contain the origin of replication of the parental plasmid, since it is stably maintained. If so, the origin of replication must be located within the *Bam*-6 fragment (Fig. III.13). This region of the *Y. pestis* plasmid shared no detectable DNA homology with the plasmid from *Y. enterocolitica*. Therefore, *Bam*-6 contains the origin of plasmid replication unique to *Y. pestis*.

Thus, it was suggested that plasmids of *Y. pestis* and *Y. enterocolitica* have different regions for the origin of replication and other sequences essential for plasmid maintenance. When *Bam*-6 was isolated from an agarose gel, labeled with [^{32}P], and hybridized to a *Bam* HI-restriction digest of plasmid DNA from *Y. pestis* 019, *Bam*-6 was found to share homology with *Bam*-13. This indicates that the plasmid from *Y. pestis* contains at least one repeated sequence.

On the basis of this analysis, Portnoy and Falkow (1981) speculated that the virulence of *Yersinia* might have been acquired when the regions encoding the Ca^{2+} dependence were integrated into several distinct ancestral plasmids and maintained during evolution of the species as a result of the selective pressure of host-parasite relationships, phagocyte-bacterium interactions, in particular. The 2.2-kb insertion sequences of *Y. pestis* is similar to the IS1 sequence found at eight copies per chromosome in *Escherichia coli* K12. This sequence shares DNA homology with approximately 13 sites within the bacterial cell. As was described earlier, Cal^+ *Y. pestis* produces Ca^{2+}-independent mutants at the frequency of 10^{-4} per cell per generation, and Portnoy and Falkow (1981) detected this unique 2.2-kb insertion in these Ca^{2+}-independent derivatives. Therefore, the frequency of insertion into the plasmid resulting in Ca^{2+} dependence is roughly 10^{-5}. Although no phenotype was found to be associated with the 2.2-kb insertion sequence except for its genetic mobility, Portnoy and Falkow (1981) tentatively designated it IS 100, as the representative IS of *Y. pestis*. IS are known to play a role in the production of deletion mutants (Part I, Klecker et al. 1977). IS 100, therefore, may have been involved in the formation of Ca^{2+}-independent *Yersinia* organisms resulting from deletion of some base sequences from this plasmid.

These extensive and elegant studies of molecular genetics, however, cannot elucidate the role of Ca^{2+} dependence in pathogenesis, and knowledge of the details of bacterial pathogenesis is required for us to determine how best to protect against infection. The authors of this book believe that the mechanism responsible for the induction of cell-mediated immunity may be tied to the role of Ca^{2+} dependence of *Yersinia*. Most specialists of bacterial infection-immunity believe that more or less limited intracellular multiplication of immunogenic living bacteria is a prerequisite for the induction of cell-mediated protective immunity against intracellular parasitic bacteria. The DNA region responsible for Ca^{2+} dependence itself might not encode the production of any antigenic determinant(s), but could, nonetheless, enable the intracellular *Yersinia* organisms to multiply. To support this concept, Charnetzky and Shuford (1985) have shown involvement of the specific *cal* region defined by Portnoy in the survival and growth of *Y. pestis* cells within mouse peritoneal cavities and within mouse peritoneal macrophages maintained in vitro using the mutant strains donated by Portnoy. Such intracellular behavior of *Yersinia* organisms could convert macrophages from an Ia-negative state to an Ia-positive (refer to Part II, Chap. 6, p. 24), due to the function of Cal phenotype. On the other hand, the antigenic epitope of V-protein binds to, i.e., recognized by, the specific T-cell receptor to produce T-cell-mediated immunity as the result of dual recognition (Part II, Chap. 11), although extensive immunization with V-antigenic epitope can produce humoral immunity without participation of the Cal phenotype, i.e., a physiological character designated calcium dependency.

In conclusion, VW and Cal function as the most critical virulence factors of *Yersinia* to show pathogenicity for the sensitive host animals, but confer the latter, the most effective protective immunity, if they function without cooperation of the other virulence factor(s) and allow the host animals to be cured from the damage due to the infection.

P1

This virulence factor unique to *Y. pestis* is encoded by a plasmid *pst* and lacking in lesser *Yersinia* species.

Actually, P1− *Y. pestis* G-32 and Dodson, equipped with F1+VW+P+Pu+, gave poorly-graded and erratic-lethal dose responses in mice and guinea pigs (Surgalla and Beesley 1971). Lower LD_{50}'s are found when i.p., s.c., or i.d. (intradermal) routes are used than when the i.v. route is used. However, all i.d. challenged guinea pigs developed visible skin lesions and bubos, although none died. These two strains killed 31 of 240 guinea pigs challenged i.p. or i.v. with 625 to 2×10^6 organisms, demonstrating the importance of P1 as a virulence factor. None of 4 animals from each of 24 groups rechallenged i.d. 30 days later with 10^6 organisms of a fully virulent strain Kim 10 died. To reconfirm these results, 4 guinea pigs from each group received i.d. inoculation with 124, 25, 5, and 1 organisms, respectively, and all animals receiving 124 or 25 organisms developed skin lesions and bubos, but 3 of 4 animals receiving 5 organisms, and 1 animal receiving 1 organism, were killed; whereas all animals receiving more than 25 organisms survived.

All animals that were recognizably infected by strain Dodson and survived were not killed upon i.d. reinfection with 9.8×10^3 organisms of strain K10, 47 days later. Therefore, P1− *Y. pestis* organisms starting from a number approaching one organism were not eliminated by nonspecific defense mechanisms, but rather multiplied enough to induce protective immunity in guinea pigs. Similar results were obtained using mice and the P1− strain G-32. Again dose-response curves were poorly graded; all mice did not survive the initial challenge, but those which did, resisted subsequent s.c. challenge with 9.3×10^3 organisms of strain K10 49 days later. Infecting doses of 50 organisms stimulated formation of resistance. Representative data is shown in Table III.5.

These results indicate that P1 is the responsible virulence factor that permits the invasive (penetrating) infections of *Y. pestis* as opposed to the generally noninvasive infections of *Y. pseudotuberculosis* and *Y. enterocolitica*.

The i.p. LD_{50} of *Y. pestis* strains of the F1+VW+Cal+P1−P+Pu+ phenotype were independently found to be as high as 10^5 organisms in the mouse and 10^6 organisms in guinea pig (Brubaker 1972).

However, the designation of this virulence factor was derived from the abbreviation of pesticin 1, a bacteriocin resembling typical group B colicins, quite irrelevant to a virulence function that permits invasion.

With regard to the P1 phentoype itself, Ben-Gurion and Hertman (1958), using the two-layer method of Fredericq (1954) with which the other bacteriocins, e.g., colicins, pyocin, and megacin had been characterized, found

Table III.5. Response of mice to Fl+VW+P+Pl−Pu+ strain G-32. (Modified from Surgalla and Beesley 1971)

Dose (i.p.)	Lethality (dead/20)	Percent survivors, immune[a]
5×10^7	7	100
5×10^6	5	100
5×10^5	3	100
5×10^4	5	100
5×10^3	5	60
5×10^2	2	83
5×10^1	2	63
5×10^0	1	16

[a] Mice that survived against G-32 were s.c. challenged with 9.3×10^3 organisms of strain Kim 10 after 49 days. No control mice survived.

that the supernatant fluid from a well-aerated 48 h broth culture of *Y. pestis* grown at 30 °C contained a substance which inhibited the growth of *Y. pseudotuberculosis*. The name "pesticin" was proposed for this material after determination of its sensitivity to proteolytic enzymes, thermolability, pH range (pH 6 to 8), and inducibility by ultraviolet (uv) irradiation. Pesticin production was greater at 27 °C than at 37 °C, while its activity was stronger at 37 °C than at 30 °C. These authors and others (Brubaker and Surgalla 1961, 1962a; Brubaker et al. 1965a; Burrows 1965) have found that most *Y. pestis* strains produce pesticin. However, the nonproducing *Y. pestis* strain TRU was not sensitive to the pesticin produced by other strains of *Y. pestis,* unlike the cases occurring with other bacteriocins which are produced by other species of bacteria. All of the authors agree that only serovar 1 of *Y. pseudotuberculosis* and certain strains of *E. coli,* including the colicin-indicator strain ∅ and a few isolates of *Y. enterocolitica* are inhibited by pesticin. The structural difference that characterizes pesticin-sensitive serovars is the presence and linkage of paratose (3,6-dideoxy-D-ribohexose) in serovar 1 organisms (Brubaker et al. 1965). One class of *E. coli* strain ∅ mutants selected for resistance to pesticin (Ben-Gurion and Hertman 1958) are no longer sensitive to the colicins B, D, I, and SI. These various bacteriocins might share a common receptor on the surface of the pesticin-sensitive *E. coli* strain ∅, but paratose has not been reported in this strain.

Extensive studies on the specificity, induction, and adsorption characteristics of P1 employing the assay method standardized by Surgalla et al. (1970) to quantitate the pesticin activity directed against sensitive cells of *Y. pseudotuberculosis,* were reported by Hu et al. (1972) and then by Hu and Brubaker (1974).

Although they started purification procedures from ammonium sulfate precipitation and assayed inhibitory activities of the calcium hydroxylapatite dialysates to the growth of cells of all four aforementioned indicator bacterial species, Hu et al. (1972) used DEAE-cellulose, while Hu and Brubaker (1974) used Sephadex G-200 and diethylaminoethyl-cellulose for column chromatography. Another divergency also existed between the starting materials they used, i.e., the crude extract of the ultrasonically disrupted *Y. pestis* cells used by Hu et al. (1972) versus the cell-free extracts of noninduced *Y. pestis* used by Hu and Brubaker (1974). However, the quite high titer of intracellular P1 concentration of noninduced organisms increased furthermore from five- to eightfold when Hu et al. (1972) induced the P1 production with UV irradiation and suggested the difference between intra- and extracellular P1 molecules. In fact, substitution by Hu and Brubaker (1974) with mitomycin-C treatment for uv irradiation resulted in the discrimination at the final step of purification between two fractions designated α- and β-pesticin showing identical activity. Their absorption spectra were typical of simple proteins and neither contained significant hexose or hexosamine. The molecular weight was about 65,000 as determined by equilibrium sedimentation and gel filtration in guanidine hydrochloride. Although isoelectric points of α- and β-form were determined to be 5.49 and 5.87, respectively, both preparations contained similar ratios of amino acids, lacked detectable cystein, and possessed a common primary structure as judged by peptide analysis of tryptic digests. They were mutually interconvertible in vitro, with equilibrium favoring formation of the α-form, although rapid conversion into a third form, termed γ-pesticin was observed. The latter eluted from calcium hydroxylapatite after the β-form and lacked significant biological activity. The identical antigenicity of these three forms was demonstrated by immunodiffusion. These results suggested that P1 can exist in conformer states, as do certain cysteine-deficient colicins.

The mode of action of P1 had been once believed to be the inhibition of DNA synthesis, degradation of RNA, and inhibition of protein synthesis in the sensitive-indicator strain (Elgat and Ben-Gurion 1969) which is similar to the action of colicin E2. However, Ferber and Brubaker (1979) showed that P1 had no primary effect on net synthesis of DNA, RNA, and protein, but rather that it interefered with cell-wall biosynthesis. Upon treatment with homogeneous α-conformer P1, *E. coli* ∅ cells increased their cell mass, but underwent conversion to osmotically stable spheroplastlike structures identical to those illustrated for P1-sensitive strains of *Yersinia* (Hall and Brubaker 1978). The balanced synthesis of DNA, RNA, and protein was shown by net incorporation of their respective radioactive precursors into trichloracetic acid-insoluble materials at ratios similar in P1-treated and untreated cells to alter the previous concept. A N-acetylglucosamindase activity was demonstrated by the release of a cell-wall specific label ([^{14}C]-diaminopimelic acid) from both resting cells of P1-sensitive bacteria and the purified mureinlipoprotein treated with P1. However, P1 hydrolysis of mureinlipoprotein purified from both P1-sensitive and P1-resistant cells occurred over a broad pH range, with an optimum of 4.7. Therefore, the presence or absence of absorption sites for the P1 molecule apparently differentiates P1-sensitive cells from P1-resistant cells.

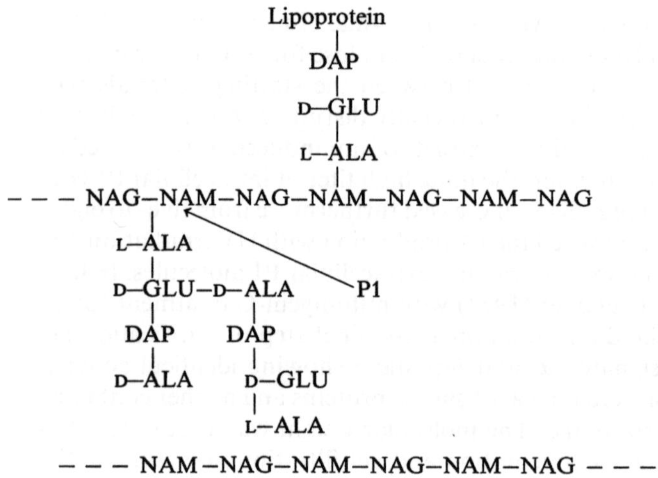

Fig. III.14. Potential cleavage site in *E. coli* murein-lipoprotein for P1 (Modified from Ferber et al. 1979)

Thus, P1 is a lysozyme that catalyzes the hydrolysis of the β-1, 4, bond between N-acetylglucosamine and N-acetylmuramic acid in the glucan backbone. The cumulative effects of pesticin-dependent hydrolysis of old murein and partial inhibition of glucan-strand elongation lead to the eventual death of sensitive bacteria. The proposed potential P1-cleavage sites in *E. coli* mureinlipoprotein (Ferber and Brubaker 1979) are shown in Fig. III.14.

In spite of the exhaustive molecular characterization and elucidation of molecular activity of P1 as a bacteriocin, the question remains how, if at all, P1 relates to the interaction of *Y. pestis* with the host-animal defense mechanisms (Part II). Burrows et al. (1964) reported that full virulence for mice was restored in a F1+VW+Cal+P1−P−Pu+ avirulent *Y. pestis* strain 014 when iron was injected into the infected mice and that iron supplementation also enhanced the virulence of a VW+P1−P+ *Y. pseudotuberculosis* strain in mice. Since the sensitivities of *E. coli, pst*$^-$ *Y. pestis,* and serovar 1 *Y. pseudotuberculosis* to P1 are enhanced by EDTA and inhibited by Fe^{3+} and hemin, the receptor for P1 resembles the receptor for group B colicin of *E. coli* which is known to be involved in iron metabolism. However, another virulence factor P of *Y. pestis* is also involved in iron metabolism as will be described in the next section. Therefore, these results indicate that in in vivo environments both P1 and P phenotypes and iron metabolism interrelate and that one or more of these influence the virulence of *Y. pestis* and *Y. pseudotuberculosis*. It is perhaps relevant that *Yersinia* organisms characteristically contain high levels of the iron-containing enzyme, catalase.

Recently, Moncla et al. (1983) proposed another possibility of *pst* plasmid involvement to virulence expression unique to *Y. pestis*. They found that *Y. pestis* organisms lacking both *pst* and *vwa*=*cal* plasmids do not take up palmitic acid, but their levels of an energy-dependent constitutive uptake system for fatty acids constituted of two enzymes of β-oxidation, i.e., β-hydroxyacyl-coenzyme A dehydrogenase and the two enzymes of the glyoxalate shunt, i.e., isocitrate and malate synthetase, are comparably high to those observed in the wild-type strain possessing both these plasmids. Because the wild-type organisms harboring *pst* and *vwa=cal* plasmids not only resist intracellular-killing activity of fatty acids (Part II, Chap. 9, p. 265), but also take up palmitic acid as a primary carbon and energy source to express virulence, the function of *pst* plasmid may be the involvement in determining the characteristics of the bacterial cell surface in cooperation with *vwa=cal* plasmid in order to maintain intraphagocytic survival of the bacteria.

However, the most obvious correlations between the P1 phenotype and the mechanisms of virulence expression by *Y. pestis* reside not in the P1 molecule itself, but rather in the associated production of "invasive enzymes" presumably controlled by linked but distinct genes located on the *pst* plasmid. Jawetz and Meyer (1944) first reported that extracts from both virulent and avirulent *Y. pestis* strains contain a factor which enhanced spreading and capillary permeability and that both avirulent and virulent *Y. pestis* organisms are able to coagulate plasma from rabbits, but not plasma from mice, guinea pigs, squirrels, horses, sheep, or man. The recognition of fibrinolysin (Madison 1936) and coagulase production by *Y. pestis,* then, preceded the finding of P1 production by 14 years. Jawetz and Meyer (1944) concluded that the serum, plasma, or other body fluids of animals immune to plague infection are unable to destroy or lyse *Y. pestis* organisms in vitro and in vivo in the absence of phagocytic cells, although these immune animals are able to restrict the bulk of the injected *Y. pestis* organisms at the site of injection.

The fibrinolysin activates plasminogen to plasmin, resulting in the proteolytic dissolution of fibrin. This can be seen as a clear liquid spot on an opalescent fibrin film in the laboratory (Beesley et al. 1967), whereas coagulase brings about coagulation of pooled rabbit plasma containing 10 mg ml^{-1} sodium citrate. However, no evidence has been obtained to show that the fibrinolysin (F) and coagulase (C) activities produced by *Y. pestis* result from the production of distinct chemical entities (Beesley et al. 1967). In contrast, crystalline trypsin, for example, activates both plasminogen and prothrombin to a fibrinolysin and coagulase, respectively (Fruton and Simons 1958). With regard to the activity, as little as 10 ng of cellular debris containing *Y. pestis* fibrinolysin is enough to convert measurable amounts of plasminogen to plasmin. On the other hand, 0.1 ml plasma is required in the qualitative assay for *Y. pestis* coagulase with decreasing amounts of bacterial-cell extract or cell debris added for the quantitative assay.

Extensive numbers of laboratory stock strains and artificially induced *pst* mutants have been examined in many laboratories all over the world; strains lacking only one or two of the these (P1 –, fibrinogenic, or coagulase) activities have yet to be encountered (Brubaker and Sargalla 1961). The difficulty in trying to dissociate these activities into two or three entities is clear. However, Beesley et al. (1967) reported that P1 activity could be discriminated from fibrinolysin (F) and coagulase (C) activities based on their intracellular localizations, difference in their stabilities towards heat, and their separation by Sephadex G-200 column chromatography.

Therefore, the reasonable conclusion would be that P1 is encoded by one gene and F and C are encoded by another gene or located on the same replicon.

With regard to the virulence determinant *pst* encoding P1, Tsukano and Wake (1982), using a procedure described under *fra* in this chapter, determined the localization of *pst* genes of *Y. pestis* strain Yreka on the plasmid of 7 Mdal size. An electron micrographic picture of it has been included in Fig. III.3.

The molecular size of this plasmid identified by Tsukano coincides with the 6 Mdal plasmid revealed by Ferber and Brubaker (1981) using agarose-gel electrophoresis. This plasmid, termed pEL 100, was unique to P1+ strains of *Y. pests* and missing in lesser *Yersinia* species. However, Ferber and Brubaker (1981) found that a particular F1+VW−P1+P−Pu+ strain A1122 does not harbor a 6 Mdal plasmid (pEL 100), but a 20 Mdal plasmid (termed pEL 101). it must be noted that neither the 6 Mdal plasmid (pEL 100) nor the 20 Mdal plasmid (pEL 101) was detected in the P−P1− mutant derived from strain A1122. *Y. pestis* strain A1122 is avirulent and is unique in its overproduction of P1 relative to other pst^+ *Y. pestis* strains. Presumably, the 20 Mdal plasmid (pEL 101) rather than the 6 Mdal plasmid (pEL 100) controls P1 production in this strain. The 20 Mdal plasmid of A1122, which may contain repeated nucleotide base sequences for P1 production or regulation of *pst,* may be altered relative to that functioning when the genes are in the 6 Mdal plasmid.

In summary, such a hypothesis could be proposed that *Y. pestis* has been established as a species causing invasive (penetrating) infection in mammalian hosts via diversification stemming from their progenitor species which has a close taxonomic relationship with *Escherichia coli.* Whereas *Y. enterocolitica* and *Y. pseudotuberculosis* harboring *vwa* plasmid, but not *pst* plasmid, have segregated into species causing localized exudative infection due to their inability to invade, *Y. pestis* has been selected undergoing a process of "degenerative" evolution analogous to that assumed to have occurred in certain obligate intracellular parasites, such as rickettsiae and chlamydiae (Moulder 1962). For such niche segregation, i.e., evolutional diversification, the host-defense mechanisms delineated in Part II must have played a role of selective pressure, and acquisition by progenitor *Yersinia* organisms of *pst* plasmid, a critical role for the segregation. In contrast, *Y. pseudotuberculosis* and *Y. enterocolitica* which are free from *pst* plasmid and, therefore, do not produce pesticin to result in P1− phenotype, are primarily transmitted orally and thus must be able to maintain themselves in a natural environment, e.g., soil and water.

P

This virulence factor is encoded by a plasmid *pgm*.

Y. pestis organisms with all of the established virulence determinants except *pgm*, phenotypically F1+VW+P1+P−Pu+ show an intraperitoneal LD_{50} for the mouse $>10^7$ organisms and for guinea pig of $>10^8$ organisms (Brubaker 1972), indicating the importance of P.

This factor was found when Jackson and Burrows (1956a) using a suitable medium under well-defined conditions, but a concentration of hemin greatly in excess of the minimum requirement for growth, observed that some, but not all, strains of *Y. pestis* produced dark-brown colonies. After prolonged incubation, nonpigmented secondary colonies developed from the pigmented ones, but the isolated nonpigmented organisms displayed reduced virulence for mice. The pigmentation resulted from the absorption of hemin by the bacteria, colonies of P+ strains were not pigmented when they were seeded on a cellophane sheet directly covering the agar surface which did not permit the perme-

Table III.6. Recipe for AGA-agar medium

Prepare stock solutionsA, B, C, D, E, and F separately. The following are for 2 liters of final medium.

			10 plates per 200 ml bottle
A. Oxoid Agar (No. 2)	40 g/1700 ml Aq. Dist.		170 ml (8 oz flat)
pH 7.0			
autoclave			
B. KH_2PO_4	8.24 g/100 ml Aq. Dist.		10 ml (½ oz bottle)
pH 8.0			
autoclave			
C. NH_4Cl	0.64 g		
$(NH_4)_2SO_4$	0.64 g		
$MgSO_4$ $7H_2O$	0.1 g	50 ml Aq. Dist.	5 ml (bijon)
$ZnSO_4$	0.0002 g		
$MnSO_4$	0.0002 g		
		Millipore filter	
D. Galactose	4.0 g		
dl-phenylalanine	0.24 g		
l-cysteine HCl	0.1 g		
dl-methionine	0.12 g		
dl-valine	0.1 g	50 ml Aq. Dist.	5 ml (bijon)
dl-isoleucine	0.11 g		
glycine	0.1 g		
Na lactate	0.2 g		
		Millipore filter	
E. Hemin	0.2 g/100 ml N/100 NaOH		10 ml (½ oz bottle)
F. l-arginine	0.2 g/10 ml Aq. Dist.		1 ml

Melt agar and add additions at 55° C.
Before pouring add 0.2/200 ml medium ferric ammonium citrate.

ation of the hemin. The grades of pigmentation varied depending on the medium constituents. The composition of a suitably-defined medium for distinguishing P+ from P−, AGA, which has been used for the studies of not only P, but also another virulence factor Pu, described in the next section, is given in Table III.6.

Simultaneously with this finding, Jackson et al. (1956a) noticed also that certain basic dyes with planar chemical configurations are absorbed by P+, but not P− cells. In addition, P+ strains grown in conventional liquid medium containing these basic dyes absorbed the latter more efficiently than P− ones, similarly to those grown in well-defined medium. Therefore, Surgalla et al. (1970) recommended the routine use of Congo red dye in conventional medium for the discrimination between P+ and P− bacteria from the practical point of view.

However, this phenomenon of Congo red dye absorption by virulent bacteria is not unique to *Y. pestis,* and Payne and Finkelstein (1977) used this method successfully to differentiate virulent strains of *Vibrio cholerae, Neisseria gonorrhoeae,* and *Neisseria meningitidis* from avirulent ones.

In this context, such a universality of the virulence factor P among pathogenic bacteria could be considered in relation to the microorganismic machinery involved in the transport and storage of iron, i.e., siderophore (refer to Part I). Since all *Yersinia* (including P− *Y. pestis*) can utilize hemin as a sole source of iron, the ability to form a pigmented colony evidently reflects the particular machinery for iron storage unique to P+ organisms. The nature of this outer membrane storage complex has not been defined; the component is extremely hydrophobic, as judged by its insolubility in aqueous solvents. Recent studies have shown that P+ cells of *Y. pestis* grown with hemin as the sole source of iron also contain a high-molecular weight cytoplasmic iron storage component, in which the cation exists as Fe^{3+}. Accordingly, P+ organisms are capable of removing the cation from hemin and storing it internally as inorganic iron (Brubaker 1984).

In addition, P1 as well as P phenotypes are involved in iron metabolism as was delineated in the previous section; loss of phenotype P results in the loss of some surface receptor for P1.

Important is the discrimination of this outer-membrane iron-storage complex expressed as P+ from the cytoplasmic iron-storage component with which P− organisms are also sharing in common. For the recognition of P as the critical virulence factor, the counterpart host-defense mechanisms which are surmounted by P should be determined. A clue to this determination was given when Jackson et al. (1956b) demonstrated that *Yersinia* organisms with all virulence factors except P restored virulence when exogenous iron was supplemented. Wake et al. (1970) concluded that the host iron-binding proteins, such as transferrin and lactoferrin, constitute the defense mechanisms to be surmounted by P (refer to Part II, Chap. 6, p. 120). In line with this complexity of the discussion, siderophore as a microorganismic machinery involved in the transport of iron will be delineated in a separate chapter.

Thus, confining the discussion to P, Perry and Brubaker (1979) employed the Congo-red agar method for the detection of the P+ character and reported that wild-type *Y. pseudotuberculosis* and *Y. enterocolitica* are *pgm*$^{-}$, whereas Burrows (1973), using AGA agar, reported that the majority of colonies from the spleen emulsion of a mouse, moribund from infection with *Y. pseudotuberculosis* strain MRE 1224 (formerly called strain PBI/+) formed colonies pigmented as seen with P+ colonies and that a small minority of isolates formed more densely pigmented colonies. Whether this discrepancy resulted from the difference in assay methods or not is unknown, but Wake (unpublished data) differentiated *Y. pestis* P++ and P+ colonies using Congo-red agar, suggesting the presence of more pigment-adsorbing material on P++ cells than P+ cells, irrespective of the assay method. Another alternative interpretation is that not only the *pgm* gene is coding for the production of this material.

Be that as it may, the association of P+ with a plasmid which regulates the production of material(s) responsible for the absorption of Congo-red dye to the cell surface of various *Y. pestis* strains is suggested by indirect evidence (Wake 1976; Wake et al. 1978; Tsukano et al. 1978; Sutoh and Wake 1979). First, the rates of spontaneous phenotypic variations of *Y. pestis* strains Yreka, 195/P, Alexander, Bryans, and MII40 from P+ to P− during maintenance at 4 °C, 27 °C, and 37 °C were too high to exclude the plasmid-associated nature of *pgm* (Sutoh et al. Fig. III.15), particularly, in F1− strains, such as Bryans and MII40. Second, all drugs known to be effective in curing bacterial plasmids, such as acriflavine, acridine orange, ethidium bromide, and sodium dodecyl sulfate (SDS) were effective in enhancing the recovery of P− mutants from fully virulent *Y. pestis* Yreka and its avirulent derivative strain 37-6e9 (Tsukano et al. 1978).

However, this evidence is only suggestive; definitive proof requires that the isolated-plasmid DNA preparation be obtained and used to transform *Yersinia* organisms from P− to P+. Recently, such definitive proof was obtained by Tsukano et al., using the same method for the demonstration of the plasmid nature of *fra* (p. 194, introduced in Wake 1984). However, difficulties in identifying the plasmid(s) associated with P may reside in the plural nature of *pgm*-bearing plasmids as postulated by Burrows (1973). Even if one could identify such a plasmid, it might represent only one of many *pgm* gene(s) present in the total plasmid complement.

To summarize, P is a virulence factor detectable by its ability to absorb hemin, ferric iron, and some basic dyes with planar chemical configuration, e.g., Congo-red dye. Although the nature of this outer-membrane storage complex conferring this ability to P+ bacteria has not been defined, it behaves in close association with P1 receptor being involved in iron metabolism of the bacteria. The fact that P− *Y. pestis* organisms with all the other established virulence factors and a high molecular weight cytoplasmic iron-storage component show restored virulence when exogenous iron is supplemented to the host animals, indicates that this outer-membrane complex takes over the ferric iron from host iron-binding proteins that constitute host-defense mech-

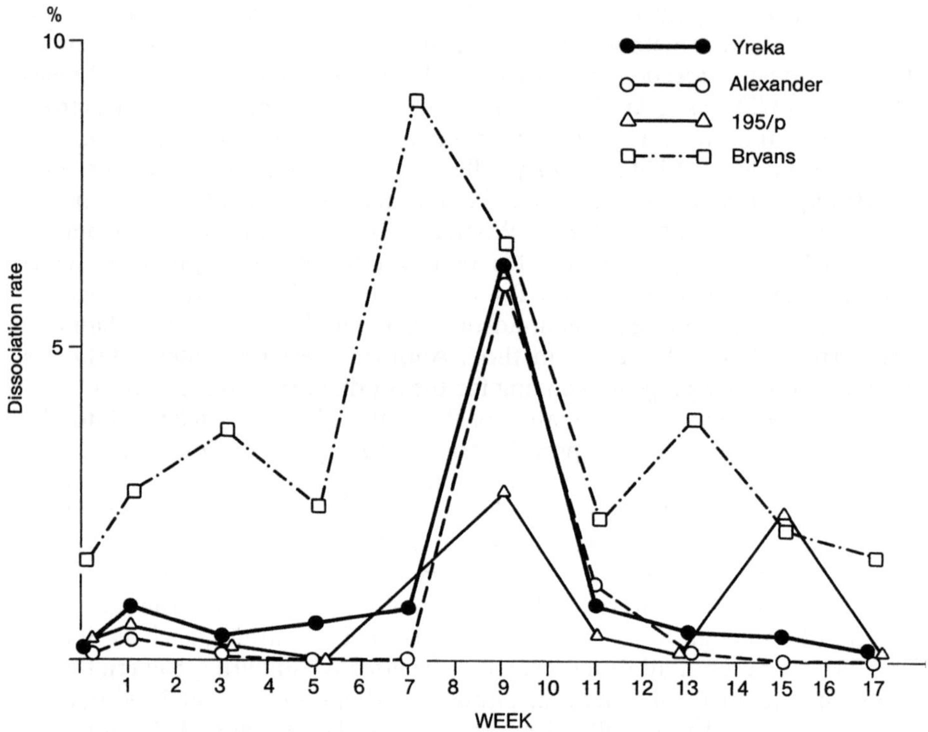

Fig. III.15. Dissociation rate of P^- colonies during long preservation of *Y. pestis* cultures under 4 °C. Four trypticase soy-agar slant cultures grown at 27 °C for 2 days were preserved in a refrigerator at 4 °C. Every week, one loopful growth of each was suspended in sterile saline and its appropriate dilution was spread over the Congo-red agar plate. After incubation for 3 days at 27 °C, the percentage of white (P^-) colonies among the whole number of grown ones was counted. Each *point* represents an average of three plates

anisms with stronger affinity to Fe^{3+}. The cytoplasmic iron storage component receives this ferric iron from in vivo environment via this machinery and utilizes it for bacterial multiplication, i.e., a critical phase of infection. Thus, being involved in iron metabolism in association with P1, *Y. pestis* has been segregated as an obligate animal-host-dependent parasite causing invasive infection from the progenitor *Yersinia* organisms. However, P character solely does not seem to be critical for this segregation, because P+ phenotypes of the other bacterial species, also, can cause infection harming the ability to take over iron from host iron-binding proteins for their multiplication, irrespective of invasive, exudative, or superficial infections.

Although the chemical nature of the *pgm* plasmid regulating P is to be determined, the segregation rate of P− organisms, sensitivity to plasmid-curing drugs, and the transfer experiment of P+ phenotype to P− recipient, indicated the location of *pgm* gene(s) on plasmid replicon(s).

Pu

This virulence factor is encoded by the chromosomal gene *pur*.

Bacon et al. (1951) isolated various auxotrophic mutants of virulent *Salmonella typhi* using a chemically defined selective minimum medium and X-ray or UV irradiation as the mutagenic agent. Those mutants which required purines, p-aminobenzoic acid (PABA), or aspartic acid were avirulent for mice when injected intraperitoneally. The addition of the respective nutrient to the peritoneal cavity at the time of i.p. injection enhanced the virulence of these mutants. They devised a similar synthetic medium for the selection of similar mutants of *Y. pestis,* i.e., AGA medium previously presented.

They plated X-ray irradiated cells of a fully virulent strain MP6 on this medium prepared either with or without selected nutrient supplements. A mutant strain, M1, requiring guanine, xanthine, or hypoxanthine, and unable to utilize adenine, was isolated. The virulence of M1 increased over 1000-fold in terms of the average lethal dose upon injection of a single dose of 4 mg of hypoxanthine into the test animals with these organisms (Burrows et al. 1954). Therefore, the M1 mutant was avirulent, not because of alterations in the other virulence factors, but because of its inherent purine dependency, in other words, the lack of endogenous purine synthesis. Burrows designated this virulence factor of *Y. pestis* as Pu (Burrows and Bacon 1958), and Brubaker gave it the genetic code designation *pur* (Brubaker 1972).

Attenuated F1+VW+P1+P+Pu− organisms like M1, when employed as live vaccines for the immunization of mice or guinea pigs elicite different responses than F1+VW+P−, F1−VW+P−, F1−VW+P+, F1−VW−P−, and F1−VW−P+ attenuated organisms. In the case of a single dose of vaccine in mice, the protection against i.p. challenge with 2000 organisms of a fully virulent strain differed significantly only between mice immunized with F1+VW+P−Pu+ and F1+VW+P+Pu− organisms. Protection provided by immunization with the former exceeded that provided by immunization with the latter. This result indicates that the lack of Pu+ in the latter effected to result in the minimum in vivo growth of the bacteria because of their inability to biosynthesize purine, whereas lack of P+ in the former allowed the limited growth of the former by the expense of internal storage of iron (refer to previous section).

To support this explanation, F1+VW+P+Pu− organisms with the minimum capability of in vivo growth produced the mice resistance more efficiently when the live vaccine was administered in two doses at 7 day intervals, while administrations in two doses of F1+VW+P−Pu+ or F1+VW−P+Pu+ organisms were much less effective to produce the mice resistance. Such results were derived from the fact that the mice whose immune systems were stimulated with nonmultiplying antigens twice could produce more effective immunity than those stimulated by living vaccine with the limited multiplication of the bacteria.

In the experiment using guinea pigs, also, F1+VW−P+Pu+ and F1+VW+P−Pu+ organisms gave the animals remarkable protection, although booster administration with F1+VW+P+Pu− organisms further enhanced the immunity.

In the context of these experiments, the conclusion is that Pu− organisms do not multiply in vivo, but their production of F1, which is a protective antigen, is greatly enhanced at the subcutaneous injection site. However, Pu+ organisms devoid of VW or P do multiply, although limitedly and form an immunity characteristic of living vaccine in which two-dose administrations are avoided due to adverse effects.

In summary, purine auxotrophs (Pu−) were found to have reduced virulence. Of these, mutants blocked at a point after the synthesis of inosine monophosphate and thus dependent on adenine or guanine per se were completely avirulent. These purines, which in bound form constitute the major nitrogenous bases in nucleotides, occur only in traces in most host cells, usually as products of the enzymatic hydrolysis of nucleic acids and nucleotides (Lehninger 1975). Therefore, *Y. pestis* organisms have to biosynthesize these bases endogenously for their in vivo multiplication to express virulence, although the other essential small molecules, e.g., amino acids, vitamins, and pyrimidines, are sufficiently provided by the in vivo microenvironment. Obviously, genes encoding the enzymes neccessary for the synthesis of essential small molecules must have been inherited from the progenitor *Yersinia* organisms and, therefore, be chromosomal. In fact, the ratio of Pu− colonies to the parent colonies on minimal-supplement agar following X-irradiation was 1:6,000 (Burrows et al. 1954), too small a frequency to assume its location on a plasmid.

Interactions Among Plasmids and Their Influence on Virulence

The idea that host-parasite relationships of unicellular pathogenic microorganisms deal with a host parasitized by microorganisms parasitized by virulence determinants: triple biology, was proposed in the Summary of Part I. Evidence supporting this idea indicates that the expressions of virulence determinants can be modified in various ways by the influence of other plasmid species at the unicellular level. Therefore, the host reactions diseases or immune responses to these microorganisms are also modified by influences from complex interactions among various plasmids with divergent biological functions: for example virulence associated plasmids, plasmids associated with drug-resistance (R) or sexuality (F-lac) in combination within a single microorganisms harboring them.

For example, Burrows (1973) observed the mutual interaction between *pgm* and *vwa*=*cal* in *Y. pseudotuberculosis* strain MRE1224. This observation forced him to discriminate one *pgm* gene from another *pgm* gene which was influenced by the presence of the *vwa*=*cal* plasmid. First, he confirmed that densely pigmented colonies (P++) are always Cal− and, second, a moderately pigmented calcium-independent (P+Cal−) colony rarely segregated into a less pigmented calcium-dependent (P±Cal+) one. These changes can be explained on the assumption that the *vwa*=*cal* plasmid had influenced one *pgm* gene to repress its expression, but the moderate pigmentation occurred when the influence from *vwa*=*cal* plasmid was blocked. These two *pgm* genes expressed their function fully by the absence of *vwa*=*cal* plasmid. When both *pgm* genes are deleted, white sectors which are the growth of Cal+ or Cal− bacteria, segregate from the less pigmented Cal+ or Cal− (P±Cal+ or P±Cal−) bacteria. In contrast, the densely pigmented (P++Cal−) colonies segregated to give a sector with pigmentation intermediate between moderate (P+) and less moderate (P±) appearance. Well-isolated colonies of this derivative grow to over 5 mm in diameter and frequently showed minute densely-pigmented spots within them. Subculturing from these spots allowed isolation of colonies indistinguishable from the parental densely-pigmented calcium-independent (P++Cal−) colony. Thus, the virulence-associated plasmids harboring in the single bacteria are not always mutually irrelevant, but sometimes regulate the phenotypic expression of another plasmid. In a particular strain of *Y. pestis,* e.g., A1122, P1 production is controlled not by the 6, but the 20 Mdal plasmid (refer to previous section). Presumably, the trimer molecule of the former is functioning in this strain, because its P1 activity is found to be extremely high (Ferber et al. 1981).

The transmissibility of virulence plasmid with the aid of mobilizing activities of transposons (refer to Part I, Chap. 3) has been frequently reported without regard to virulence expression to elucidate the plasmid association of each phenotype. Since Ginoza and Matney (1963) transferred an R-plasmid from *E. coli* B into five strains of avirulent *Y. pestis* (TS, A4, TRU, Yokohama, and EV76) and the *Y. pseudotuberculosis* strain 4C, Kol'tsova et al. (1973) transferred pst^+ gene from pst^+ *Y. pestis* to pst^- *Y. pestis* with the aid of R-plasmid equipping *tra*-operon. The successes of Tsukano (introduced in Wake 1984) who transferred each fra^+, pst^+, and pgm^+ from Fl+Cal−P1+P+ $Tc^rKm^rApc^r$(RP4) *Y. pestis* organisms to F1−Cal−Pl−P−$Su^rSm^rTc^r$(N-3) were individually described in the previous section.

These R-plasmids can be divided into fi^+ and fi^- types depending on whether the coexistence of the plasmid with the F-type plasmid inhibits the formation of F-type pilli by the host bacteria. On the other hand, plasmids can be classified into many incompatibility groups on the principle that two species of plasmid belonging to the same incompatibility group cannot coexist stably within the same host bacteria.

Taking these classifications of R-plasmids and their possible interaction with the virulence-associated plasmids into consideration, Wake (1976) and Wake et al. (1977) investigated the effects on virulence of the transfer of R-plasmids into fully virulent (F1+VW+Cal+P1+P+Pu+) *Y. pestis* strains. The procedure, depicted in Fig. III.16, yielded five colonies of *Y. pestis* harboring R-plasmid NR1 ($Sm^rCp^rTc^r$) from the strain Yreka, four colonies of a similar drug-resistant strain from the strain Bryans, but no such colony from

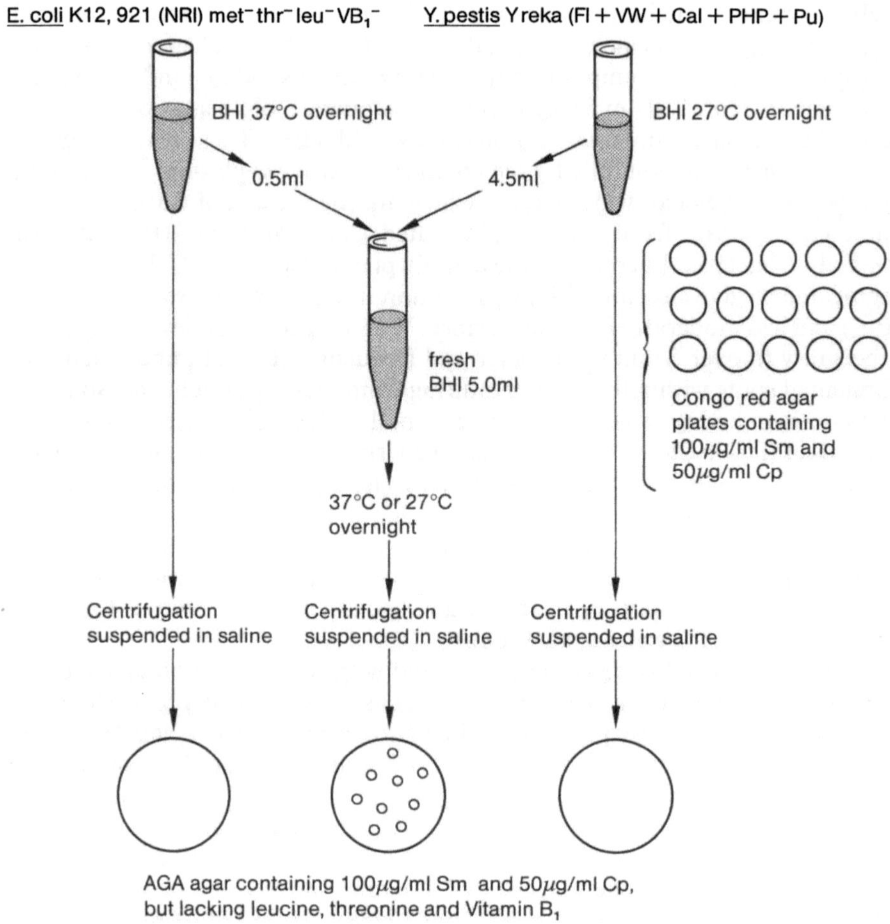

Fig. III.16. The transfer experiment of a Sm^rCp^r R-plasmid NR1 from *E. coli* K12 into *Y. pestis* Yreka by conjugation

strain Alexander. These results indicate that the transmissibility of R-factor into *Y. pestis* organisms are divergently controlled by the unique physiological conditions inherent to each recipient strain owing to their divergent derivative histories, in addition to the inadequacy of the synthetic medium used to enhance the growth of viable *Y. pestis* organisms. Therefore, a conventional agar medium, i.e., brain-heart infusion agar (Difco), which has been shown to enhance full growth of *Y. pestis* organisms, was used with the addition of Congo-red dye to discriminate the pigmented (P+) *Y. pestis* colonies from the nonpigmented (P−) *E. coli,* i.e., the donor of the R-plasmid. The appropriate antibiotics were added to these agar media for the selection of drug-resistant bacteria, irrespective of recipient *Y. pestis* or *E. coli.* Figure III.17 shows the effectiveness of this method.

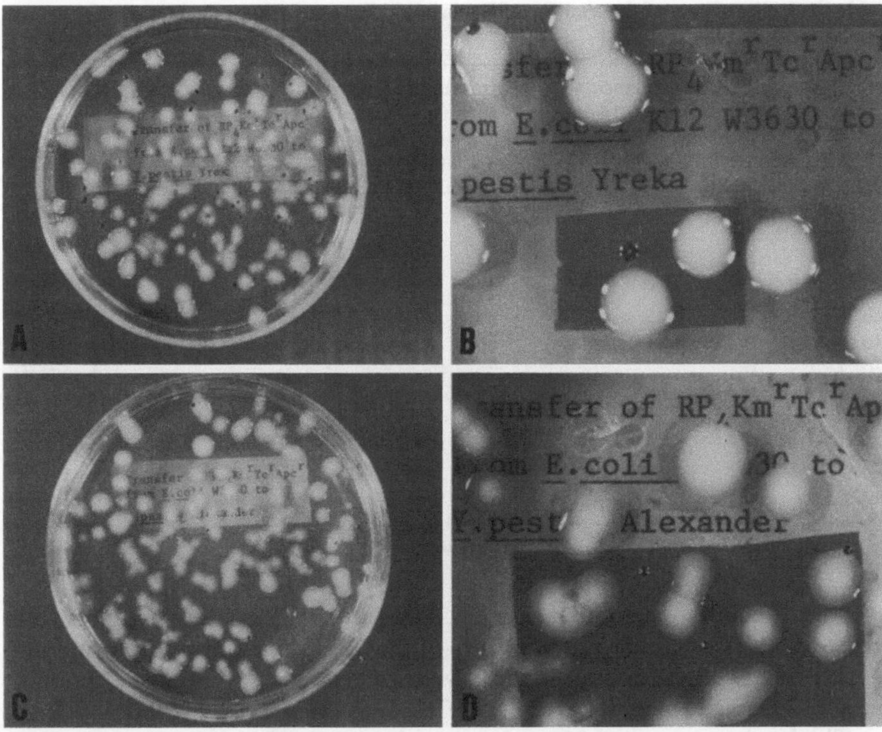

Fig. III.17. The transfer experiment of an R-plasmid RP4 (Km^r Tc^r Apc^r) from *E. coli* K12,W3630 to two virulent *Y. pestis* strains Yreka and Alexander, utilizing the pigmentation of P+ *Y. pestis*, but nonpigmentation of the donor *E. coli* K12 on Congo-red agar plates containing antibiotics. **A** General view of the Congo-red agar plate after spreading over the 0.1 ml appropriate saline dilution of the donor *E. coli* K12-recipient *Y. pestis* Yreka mixture that had been incubated overnight at 37 °C, followed by incubating for 3 days at 27 °C; **B** enlarged picture of **A**; **C** general view of the Congo-red agar plate of a similar experiment, but using *Y. pestis* Alexander as the recipient organisms; **D** enlarged picture of **C**. Dark-pigmented small *spots* are the P+ *Y. pestis* colonies that became drug-resistant

By the use of this method, a total of 12 P+ *Y. pestis* colonies harboring the R-plasmid NR1 were isolated and assayed for their virulence by s.c. injection of 10^5 organisms into each of ten mice of an outbred strain ddY/S. Simultaneously, the presence or absence of each of the established virulence factors F1, Cal, P1, and P was determined. All P+ isolates were avirulent and Cal− after receiving transfer of NR1. Furthermore, all mice surviving this virulent challenge were challenged s.c. with 3.4×10^3 organisms of strain Yreka in order to determine whether the 10^5 organisms that failed killing mice had multiplied in vivo and conferred the mice protective immunity or not. The results are summarized in Table III.7.

In order to clarify what influence the plasmid NR1, whose genetic structure is diagrammatically shown in Fig. III.18, gave to *vwa* = *cal* plasmid, the following experiments were performed.

Table III.7. The effects of the transfer of R-plasmid NR1 into a fully virulent *Y. pestis* strain Yreka on its virulence

Substrain number	Mouse virulence; dead/10 mice 14 days after s.c. infection with 10^5 organisms	Immunogenicity; survival/10 mice after s.c. infection with 3.4×10^3 Yreka	Assay of virulence factors			
			Fl	Cal	Pl	P
No. 1	0	7	+	−	+	+
No. 2	0	0	+	−	n.d.[a]	+
No. 3	0	8	+	−	+	+
No. 4	0	4	+	−	+	+
No. 5	0	5	+	−	+	+
No. 6	0	6	+	−	+	+
No. 7	0	7	+	−	+	+
No. 8	0	7	+	−	+	+
No. 9	0	0	+	−	−	+
No. 10	0	5	+	−	+	+
No. 11	0	3	+	−	+	+
No. 12	0	7	+	−	+	+

[a] n.d.: not done, substrain No. 2 did not grow on Pl-assay agar.

One of such isolates (substrain No. 3 in Table III.7), designated SCY, was treated with ethidium bromide to cure of the NR1; and a resulting strain designated 376e9 was found to be sensitive to Sm (streptomycin) and Cp (chloramphenicol) and calcium-dependent (Cal+) as is shown in Fig. III.4.

The in vitro assays of 376e9 for the established virulence factors demonstrated F1+VW+Cal+P1+P+Pu+ (see Fig. III.12). However, it was avirulent in the mouse-virulence assay, contradictory to the previous experiences. For evidence that may give solution to this problem, cells recovered from the slant on which 376e9 organisms had been maintained at 4 °C for more than a year, were reexamined and all were found to be sensitive to tetracycline (Tc^s) and calcium-independent (Cal−). On the other hand, SCY segregated spontaneous tetracycline-sensitive P− mutants, i.e., $Sm^rCp^rTc^s$ F1+VW+Cal−P1+P−Pu+ designated NSY, after prolonged storage at 4 °C (Sutoh and Wake 1979).

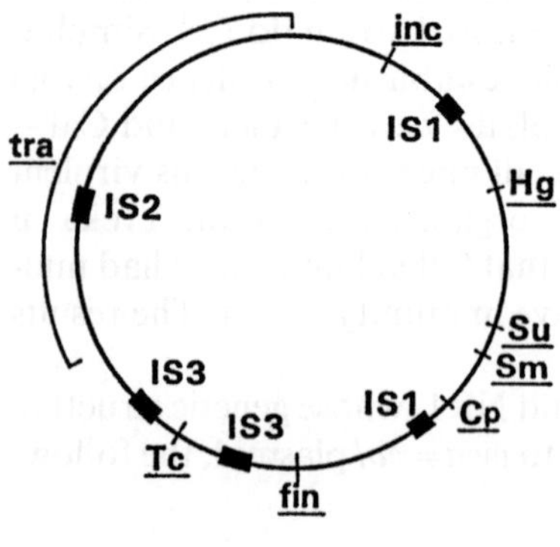

Fig. III.18. Genetic structure of R-plasmid NR1. This plasmid is sometimes synonymously designated R100 or 222

To summarize these results with reference to the genetic structure of NR1 plasmid shown in Fig. III.18, it can be concluded that (1) the presence of whole NR1 plasmid within the fully virulent *Y. pestis* Yreka organisms repressed the expression of the *cal*+ region of the *vwa*−*cal* plasmid; (2) the treatment of SCY with ethidium bromide was effective to delete the DNA segment of NR1 plasmid between the two IS1 and release the *cal*+ gene from the repression; (3) the prolonged storage at 4 °C of both 376e9 and SCY were effective to delete the DNA segment regulating tetracycline resistance (Tc) together with the *cal*+ region of the *vwa*−*cal* plasmid.

Because these experiments indicated that the transfer of this R-plasmid may alter the expression of the virulence factor Cal and alter the virulence of organisms in mice, Wake (unpublished) similarly transferred R-plasmids of various incompatibility groups and of both fi^+ or fi^- types and measured the transfer rate, inhibition of expression of cal^+ genes, and mouse s.c. virulence by the use of five animals in one group to be infected with the recipient bacteria. R-plasmids other than NR1 did not inhibit the expression of cal^+ genes, and the virulence of the transformed recipients was variable (Table III.8). R-plasmid RP4 was unique in its high transfer rate, irrespective of the recipient *Y. pestis* strains (8×10^{-2} in Yreka and 3×10^{-2} in Alexander). The transfer rate of RP4 from *E. coli* K12.3630 (RP4) into *Y. pestis* Yreka at 27 °C was tenfold higher than at 37 °C. The assay for the calcium dependency of the recipient *Y. pestis* showed that the harboring of RP4 plasmid did not alter the Cal+ phenotype, and s.c. challenge with 10^2 to 10^3 organisms of this $Tc^rKm^rApc^r$ F1+Cal+P1+P+ *Y. pestis* Yreka (RP4) did not kill all five mice, i.e., one to three mice survived to the 14th day. The mice which survived for 14 days after challenge developed resistance to s.c. challenge with similar doses of the fully virulent *Y. pestis* Yreka. In this context, the intracellular coexistence of RP4 plasmid does not influence the DNA segment responsible for the expression of calcium dependency in contrast to NR1 plasmid, which does. However, the reduced virulence accompanying immunogenicity of the strain *Y. pestis* Yreka

Table III.8. Transfer of various R-plasmids from an *E. coli* K12 donor to fully virulent *Y. pestis* recipients and its effects on virulence

R-plasmid	Incompatibility group	fi Type	Transfer rate	Inhibition to Cal	Mouse virulence	Immunogenicity
R38B	F I	+	1×10^{-4}	−	+	±
NR1 (R100)	F II	+	1×10^{-4}	+	−	+
RI	F III	+	1×10^{-4}	N.T.[a]	N.T.	N.T.
Co1B	F IV	+	N.T.	N.T.	N.T.	N.T.
R124	N	+	1×10^{-4}	N.T.	N.T.	N.T.
N-3	P I	−	1×10^{-4}	−	+	N.T.
RP4	W	−	1×10^{-1}	−	+	+
S-a		−	1×10^{-4}	N.T.	N.T.	N.T.

[a] N.T.: not tested.

(RP4) indicates the influence of RP4 plasmid to some unknown stage of full virulence expression which is not related to immunogenesis, i.e., in vivo multiplication.

Independently to above studies, Zsigray et al. (1983) transferred another category of plasmid, i.e., F-lac from *E. coli* 23, 10_s to a virulent $fra^+ vwa^+ cal^+ pst^+ pgm^+ pur^+$ *Y. pestis* strain, 360. Equal volumes of 37 °C grown *E. coli* (F-lac) and 26 °C grown *Y. pestis* 360 were mixed and cultivated at 37 °C for 4 h. Dilutions of the mixed cultures were spread on a minimal-agar medium on which only lactose-positive Y. pestis organisms can grow. The resultant *Y. pestis* (F-lac) clones retained all of the known virulence traits, but were avirulent (mouse LD_{50} was 4×10^8 organisms). In contrast, two segregants from ten transconjugants showed mouse LD_{50}'s of 27 ± 10 organisms or 28 ± 27 organisms and were calcium-independent after loss of the F-lac plasmid.

However, Zsigray et al. (1983) found that the virulence and calcium dependency of *Y. pestis* 360 were not affected by the transfer with F°-lac (a variant of the sexuality plasmid) or R100 (synonymous with NR1). Therefore, it is possible that some unique nucleotide base sequences repressing *cal* genes are present in the NR1 used by Wake (1976) and the F-lac used by Zsigray et al. (1983), but not in the F°-lac and R100 used by Zsigray et al. (1983).

In conclusion, the idea that was proposed in the Summary of Part I provides us new scopes for study of *Yersinia* and host-parasite relationships at the level of molecular genetics. The interactions among variable or the same species of plasmids, irrespective of whether inherently or exogenously infected, can repress or enhance the expression of the virulence-associated phenotypes. Thus, the virulence plasmids of *Yersinia* organisms the phenotypic expressions of which are detectable individually are furthermore feasible for dissection by the use of other species of plasmid with known genetic structures. In addition, it would be unsurprising if we found that a certain virulence factor of a particular *Yersinia* strain, isolated from a different origin, was associated with a peculiar plasmid of particular molecular sizes. These rare cases can occur as the result of unique derivative histories of each strain in nature.

Chapter 3 Other Mechanisms of *Yersinia* Organism Functioning for Expression of Their Virulence

As a supplement of those established virulence factors, whose expressions and target host-defense mechanisms for the conquest have been confirmed experimentally, some other mechanisms relating to the virulence expression of *Y. pestis* are delineated in this chapter, comprising all possible mechanisms that are accessory for perfection of that dreadful virulence. In detail, the first two sections deal with toxic substances that injure host-animal organisms pharmacologically, the latter two sections, the metabolisms of individual components, nutritionally required for the bacterial in vivo growths.

The discrimination of toxicity from virulence is one of the most important points in understanding the concept of pathogenicity as expressed in the "disease formula" presented in Part I (p. 11).

For example, challenge doses of 10^8 avirulent *Y. pestis* organisms killed most mice within 14 to 74 h, but the surviving mice continued to live thereafter, while those which survived similar challenge doses of virulent *Y. pestis* organisms finally died after a recognizable period without deaths as shown in Fig. III.19. This discriminates toxic deaths due to large doses from deaths due

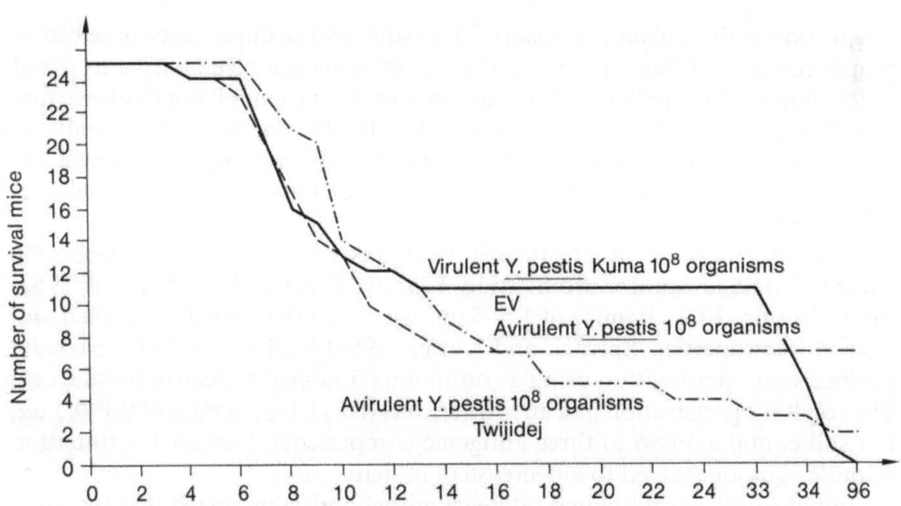

Fig. III.19. Death survival time courses of ddY/S mice after subcutaneous infection with a large dose of virulent and avirulent *Y. pestis* strains

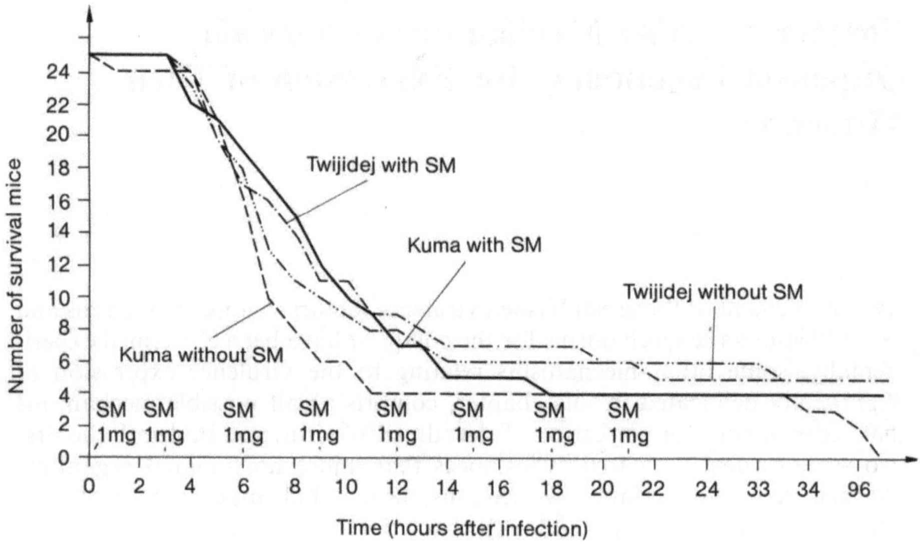

Fig. III.20. Time course of ddY/S mice treated with repeated i.p. injections of streptomycin (SM) after subcutaneous infection with large doses of virulent or avirulent *Y. pestis* cells. ———— Injected s.c. with 10^8 virulent *Y. pestis* organisms strain Kuma followed by repeated i.p. injections of 0.1 mg each SM. – – – – – – The same as above, without SM treatment. —·—·—·— Injected s.c. with 10^8 avirulent *Y. pestis* strain Twijidej organisms, followed by repeated injections with 1.0 mg each SM. —··—··— The same as above, without SM treatment

to infection with virulent organisms. This indicates the importance of selecting proper infective doses for the elucidation of virulence mechanisms to avoid deaths due to direct toxicity of the microorganisms not involving virulence factors. The experimental results shown in Fig. III.20 support this and indicate the dissociation of infection per se from the *disease-onset due to bacterial virulence:* an important concept for the proper understanding of host-parasite relationships.

Various groups of investigators have expressed divergent opinions on the nature of toxic substances produced by *Y. pestis*. Fraction II (Baker et al. 1952) showed an i.p. LD_{50} in mice of 1.5–5 µg. Ajl et al. (1955) further purified and studied its properties. Spivack and Karler (1958) further purified a partially purified toxin preparation using a continuous flow-paper electrophoretic cell. The resultant preparation had an intravenous (i.v.) LD_{50} in mice of 0.1–0.3 µg, but still contained two to three antigenic components. Further fractionation of these components led to a decrease in toxicity.

Initial studies with isolated rat heart mitochondria indicated that the toxin was interfering with the oxidation of alpha-ketoglutarate, pyruvate, beta-hydroxybutyrate, malate, and succinate (Packer et al. 1959). The biochemical mechanism for this interference was blockage of the reduced nicotinamide adenine dinucleotide-coenzyme Q reductase in the terminal electron-transfer chain (Kadis et al. 1965). The lethal effect of murine toxin is believed to result from heart malfunction (Kadis and Ajl 1970) or circulatory failure after peripheral vascular collapse (Hildebrand et al. 1966).

However, Brown and Montie (1977) recognized a pitfall for the investigators hidden within the experimental design which led to these conclusions. The toxin concentrations used in the majority of these studies were between 0.5 and 5.0 mg ml^{-1}, 1,000 times the normal mouse LD_{50}, and the concentration of toxin used frequently exceeded the concentration of mitochondrial protein used in these in vitro tests.

When they examined the effect of murine toxin in the living mouse, doses corresponding to 1–2 LD_{50} (for the mouse) could not be detected in an in vitro test with mitochondria. Starved mice gave decreased LD_{50}'s (0.5 to 1.2 μg per mouse) and mice kept at 37 °C were not killed by the toxin.

Thus, the effect of murine toxin was found to be related to the physiological conditions of the living mouse whole body, rather than to the intracellular respiratory enzymatic system of a particular target organ. Therefore, they assumed the activity of murine toxin as resembling a hormone that binds to some receptor distributed on the cell surfaces of various tissues and regulates the cell function. Analogous mechanisms have been established with cholera toxin (CT; refer to Part I, Chap. 4, p. 25), which is known to stimulate adenyl cyclase activity of the cell after binding to the receptor. When CT was injected i.p. and then plague murine toxin administered intramuscularly (i.m.), 1 to 2 LD_{50} of the latter did not show toxicity. Similarly, repeated administrations of glucagon, but not epinephrine, protected mice against the injection of murine toxin from death or prolonged survival time of those mice which were killed.

All these experimental results coincide with the toxic effects of catecholamines, which were established as the beta-adrenergic blocking agents that compete for the binding to the beta-adrenergic receptors as follows: In the normal physiology of a mouse body, the hormones, such as epinephrine, bind to these receptors and send intracellular signals to mobilize free fatty acids to keep the body temperatures of sensitive animals constant. When these receptors are occupied by the competitive bindings of plague murine toxin, i.e., the receptors are blocked, thermogenesis is inhibited resulting in hypothermia and death of sensitive animals. If the murine toxin activity functions in this way, i.e., as the beta-adrenergic blocking activity, no in vitro assay system is available for its quantitative measurement.

For the chemical nature of plague murine toxin, a protein consisting of active subunits of the molecular weight 12,000 to 24,000 is reported (I. C. Montie and D. B. Montie 1971).

With regard to the genetic trait controlling the murine-toxin production, treatments of *Y. pestis* strain SCY with ethidium bromide, acriflavine, acridine orange, and the aerated cultivation at 41 °C for 48 h, have failed to yield murine-toxin-nonproducing derivatives in contrast to the other plasmid-mediated virulence factors, i.e. F1, VW, P1, P which were lost at least in some of these derivatives (Wake et al. 1977). While these results support the chromosomal nature of the gene controlling the murine-toxin production, Ferber and Brubaker (1981) described that all, but one *Y. pestis* strain which produce the murine toxin, were found to harbor a 65 Mdal plasmid species. An exceptional murine-toxin-producing strain A1122 being devoid of the 65 Mdal plasmid, is devoid of the 6 Mdal plasmid, but expresses P1 with enhanced activity presumably by the presence of the 20 Mdal plasmid (see previous section). In conclusion, the genetic control of murine-toxin production is to be studied in future.

Endotoxin, i.e., Lipopolysaccharides of *Yersinia* Organisms

The autopsy findings of plague-sensitive, but murine-toxin-resistant animals, e.g., guinea pigs and humans, killed by bubonic plague are characterized by the presence of visible and microscopic abscesses and granuloma formation in the liver and spleen (Chap. 2), with sepsis the final outcome in most of the cases. However, marked subcutaneous hyperemia and hemorrhage giving a darkened appearance due to the change of hemoglobin into hemosiderin also constitute characteristic features of plague death (hence the origin of the term "black death").

Figure III.21 shows three nude mice: one normal, one hyperemic due to s.c. challenge with a virulent *Y. pestis* Yreka organism (pink in natural color), and one darkened in color after death had occurred.

In order to identify the toxic substances of *Y. pestis* related to its pathogenesis, many investigators have isolated and examined the endotoxin, i.e., lipopolysaccharide (Part I, Fig. I.3) from *Y. pestis,* although most classical methods for lipopolysaccharide purification were unsuccessful.

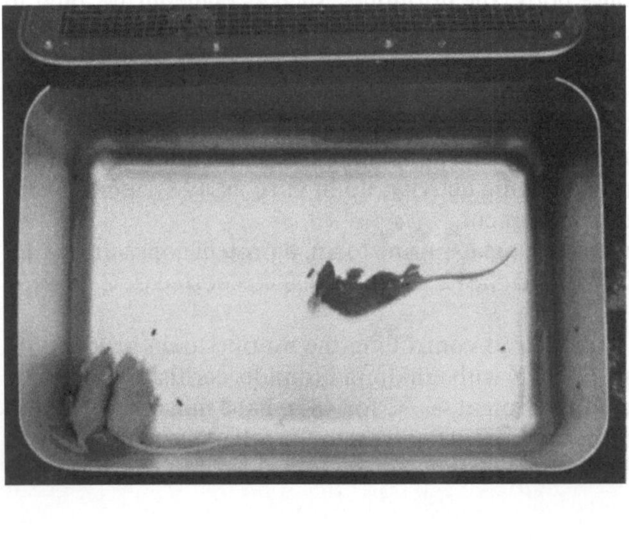

Fig. III.21. Black appearance of plague-killed nude mouse. Two of three nu/nu mice deriving BALB/C strain were infected s.c. with 100 organisms of a fully virulent *Y. pestis* strain Yreka. One dead mouse with black appearance (right), another infected mouse with reddish skin color (left), and one normal mouse (middle) are seen

An endotoxin preparation prepared by Walker et al. (1966) starting from the acetone-dried powder of three different *Y. pestis* strains, by the use of hot phenol extraction followed by pronase digestions, ethylether extraction, and alcohol precipitation, had a mouse LD_{50} of 3 mg and a guinea pig LD_{50} of 25 mg. When they started by disrupting the bacterial cells using ball-mills (i.e. grinding-mills in which the materials to be ground is tumbled in drums with heavy balls of iron, steel or stone), the mouse LD_{50} decreased from 20 mg to 7.5 mg upon purification. However, both preparations contained one or two impurity antigens demonstrable in Ouchterlony gel-diffusion plates. All preparations showed an absorption at 510 mμ and the presence of 7-carbon sugars, lecithin, cephalin, and ethanolamine. Endotoxin from ball-mill-ground material decreased liver glycogen and plasma glucose, and increased blood urea nitrogen in mice within 21 h of i.v. injection of one-half a LD_{50} dose. Phenol-extracted endotoxin, introduced into mice and guinea pigs, and ball-mill prepared endotoxin, introduced into mice, induced petechial hemorrhages, diffuse hemoglobin imbibition by connective tissues and enlarged congested capillaries or vessels in various organs and tissues, and acute renal tubular necrosis. Vacuolization of liver parenchymal cells resembling fatty metamorphosis was seen in both mice and guinea pigs with both preparations. In mice, brain, heart, and liver tissues were most frequently involved. In guinea pigs, the intestinal tract, subcutaneous tissues, peritoneal wall, and adrenals were most frequently congested or hemorrhagic. The signs of peripheral vascular collapse were consistent with the effects of endotoxins isolated from other Gram-negative bacteria and were most striking in mice killed 48 h after inoculation.

Later, Albizo and Surgalla (1970) isolated a *Y. pestis* endotoxin containing 2.1% nitrogen, 1.6% phosphorus, 22.5% neutral hexoses, 1.5% hexosamine, 25% esterified and amide-linked fatty acids, and 1.4% protein which also exhibited toxicity in mice (i.p. $LD_{50} = 537$ μg) and rabbits (i.v. $LD_{50} = 32$ μg). They employed the phenol-water extraction method of Westphal as modified by Tauber and Russel (1961). The lipopolysaccharide preparation exhibited antigenicity in rabbits, evoked a biphasic pyrogenic response in rabbits, stimulated resistance (in mice) to bacterial infection, and produced a localized and generalized Schwartzman reaction in rabbits. Although they did not eliminate the possibility that the lipopolysaccharide may act synergistically with proteins or other toxins of *Y. pestis,* this study provided new evidence that endotoxin could contribute to plague death. Cocking et al. (1960) reported that guinea pigs dying of plague contain approximately 1.6×10^{11} organisms. Since Albizo and Surgalla (1970) could extract two guinea pig LD_{50} (1 mg) of endotoxin from 1.8×10^{11} *Y. pestis* organisms, the lethalities in murine-toxin-resistant guinea pigs could be explained by endotoxemia.

Thus, the lethal toxicity of *Y. pestis* cells, irrespective of virulence, to plague-murine-toxin-insensitive animals, e.g., guinea pigs, could be attributed to LPS, whose structure is diagrammatically shown in Fig. I.3 using *E. coli* as the example. Within such structure of LPS, it is the lipid A portion that exerts toxicity. Another portion, O-polysaccharide, does not exert toxicity, but its role for directing the preferential infection route must be delineated on the assumption that *Y. pestis, Y. pseudotuberculosis,* and *Y. enterocolitica* have divergently segregated, i.e., evolved stemming from the common progenitor *Yersinia* organisms closely related with *Escherichia coli* (Brubaker 1984). The *Y. pestis* organism, which is the obligatory parasite being normally contained within the rigidly fixed, nutritional environments comprising its flea vector and mammalian host, is not classified into serovars by the use of O-antigen specific serum, i.e., it belongs to a single serovar.

On the other hand, *Y. pseudotuberculosis* and *Y. enterocolitica* are primarily transmitted orally and thus must be able to maintain themselves in the natural environment as *E. coli* does. Accordingly, the selective pressures generated by the need to compete with the normal flora of soil and water are evidently sufficient to ensure the preservation of full or certain lengths of O-polysaccharide chains. In fact, *Y. pseudotuberculosis* is classified into 10 serovars, and *Y. enterocolitica* into 57 serovars. One can assume that the more the number of the monosaccharides as the repeating units of O-polysaccharide, the more varieties of the former are involved, thus resulting in more varieties of serovars. In accord with this assumption, Samuelson et al. (1974) reported that the lengths of *Y. pseudotuberculosis* O-polysaccharides seem to be rather short.

Further evidence supporting this assumption was obtained by Wake (1984) at the conclusion of the following experiment.

Characterization of deep rough *Sal. typhimurium* mutants devoid of O-polysaccharide chains, selected by the use of phage C21, and semirough mutants with shorter chains, selected using phage FO is shown in Fig. I.16 (Part I). The T4 phage was available for the identification of the smooth type (with complete O side chains). In spite of species differences between *Y. pestis* and *Sal. typhimurium,* wild-type *Y. pestis* Yreka organisms were infected with T4, FO, and C21 phages. The number of plaques evident on treated lawns was 1 with T4, none with FO, and 20 with C21 in a typical experiment. This result indicates that O-polysaccharide chains present could not hide the receptors for C21 and that almost all receptors specific for T4 and FO are absent on the cell surface of wild-type *Y. pestis,* as are (presumably) O-polysaccharide chains.

Based on this assumption, a relatedness of O-polysaccharide composition to the adaptation of *Yersinia* organisms for the oral infection route was examined.

Genetic differences exist in the sensitivity of mice to oral challenge with *Yersinia* organisms. When the inbred mouse strain BALB/CN is challenged orally with 10^9 *Y. pestis* Yreka cells, 40% survive, but all are killed within 12 days by a similar dose of *Y. enterocolitica* WA2-1. In contrast, with the inbred-mouse strain C57BL/6N all animals were killed within 4 days by the same dose of *Y. pestis* Yreka, and 50% survived *Y. enterocolitica* oral challenge (Fig. III.22A). The four surviving BALB/CN mice and five surviving C57BL/6N mice were challenged s.c. with 10^3 *Y. pestis* Yreka cells. All of the BALB/CN that had survived the oral challenge with *Y. pestis* were killed, indicating

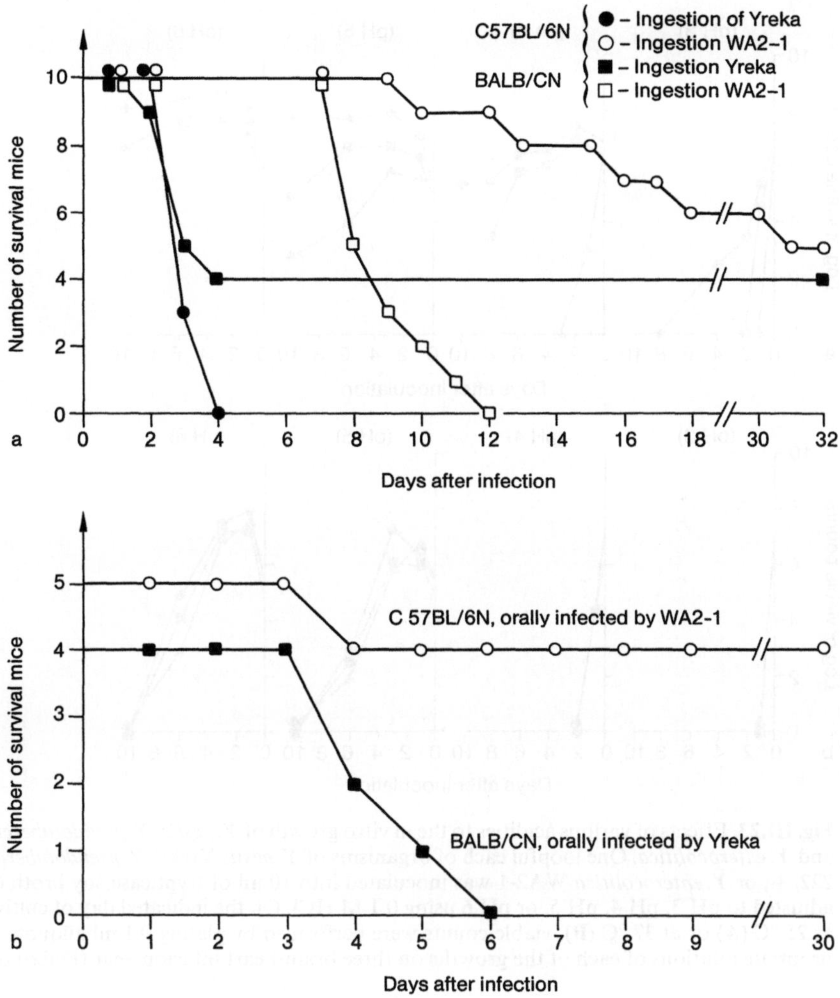

Formation of immunity after oral infection

Fig. III.22. The lethal pathogenicities of VW+ Cal+ *Y. enterocolitica* (08) and Fl+ VW+ Cal+ Pl+ P+ *Y. pestis* Yreka after peroral infection to C57BL/6N and BALB/CN mice and their potentialities to confer protective immunity against *Y. pestis* on the survivor mice. Each group of ten mice received 3.3×10^9/ml of WA2-1 or 1.3×10^9/ml of Yreka organisms in water suspensions at 48 h water deprivation **(A)**. The five survivor C57BL/6N and the four survivor BALB/CN mice were s.c. challenged with 7.6×10^2 organisms of *Y. pestis* Yreka **(B)**. The number of surviving mice in each group was depicted as follows: C57BL/6N, perorally (p.o.) challenged with WA2-1 (–○–); C57BL/6N, p.o. challenged with Yreka (–●–); BALB/C, p.o. challenged with WA2-1 (–□–); and BALB/C, p.o. challenged with Yreka (–■–)

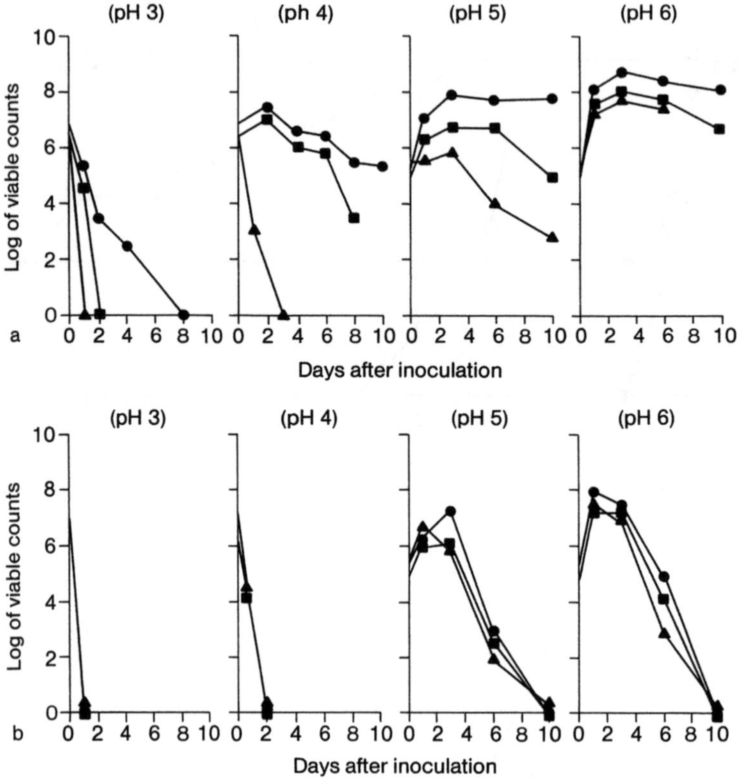

Fig. III.23. Effects of various acidities to the in vitro growth of *Y. pestis, Y. pseudotuberculosis,* and *Y. enterocolitica.* One loopful each of organisms of *Y. pestis* Yreka, *Y. pseudotuberculosis* 222/+, or *Y. enterocolitica* WA2-1 was inoculated into 10 ml of trypticase soy broth (BBL) adjusted to pH 3, pH 4, pH 5, or pH 6 using 0.1 M HCl. On the indicated day of cultivation at 25 °C (**A**) or at 37 °C (**B**), viable counts were performed by plating 0.1 ml aliquots of appropriate dilutions of each of the growths on three brain-heart infusion agar (Difco) plates

that no immunity was developed following the oral infection. In contrast, four of five C57BL/6N mice that had survived oral challenge with *Y. enterocolitica* organisms survived the s.c. challenge with a lethal dose (7.6×10^2 organisms) of *Y. pestis* Yreka (Fig. III.22 B). This indicates that the orally-introduced *Y. enterocolitica* organisms remained viable in the mice and stimulated production of a protective immunity, whereas the *Y. pestis* organisms were killed without stimulating the development of a protective immunity.

In order to demonstrate the role of O-polysaccharide in directing the preferential infection route of these three *Yersinia* organisms in comparison, a novel facet of host-defense mechanisms unique to the oral infection route must be pointed out, i.e., perorally-infected microorganisms are confronted with the acidity of mouse stomach juice (about pH 2.6 according to Nakagawa, private communication).

Table III.9. Intagastric infection of mice with virulent or attenuated *Y. pestis*

Infection dose (CFU)	Strains infected and their virulence				
	Yreka, virulent			EV, attenuated	
	Death/ infected	Death (%)	Day of the last death	Death/ infected	Death (%)
3×10^7	6/ 6	100	3	0/ 8	0
3×10^6	3/ 8	37.5	4	0/ 9	0
3×10^5	2/ 7	28.5	4	0/ 8	0
3×10^4	0/10	0		0/ 9	0
3×10^3	0/10	0		0/10	0
3×10^2	0/ 9	0		0/ 9	0
3×10^1	0/ 7	0		0/10	0
3×10^0	0/10	0		0/ 9	0

Based on the possibility that a biological function of the O-polysaccharide chains of Gram-negative bacteria is to provide protection against environmental stresses, including those presented by acid pH, the viabilities of *Y. pestis, Y. pseudotuberculosis,* and *Y. enterocolitica* were determined following incubation at pH 3, pH 4, pH 5, and pH 6 at temperatures of 25 °C or 37 °C (Wake 1984). As shown in Fig. III.23, *Y. enterocolitica* is most resistant, *Y. pseudotuberculosis* less resistant, and *Y. pestis* more sensitive to the killing effect of acidic pH. Presumably, the O-polysaccharide chains of *Y. enterocolitica* and *Y. pseudotuberculosis* provide a survival value at an acidic pH. From the viewpoint of biological evolution, these microorganisms have been selected as orally-transmitted pathogens in part due to their wider acceptable pH range.

However, this does not deny the pathogenicity of orally-transmitted plague infection reported by Christie et al. (1980). Butler et al. (1982) reported 50% lethal doses of bacteria by intragastric inoculation of: *Mus musculus* $\log_{10}$ 6.32, *Zygodontomys pixuna* 5.46, and *Rattus rattus* 9.62. The infection microorganisms, however, induced generalized infection and were not found in feces, in clear contrast to orally-transmitted *Y. enterocolitica* which survive for a longer time, are found in feces, and stimulate the host animals to develop cell-mediated immunity (Alonso et al. 1978; Alonso et al. 1980; Kaneko and Hashimoto 1983; Wake et al. 1983).

Wake (unpublished) also intragastrically infected an outbred-mouse strain with a fully virulent *Y. pestis* Yreka and an attenuated-strain EV. (Here, the difference between the aforementioned oral infection with the intragastric one, is that the animals receive the microorganisms contaminated with food or water in the oral infection, whereas organisms are introduced through intragastric tubes in the intragastric infections.) The results are shown in Table III.9. The attenuated strain did not kill the mice and only large doses of the virulent *Y. pestis* initiated fatal infections. However, fewer numbers of both or-

Table III.10. Protection of mice that had survived intragastric challenge with a fully virulent Y. pestis strain Yreka against a second intragastric challenge with the same strain[a]

Dose (CFU) of the first infection	Survivals/ infected	Survivals (%)	Day of the last death
3×10^6	1/ 5	20	4
3×10^5	1/ 5	20	4
3×10^4	4/ 8	50	4
3×10^3	2/10	20	4
3×10^2	1/ 8	12.5	3
3×10^1	1/ 7	14.2	4
3×10^0	3/ 6	50	3
0	0/ 2	0	3

[a] Challenge dose: 4.8×10^7 CFU.

Table III.11. Protection of mice intragastrically immunized with attenuated *Y. pestis* strain EV against subcutaneous challenge with the fully virulent strain Yreka organisms[a]

Immunizing dose (CFU)	Survivals/infected	Survivals (%)
3×10^7	5/ 8	52.5
3×10^6	3/ 9	33.3
3×10^5	1/ 8	12.5
3×10^4	1/ 9	11.1
3×10^3	6/ 9	60.0
3×10^2	1/ 9	11.1
3×10^1	0/10	0
3×10^0	0/ 9	0
0	0/10	0

[a] Challenge dose: 4.8×10^3 CFU.

ganisms than showed lethality had multiplied within the mouse bodies because. Table III.10 and Table III.11 show that both organisms conferred protective immunity to some of the host mice against lethal doses of the *Y. pestis* Yreka as judged by subsequent s.c. challenge with Yreka cells 14 days later.

How *Yersinia* Organisms Transport Iron that Enhances Virulence

As was exemplified by some pathogenic bacteria in Part I, microorganisms use the same general approach for transporting iron into the cell at anything more than a residual rate. It has been recognized that the machinery working for this approach is the final accomplishment of microorganismic evolution since photosynthetic organisms started pumping oxygen into the atmosphere 2 b. y. ago, because available iron has been converted to its ferric form, which is virtually insoluble. Also, it has been elucidated that this high-affinity system has two parts: the production of an iron-binding molecule, which is excreted from the cell; and the elaboration of a membrane-receptor molecule, which binds the iron complex and transports the metal to the cell's interior (refer to Fig. I.15). Furthermore, e.g., in *E. coli,* the membrane receptor and the enzymes that synthesize the iron-binding molecule are coded for by five genes in one operon, which is turned off when iron levels in the cell have been satisfied and the switch off appears to be a complex formed between iron and a polypeptide (a repressor protein), which is coded for at a separate location from the operon. The term, siderophore, Greek for "iron bearer", is applied to molecules that can bind iron at very high affinities. The several hundred different siderophores now known are classifiable into two groups: those based on phenol catechols and those based on hydroxamates (see Neilands, J. cited by Roger Lewin in Science, 225:401–402, 1984).

Based on such background knowledge, indirect evidence for siderophore production by *Y. pestis* (Wake et al. 1975) is validated, because 3×10^5 *Y. pestis* organisms overlaid on the deferrated media allowed a limited number of "early colonies" to grow after incubation at 27 °C for 3 days, but these became surrounded by satellite colonies after prolonged incubation. As the control, the paper disks put on the deferrated agar medium after immersing into $FeCl_3$ solution, also allowed the growths of satellite colonies after 3-day-incubation at 27 °C. Another remarkable finding was that five P+ *Y. pestis* strains formed 15– to 33-fold more, and a P++ *Y. pestis* MII40 strain, 94-fold as many early colonies in comparison with three P– *Y. pestis* colonies (Table III.12 and Fig. III.24), indicating a close relationship of the virulence factor P (refer to Part III, Chap. 2, p. 223) with the potentiality of siderophore production, although the latter is demonstrated to be present, although weakly, in P– *Y. pestis,* also. Presumably this difference in the potentialities for siderophore production between P+ and P– *Y. pestis* organisms is caused not only by the presence of the outer-membrane iron-storage complex but also of the cytoplasmic iron-storage component in P+. The P– *Y. pestis* organisms lack the outer membrane iron-storage complex although the cytoplasmic iron-storage component is present (refer to Part III, Chap. 2, p. 223).

Contrary to these data, however, Perry and Brubaker (1979) spotted iron-starved cells of *Yersinia* and control organisms capable of producing known siderophores on the surface of iron-limited agar (containing citrate or nitrilotriacetate) seeded with *Salmonella typhimurium* LT2 enb-7. After incubation for 16 h (37 °C) or 24 h (26 °C), distinct zones of growth caused by iron-depletion release and diffusion of siderophores surrounded colonies of the positive-control organisms, but not *Yersinia* colonies. The method employed was derived from Miles et al. (1975), who concluded that the capacity to synthesize iron chelators is, by itself, not an index of the capacity of a strain to proliferate

Table III.12. Interrelationships among SID-producer ratio, percent of pigmented colonies, virulence, and toxicity of 9 *Y. pestis*, 2 *Y. pseudotuberculosis*, and 1 *Y. enterocolitica* strains

Strain designation[a]	SID-producer ratio[b]	Subcutaneous virulence (LD_{50})[c]		Intravenous mouse toxicity (LD_{50}, g/mouse)[d]
		Mouse	Guinea pig	
Y. pestis				
Yreka Fl+VW+Pl+P+ (96.2%)	17.0	13	110	35.4
Alexander Fl+VW+Pl+P+ (55.3%)	15.0	34	68	26.8
Bryans Fl+VW+Pl+P+ (69.1%)	18.5	16	5200	30.8
195/P Fl+VW+Pl+P+ (91.4%)	25.9	7900	150	70.0
MP6 Fl+VW+Pl+P+ (92.4%)	32.5	21	120	17.7
EV Fl+VW+Pl+P− (0%)	1.0	10^6	10^6	21.8
A1122 Fl+VW−Pl+P− (0%)	2.8	10^6	10^6	20.4
Tjiwidej Fl+VW−Pl+P− (0%)[e]	2.6	10^6	10^6	28.7
MII40 Fl−VW−Pl−P+ (99.4%)	93.6	10^6	10^6	40.5
Y. pseudotuberculosis				
14I Fl−VW−Pl−P− (0%)	1000	10^6	10^6	400.0
222/+ Fl−VW+Pl−P− (0%)	1000	10^6	10^6	400.0
Y. enterocolitica				
03 Fl−VW−Pl−P− (0%)	1000	n.d.[f]	n.d.	n.d.

[a] All established virulence factors except Pu were assayed according to Surgalla et al. (1970): percents in parentheses indicate those of pigmented colonies among 71–7771 (mean 1823) isolated from the respective strains.
[b] The number of early colonies of the respective strain, formed before third day of incubation at 27° C on DBHIA plate, divided by those of EV (1 or 2) in parallel experiments, using same lot of DBHIA. Results of two experiments are combined in this table.
[c] Mean lethal dose (LD_{50}) was calculated from deaths per ten mice or six guinea pigs on day 14 according to the method of Reed and Muench.
[d] Calculated from deaths per ten injected mice on day 2 after receiving appropriate dilutions of acetone dried-powder saline suspension according to the method of Reed and Muench.
[e] Presumably, an accidental P− isolate in our laboratory.
[f] n.d., not done.

in vivo after assaying randomly selected avirulent and virulent strains of *Klebsiella* species, *E. coli,* paired S and R (avirulent) *Escherichia* (one pair), *Salmonella* (four species, six pairs), and of *Shigella* (three species, six pairs). In these experiments, the organisms assayed were tested as a *population,* rather than as individual bacteria as assayed by Wake et al. (1975). If any individual siderophore-producing bacteria in the spotted *population* excretes diffusible siderophore, one might erroneously judge that the *population* as a whole was producing siderophores or that no siderophores were produced, or one might see limited siderophore effects, if the predominant cells in the *population* are still repressed for siderophore-production by the residual iron contained in the agar. In fact, Perry and Brubaker (1979) described a narrow (1 to 2 mm) dense

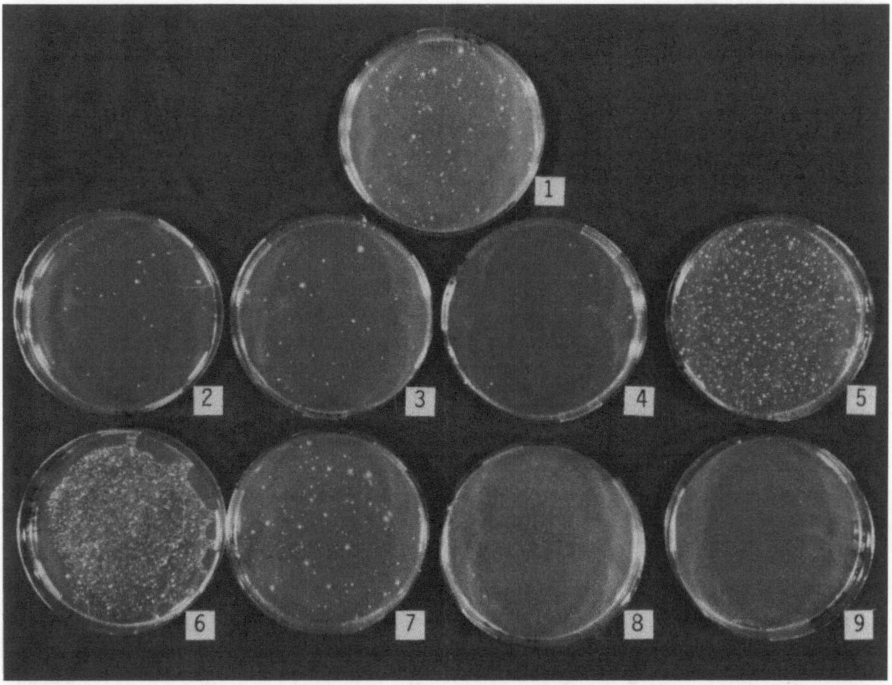

Fig. III.24. General view of the plates (one of each strain), showing the numbers of SID-producers in the experiment of Table III.7. All plates received inoculations of 3×10^5 organisms of the respective strain, onto the DBHIA, and incubated at 27 °C for 3 days. (*1*) Yreka, virulent-plague strain; (*2*) Tjiwidej, attenuated-plague strain; (*3*) A1122, attenuated-plague strain; (*4*) EV, attenuated-plague strain; (*5*) MII40, attenuated-plague strain; (*6*) MP6, virulent-plague strain; (*7*) 195/P, virulent-plague strain; (*8*) 14I, pseudotuberculosis strain; (*9*) 222/+, pseudotuberculosis strain. Because of confluent growths of pseudotuberculosis organisms, no colony is recognizable on plates (*8*) and (*9*)

zone of growth appearing around colonies of all species of *Yersinia* if the plates were incubated for an additional 24 h. This may indicate that the production of siderophores by *Yersinia* was delayed or that only some of the cells produced siderophore, in accordance with the observations by Wake et al. (1975). Siderophore as the cytoplasmic iron-storage component and the P+ substance responsible for pigmentation may not be the same (Wake et al. 1975). P+ seems to be controlled by plasmid-encoded factors (Chap. 3), and siderophore by the chromosome gene(s).

Furthermore, the conclusion of Perry and Brubaker (1979) that wild-type *Y. pseudotuberculosis* and *Y. enterocolitica* are *pgm*$^-$, is in conflict with Burrows (1973), who described pigmentation of *Y. pseudotuberculosis* in detail. This discrepancy may due to a high frequency of spontaneous loss of the *pgm*$^+$ gene after isolation from the infected animals. This would not be unique since *Shigella dysenteriae,* quickly loses the larger plasmids controlling invasiveness

Table III.13. Virulence-enhancing effect of SID on the P− attenuated strain EV

Pretreatment	EV dose	Lethality (deaths/10)	Survivors, immune[e] (%)
SID with arachis oil[a]	3.8×10^7	10	n.d.[f]
SID with arachis oil[a]	3.8×10^6	6*[d]	100
SID with arachis oil[a]	3.8×10^5	5*	100
SID with arachis oil[a]	3.8×10^4	5	100
SID with arachis oil[a]	3.8×10^3	3	57
SID with arachis oil[a]	3.8×10^2	4	17
SID with arachis oil[a]	3.8×10^1	4	33
Saline with arachis oil[b]	3.8×10^7	10	n.d.
Saline with arachis oil[b]	3.8×10^6	1*	100
Saline with arachis oil[b]	3.8×10^5	0*	100
SID without arachis oil[c]	3.8×10^6	0*	100
SID, i.v. injection	0	0	n.d.
SID, i.p. injection	0	0	n.d.

[a] One-fourth ml SID (titer 16) was emulsified with same volume of arachis oil and injected i.p. into ten mice.
[b] One-fourth ml saline was emulsified with same volume of arachis oil and injected i.p. into ten mice.
[c] Two-tenths ml SID (titer 16) was used to suspend and inject i.p. EV organisms into ten mice.
[d] Significant difference ($P < 0.05$), according to Fisher's exact test, between SID-oilless group.
[e] Percent of mice surviving 14 days after challenge with 10^5 Yreka organisms (7000 LD_{50}).
[f] Not done.
* Number of deaths on day 14 of the infection.

(Part I). A siderophore preparation prepared from *pgm*$^-$ EV strain was able to stimulate the in vivo multiplication of *Y. pestis* (Wake et al. 1975, Table III.13) if administered in an appropriate form.

Sutoh and Wake (unpublished) tried to isolate the siderophore from the P+ *Y. pestis* strain MII40 and to obtain direct evidence of siderophore production by *pgm*$^+$ *Y. pestis* using the method successful with *Vibrio cholerae* (Payne and Finkelstein 1978). No production of phenolate compounds were detected using the Arnow assay at 510 nm (Arnow 1937). The ethyl acetate extract or another culture-supernatant sample passed through Dowex 1 at pH 7.0 with 1 N sodium formate, followed by passage through Dowex 1 at pH 9.0, and concentrated by freeze-drying were measured for hydroxamates at 540 nm with the Csáky procedure (Csáky 1948) using hydroxylamine hydrochloride standards (Table III.14).

The growth-stimulating effect of the ethyl acetate extract was compared to that of iron, but no significant differences were found, perhaps because the synthetic medium contained 1 μM iron, allowing the growth-stimulating effect of the *Y. pestis* siderophore to stimulate some *Y. pestis* organisms to *initiate* multiplication. To confirm this hypothesis, a suspension of EV organisms grown on trypticase soy agar (BBL) at 27 °C for 48 h was diluted to 5.8×10^3,

Table III.14. Paper chromatography[a] of hydroxamate-type siderophore produced by *Y. pestis* MII40

Method of detection	Color	RF value
Ultraviolet light	Fluorescent	0.56 ± 0.07
$FeCl_3$	Red	0.55 ± 0.05
	Black	0.68 ± 0.06
Ferric ammonium citrate	Blue	0.57 ± 0.06
$K_3Fe(CN)_6$	Blue-violet	0.68 ± 0.04

[a] The solvent contained 5% ammonium formate and 0.5% formic acid in water.

Table III.15. The effect of siderophore produced by *Y. pestis* MII40 the growth of *Y. pestis* EV

Number of EV Inoculated[a]	Synthetic medium (control)	Synthetic medium added with Blutal[b] (20 MFe)	Synthetic medium added with siderophore (5%)[c]	Synthetic medium added with siderophore (10%)[c]	Synthetic medium added with siderophore (10%)[d]
5.8×10^3	1.4×10^7	1.4×10^7	9.2×10^6	1.8×10^7	1.2×10^7
5.8×10^2	2.8×10^5	3.0×10^6	6.9×10^6	6.6×10^6	8.5×10^6
5.8×10	0	0	2.0×10^6	1.8×10^6	3.4×10^6

[a] EV organisms were inoculated into 10 ml synthetic medium with or without addition of siderophore or iron to yield the indicated concentration of ethyl acetate of the MII40 culture supernatant.
[b] Chondroitin sulfate, colloidal $FeCl_3$.
[c] Lot 1 of ethyl acetate extract of MII40 culture supernatant.
[d] Lot 2 of ethyl acetate extract of MII40 culture supernatant.

5.8×10^2, 5.8×10 organisms using redistilled water, and incubated at 27 °C with or without addition of the ethyl acetate extract.

As shown in Table III.15, ethyl acetate extracts of MII40 culture in synthetic medium allowed EV organisms to proliferate to 10^{6-7} organisms ml^{-1}.

Based on the assumption that the in vivo multiplication of P– *Y. pestis* EV organisms might be stimulated by supplementation with the siderophore, 0.2 ml siderophore samples were injected i.p. with EV organisms. All mice receiving 1.2×10^7 organisms survived for 14 days, but 7 out of 10 mice were killed when challenged with 1.2×10^8 organisms. In the control group, without siderophore injection, the same proportion (seven of ten mice) were killed only when challenged with at least 3.0×10^8 organisms. This does not prove that there is a virulence-enhancing effect of siderophore on EV organisms, but clearly the siderophore is nontoxic. To confirm nontoxicity of the siderophore preparation, 0.25 ml of siderophore was injected i.p. into each of ten mice. No mice were killed. Siderophore samples mixed with an equal volume of steril-

Table III.16. Virulence-enhancing effect of Blutal on EV[a] in mice

Blutal dose[b] (ml day^{-1})	Fe contents (mg)	Number of Blutal injections[d]	Cumulative Fe dose (mg)	Dead mice / Injected mice in 14 days
0.01[c]	0.04	1	0.04	1/10
0.1	0.4	1	0.4	1/10
0.1	0.4	2	0.8	1/10
0.1	0.4	3	1.2	4/10
0.1	0.4	4	1.6	4/10
0.1	0.4	5	2.0	9/10
0.1	0.4	6	2.4	10/10
0.2	0.8	1	0.8	0/10
0.2	0.8	2	1.6	3/10
0.2	0.8	3	2.4	6/10
0.2	0.8	4	3.2	8/10
0.2	0.8	5	4.0	10/10
0.4	1.6	1	1.6	1/10
0.5	2.0	1	2.0	4/10
0	0	0	0	0/13

[a] EV-challenge dose: 2×10^4 viable cells of EV were injected subcutaneously into each mouse.
[b] Indicated volume of Blutal was injected intraperitoneally.
[c] One-tenth ml of 10^{-1} dilution of Blutal in saline.
[d] Blutal injection was repeated daily for the periods indicated after the day of EV challenge.

Table III.17. "Virulence-enhancing effect" of intraperitoneally injected Blutal on intraperitoneally injected EV[a] in mice

Blutal injected (ml)	Fe contents (mg)	Number of Blutal injections[b]	Cumulative Fe dose (mg)	Dead mice / Injected mice in 14 days
0.1[c]	0.04	1	0.04	2/ 9
0.1	0.4	1	0.4	3/10
0.1	0.4	2	0.8	10/10
0.1	0.4	3	1.2	10/10
0.1	0.4	4	1.6	10/10
Control	0	0	0	0/10

[a] EV challenge dose was 2×10^4 colony-forming units in 0.2 ml saline.
[b] I.p. Blutal injections were started on the day of EV challenge and repeated daily at specified times.
[c] One-tenth ml of a ten fold saline dilution of Blutal was injected on the day of EV challenge.

ized arachis oil was also tested. The results, shown in Table III.13 show statistically significant differences ($P<0.05$, according to Fisher's exact test) in its virulence-enhancing effect between the mouse groups challenged with EV organisms in siderophore-arachis oil emulsion and the control mouse groups which were challenged with EV organisms in saline-arachis oil emulsion. These results are similar to the experiments of Jackson and Burrows (1956b) who

Table III.18. "Virulence-enhancing effect" of intraperitoneally-injected Blutal[a] on various subcutaneous-challenge doses of EV in mice

EV-challenge dose[b] (CFU)	Fe contents[c] (mg)	Number of Blutal injections	Cumulative Fe dose (mg)	Dead mice / Injection mice in 14 days
1.1×10^7	0.4	6	2.4	10/10
1.1×10^6	0.4	6	2.4	15/20
1.1×10^5	0.4	6	2.4	18/20
1.1×10^4	0.4	6	2.4	14/20
1.1×10^3	0.4	6	2.4	12/20
1.1×10^2	0.4	6	2.4	15/20
1.1×10^1	0.4	6	2.4	9/20
1.1×10^0	0.4	6	2.4	0/20

[a] One-tenth ml Blutal was injected intraperitoneally into mice everyday for six successive days after the day of EV challenge.
[b] EV-challenge dose: determined from average colony-forming units on five BHIA plates each spread with 0.1 ml of one of serial ten fold dilutions and incubated for 3 days at 27°C.
[c] Quantity of Fe contained in a 0.1 ml dose of Blutal.

Table III.19. "Virulence-enhancing effect" of subcutaneously injected Blutal on EV[a] suspended in Blutal

EV-challenge dose (CFU)	Blutal injected (ml)	Fe contents (mg)	Dead mice / Injected mice in 14 days
2.0×10^4	0.2	0.8	10/10
2.0×10^3	0.2	0.8	6/10
2.0×10^2	0.2	0.8	0/10
2.0×10^1	0.2	0.8	1/10
2.0×10^0	0.2	0.8	0/10
2.0×10^4	0[b]	0	0/10

[a] Colony-forming units were determined as in Table III.20. Two-tenths ml serial dilutions of EV suspended in Blutal was were injected subcutaneously into each mouse.
[b] Colony-forming units of EV injected, suspended in 0.2 ml saline.

were also successful in enhancing the virulence of their P− attenuated strains by injecting i.p. hemin or such iron salts as ferrous sulfate or ferric iron emulsified within arachis oil. Thus, siderophore emulsified in arachis oil seemed to have transported iron to P− *Y. pestis* organisms. When the surviving mice were tested for immunity, the results suggested that multiplication of EV bacteria occurred in some animals even when the EV doses were as small as 3.8×10^1 organisms (Table III.13, right column).

However, Kaplan and Basford (1979) have given an alternative explanation of such apparent enhanced virulence of P− *Y. pestis* organisms, i.e., iron

Table III.20. Days of deaths of the mice which received EV abd Blutal through different combinations of routes

Day of death	1	2	3	4	5	6	7	8	9	10	11–14	Total deaths	Average day of death	Variance (log 10)
Group I[a]	0	30	26	12	8	10	3	1	0	0	0	90	3.20	0.329
Group II[b]	0	1	4	1	2	9	2	2	2	5	0	28	5.21	0.363
Group III[c]	0	1	2	0	27	30	11	1	9	3	0	84	5.98	0.128
Group IV[d]	0	3	5	4	2	2	2	2	0	0	0	20	4.05	0.383
Group V[e]	0	7	1	3	0	2	2	1	0	0	0	16	4.52	0.343

[a] Received intraperitoneally 4.3×10^4 CFU of EV and two daily i.p. injections of 0.1 ml Blutal from the day of EV challenge.
[b] Received subcutaneously 4.3×10^4 CFU of EV and six successive daily i.p. injections of 0.1 ml Blutal from the day of EV challenge.
[c] Received subcutaneously 5.8×10^4 CFU of EV suspended in 0.2 ml Blutal.
[d] Received 4.2×10^4 CFU of Yreka strain.
[e] Received 3.6×10^3 CFU of Yreka strain.

neutralizes nonspecific cellular components of defense mechanisms (Part II). Iron and hematin, but not hemoglobin, inhibit the intraleukocytic killing of bacteria by inactivating cationic proteins present in the leukocyte lysosome. This inhibition is dose-dependent, with nearly complete inhibition at 1.84 mM $FeSO_4$ and no inhibition at 0.01 mM $FeSO_4$. Injected iron also probably inhibits the H_2O_2-dependent killing mechanisms of phagocytes by interfering

[a] Doses and procedures of EV and Blutal administration are shown in Table III.19. Among the mice dead from the infection with *Y. pestis* EV organisms, which were enhanced for their virulence by the activity of ferric iron (Blutal), only those autopsied just after death were considered. Five different tissues, were cutaway: liver (*Li*), spleen (*Sp*), lung (Lu), heart (*He*), and the peritoneal side of the local skin (*Lo*) where the EV organisms had been injected. To each cut surface, a brain heart infusion agar (Difco) was rubbed onto all surfaces extensively; all plates were incubated for 3 days at 27°C. The growths of EV organisms were expressed as in Footnotes e–j.
[b] Intraperitoneal injection.
[c] Subcutaneous injection.
[d] Each dead mouse found in each group of mice receiving iron and infection with EV organisms through three different combinations of administration routes was numbered in the order in which it was picked up from the cage containing the respective group.
[e] Liver (Li).
[f] Spleen (Sp).
[g] Lung (Lu).
[h] Cardiac blood (He).
[i] Local site of EV injection (Lo).
[j] Growth of EV on the plate smeared with each organ was graded as follows: $\begin{smallmatrix}++\\++\end{smallmatrix}$ Confluent growth; $\begin{smallmatrix}++\\+\end{smallmatrix}$ confluent, but in the form of attached colonies; ++ several thousands of isolated colonies; + several hundreds of isolated colonies. Figures expressed as Arabic number denote colony numbers counted whenever possible instead of using symbols.

Table III.21. Distributions of *Y. pestis* EV organisms in the mouse bodies immediately after deaths from the virulence enhanced by iron

lutal EV	Routes of administration[a]														
	i.p.[b]					i.p.					s.c.[c]				
	i.p.					s.c.					s.c.				
Iouse[d] umber	Li[e]	Sp[f]	Lu[g]	He[h]	Lo[i]	Li	Sp	Lu	He	Lo	Li	Sp	Lu	He	Lo
Io. 1	++++[j]	++++	++++	++++	++++	++++	++++	++++	++++	+	++++	++++	++++	++++	++++
Io. 2	++++	++++	++++	++++	++++	++++	++++	+++	++	++	3	0	5	0	++++
Io. 3	++++	++++	++++	++++	++++	++++	++++	++	++	27	+	84	54	14	++++
Io. 4	++++	++++	++++	++++	++++	++++	++++	++++	++++	13	55	3	0	1	+++
Io. 5	++++	++++	++++	++++	++++						+	54	9	6	+++
Io. 6	++++	++++	++++	++++	++++						+	+	++	++	+++
Io. 7	++++	++++	++++	++++	++++						2	0	+	44	+++
No. 8	+++	++++	++++	++++	++++	11	1	9	4						+++
No. 9											0	70	0	1	+++
No. 10											++	++	+	6	+++
No. 11											+	28	35	16	+++
No. 12											0	0	0	0	+++
No. 13											+	+	+	38	+++
No. 14											0	0	0	0	+++
No. 15											71	30	1	0	+++
No. 16											1	11	1	0	+++
No. 17											+	+	36	7	+++
No. 18											22	+	8	2	+++
No. 19											+	82	31	4	+++
No. 20											0	0	0	0	+++

with the phagocytosis-associated activation of the hexose monophosphate shunt (HMPS) pathway of glucose metabolism, thus, interfering with the production of highly reactive oxygen and hydroxyl radicals. According to Kaplan and Basford (1979), the presence of 0.37 mM or 0.18 mM $FeSO_4$, reduced the production of H_2O_2 to one-seventh or to one-quarter of normal levels, respectively. Further experimentation showed that H_2O_2 in the absence of leukocytes is inactivated by $FeSO_4$.

However, the mode of action through which iron enhances the in vivo growth of P− *Yersinia* could vary depending on the combinations of administration routes of EV organisms and Blutal, due to the diverse grades of participation of circulating iron-binding plasma proteins, e.g., transferrin, that interact with free iron. This diversity could determine whether the iron is assimilated directly by *Yersinia* organisms or is utilized for the neutralization of humoral or cellular defense mechanisms. The combination of s.c.-s.c. routes, i.e., the administration of EV organisms suspended in a small amount of ferric chloride emulsified in chondroitin sulfate (Blutal) could facilitate the direct action of iron with *Yersinia* organisms, but the neutralization of the whole-body defense mechanisms might be minimal when the P− character of the *Yersinia* organisms which is responsible for the invasiveness of the bacteria is supplemented directly by coinjection with iron. The results shown in Tables III.16–20 and Fig. III.25 demonstrate that this is in fact the case. As summarized in Table III.21, the multiplication of EV organisms was, to some extent, concentrated in the local site of s.c. injection of both EV organisms and Blutal. The different slopes among the cumulative-death curves depicted in Fig. III.25 indicate that differences among the effective mechanisms of iron in enhancing the virulence of EV organisms (Wake et al. 1970).

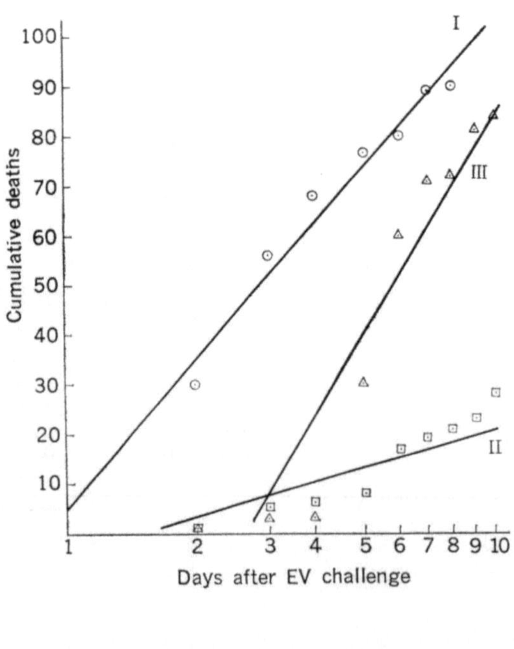

Fig. III.25. The different slopes among the cumulative-death curves of three mouse groups i.e. those mice receiving i.p. injection of 5.1×10^4 organisms of *Y. pestis* EV and i.p. injection of 0.1 g iron in the form of chondroitin sulfate ferric iron (○); those mice receiving s.c. injection of the same dosis of EV organisms and i.p. injection of 0.1 g iron (□); and those mice receiving s.c. injection of the same dosis of EV organisms suspended in 0.1 ml solution of 0.1 g iron (△)

Irrespective of whether the bacteria take up the injected ionic iron or the free ionic iron found and subsequently released from the transferrin, the bacterial mechanism responsible for the sequestering iron from the microenvironment is probably the siderophore.

Siderophore production and siderophore genetics, as they relate to the P+ phenotype, were discussed in Chap. 2.

In conclusion, *Y. pestis* uses the same general approach for transporting iron into the cell as *E. coli* with which it is closely related taxonomically (Brubaker 1984). The iron-binding molecule, which can be excreted even from the P− *Y. pestis* cell is released from repression under the condition of strict iron starvation. Therefore, these P− organisms cannot grow in vivo to express virulence, but supplementation with exogenous iron or exogenous siderophore which transports iron into the cell interior, enhances the in vivo multiplication, and therefore, the virulence.

In contrast, the outer-membrane iron-storage complex unique to P+ organisms, which can take over the iron from hemin, transferrin, or lactoferrin with higher affinity to iron, plays the role of an established virulence factor, as was delineated in Chap. 2 of Part III.

The Relationship of Arginine Metabolism with *Y. pestis* Virulence

Brubaker (1984) stated that those accepting the notion that the modern version of *Y. pestis* organism has somehow become attenuated in its virulence will probably change that opinion after reading Thomas Butler's recent account of plague in Vietnam (Butler 1983), and that highly qualified physician, Butler was nevertheless "humbled by the swift progression of this disease which carried healthy individuals to death within 3 days."

However, it has been noted over a long period that *Y. pestis* strains isolated in South America are prone to show rather weaker virulence (Pollitzer 1954). This section is devoted to include such exceptional *Y. pestis* strains in this book.

Y. pestis strains with reduced virulence have been isolated that show higher LD_{50}'s in guinea pigs, which contain 1-asparaginase activity in their serum (Broom 1961). M. Baltazard communicated to T. W. Burrows that nearly 200 strains isolated in an epizootic focus of plague at Exu, Pernambo, Brazil showed high pathogenicity for mice, but were innocuous for guinea pigs. When Burrows and Gillet (1971) compared 11 of these strains with the fully virulent *Y. pestis* strain MP6, he found that Exu strains showed higher growth rates on defined minimal agar, if the medium was supplemented with 1-asparagine. Recognition of growth stimulation by asparagine suggested an explanation for the low pathogenicity of Exu organisms for the guinea pig, which would present Exu organisms with an asparagine-free microenvironment. The resultant slowed growth rate could allow the guinea pig to limit the infection long enough for the development of an active immunity. Since 20 $\mu g\ ml^{-1}$ of asparagine in the blood was estimated to be required to permit these *Y. pestis* strains to grow as rapidly as a fully virulent strain, and since human blood plasma contains only 6 µg of asparagine, these semi-asparagine-dependent organisms cause nonacute human infections. Thus, this is an example of a bacterial pathogen with host specificity determined by its nutritional requirements and nutrients provided by the host (Burrows and Gillet 1971).

In connection with the asparagine metabolism, following reports, also, deserve pointing out: Some of the *Y. pestis* strains of Russian origin produce aspartase which disrupts the bioenergetics and nitrogen metabolism of mammalian cells. Perhaps the deficiency in this character in typical *Y. pestis* serves to ensure or at least prolong a cozy intracellular environment by providing enough asparagine to the pathogens. Furthermore, the progression of the disease caused by these Russian strains is characterized by the extracellular *Y. pestis* multiplication leading either to accelerated septicemia or recovery from the disease (Brubaker 1979, Dreyfus and Brubaker 1978).

Chapter 4 Host Resistances to *Yersinia* Organisms – Immunology in *Yersinia* Infection

Among the previously announced contents of Part III describing the interactions between both types of factors which are expressed in a particular field of host-parasite relationships, i.e., *Yersinia infection* (Introduction to Part I, p. 1), the host factors constituting the individual counterpart facet in the integrated host-defense mechanisms are the target for the function of each virulence factor. Therefore, they were delineated in the sections for each virulence factor in Chap. 2, Part III. In this chapter the changes of the host components caused by *Yersinia* infection are to be described.

Until recently, this knowledge has been accumulated on the basis of classical immunology, deriving from the efforts to develop effective vaccines for prophylactic purposes. Actually, the origin of classical immunology could be traced back to the description that a small number individuals in an epidemic area survived these epidemics and that these people took care of the dying patients during the next wave of plague epidemics (Pollitzer 1954). If artificial infection without any hazardous results is realizable, the principle for a live-plague vaccine is established according to the aforementioned human experience. Furthermore, the principle of the killed-plague vaccine has been established since the finding of "antibodies" (Part II, Chap. 7, p. 131). Much has been disputed as to the superiority of either of these two categories of vaccines. However, an objective evaluation method of the efficacy of either category using modern statistics to compare vaccine-administered and placebo-administered groups was still not available when plague epidemics prevailed in some areas of the world that could have served as fields for such trials.

On the other hand, "the swift progression of plague which carried healthy individuals to death within 3 days" (refer to previous chapter, p. 255) is a remarkable characteristic of *Y. pestis* infection. This means that *Yersinia pestis* organisms kill the host organisms before the acquired-immune mechanisms exert their effective function, because the formation of immunity requires several days. (Actually, the antibody production by tissue-cultured lymphocytes with the aid of macrophages begins, at earliest, after 2 days of antigenic stimulation.) Therefore, the changes of the host components caused by *Yersinia* infection in nonimmunized animals can be regarded as those of the natural, i.e., innate or "nonspecific" resistance mechanisms. In contrast, the modifications occurring in the host animals after the critical point of death or survival, can be regarded as involving acquired or "specific" immunity.

This dissociability of natural and acquired resistances provides the experimental host-parasite relationship system of *Y. pestis* infection, enabling medical immunologists to resolve a variety of important mechanisms of antibacterial host-resistance mechanisms at the level of modern immunology. Therefore, the functions of host-resistance mechanisms themselves are feasible to study independently from the side of bacterial virulence, although these two components are mutually interacting at all molecular-, cellular-, and individual-host levels throughout all progression of the process.

In this context, this chapter deals with: (1) natural resistance to *Y. pestis* infection expressed in the strain differences in mice and guinea pigs; (2) humoral aspects of anti-*Y. pestis* acquired immunity; (3) cell-mediated immunity as anti-*Y. pestis* bactericidal mechanisms; (4) cooperative protection provided by humoral- and cell-mediated immune responses against *Y. pestis* infection; (5) strain differences of anti-*Y. pestis* acquired resistances in mice and guinea pigs.

Natural Resistance to *Y. pestis* Infection Expressed in the Strain Differences in Mice and Guinea Pigs

Among the multifaceted, integrated host-defense mechanisms delineated in Part II, those without involvement of "specific interaction between the epitope and the combining site" (conclusion of Chap. 1, Part II, p. 85) are called natural or innate defense mechanisms and their activities, natural resistance.

These activities are exerted in time course starting from the mutual contact of host with the pathogen, e.g., *Y. pestis*. Because bubonic plague is mediated by flea bite, the skin barrier is surmounted by physical force, and bacteria cause the chemotactic mobilization of cellular components with defensive functions (Chap. 8, Part II) as the first response of the host. However, reports on this leukocyte-*Y. pestis* first contact at the intra- or subcutaneous site are lacking except for unpublished data obtained by Wake, who intracutaneously injected a virulent *Y. pestis* strain into athymic nude mice and microscopically examined the Giemsa-stained smear preparations stamped from the skin piece cut out in time course. These data showed that virulent *Y. pestis* bacilli parasitize both polymorphonuclear cells (PMNs) and monocyte-macrophages even after 1 day of infection and that PMNs discharge the intracellularly multiplied *Y. pestis* organisms to be phagocytized by monocytes (Fig. III.26). In contrast to this study, Burrows et al. (1954) studied the leukocyte-*Y. pestis* interaction by *intraperitoneally* injecting virulent *Y. pestis* organisms into mice; and Cavanaugh et al. (1959), into the mice previously injected i.p. with a starch solution or a mineral oil Bayol F. The latter, furthermore, studied the interaction of *Y. pestis* organisms with the tissue-cultured macrophages which had been elicited with the oil Bayol F. Recently, Straley and Harmon (1984) observed the growth in mouse peritoneal macrophages of *Y. pestis* lacking established virulence determinants using unstimulated, but tissue-cultured macrophages. Re-

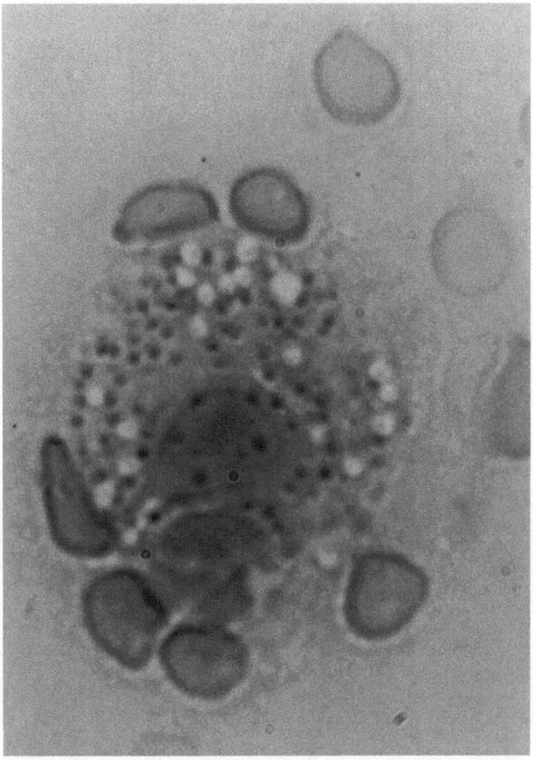

Fig. III.26. A polymorphonuclear cell is being phagocytosed by a mononuclear cell containing many *Y. pestis* cells which were intracutaneously injected 1 day before into an athymic nude mouse. Some of these bacterial cells are stained faintly, as if they are being digested within the white phagolysosomal spaces. Giema stain ×1,500

markable is that all of these authors, except Wake, used more or less artificially altered sites of leukocyte-*Y. pestis* first contact, and care must be taken when considering the influences of such artifacts before exact interpretation of the experimental results can be made. Reliance on tissue-culture system is inadvisable because of the de-differentiation and other changes that occur in the cultured cells (Smith 1972, refer to Chap. 6, Part I, p. 52).

With regard to this remark, not the interaction between each isolated facet of the defense mechanisms and *Y. pestis*, but survival times of the sensitive animals after the percutaneous infection with minimal lethal dose (MLD) of *Y. pestis* organisms is thought to serve as the indicator of the integrated effects of natural-resistance mechanisms as a whole.

Based on this principle, various inbred-mouse strains were challenged s.c. with an approximate MLD of a virulent *Y. pestis* strain Yreka (Wake 1980). Against the challenge dose of 62 organisms, 10 BALB/C mice showed the longest mean survival time (5.1 days); 10 C57BL/6J, the shortest (4.3 days), and their 20 hybrid, intermediates (4.4 days). Presumably, the deficiency in mature T-cells is not without influence on natural resistance, because 19 athymic nu/nu mice derived from BALB/C, showed an abnormally short survival

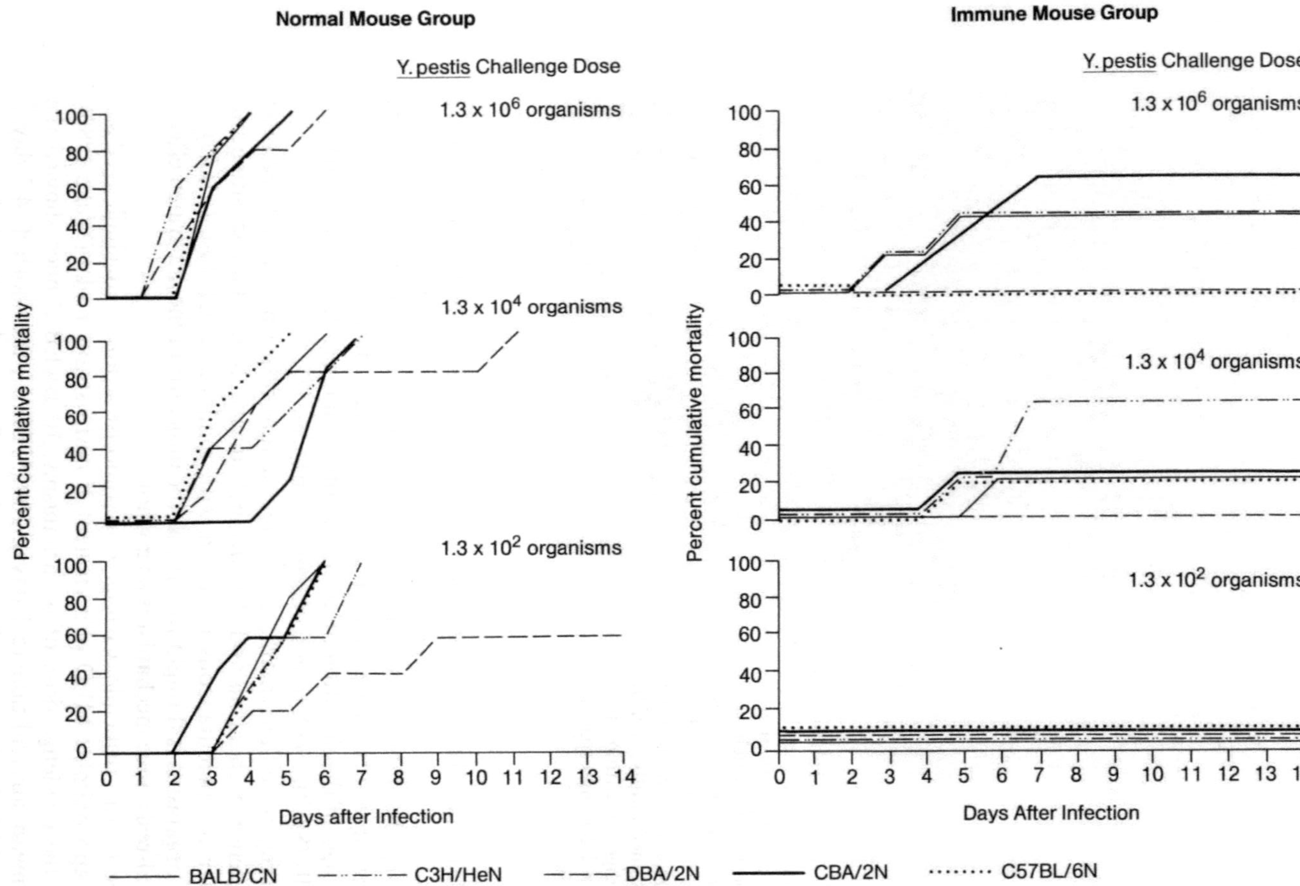

Fig. III.27. Nonrelationship between H-2 haplotype and natural or cell-mediated immune resistance to *Y. pestis* infection. Haplotypes of inbred mice used are as follows:

Haplotype	Inbred mice strains	
H-2[b]	C57BL/6N · — · — · — · —	
H-2[d]	BALB/CN ————	DBN/2N — — — — — — —
H-2[k]	C3H/HeN · · · · · · · · ·	CBA/2N · · · · · · · · · · · · ·

time (3.6 days). In addition, the order from the longest to the shortest survivors has never changed among these strains in four subsequent experiments. Thus, it was concluded that natural resistance in mice against plague is under genetic control. However, the relevance of this gene(s) with the H-2 haplotype (Part II, Chap. 2, p. 88) located on chromosome 17 was denied by the following experiments (Wake 1985).

All male BALB/CN and DBA/2N mice of $H\text{-}2^d$ haplotype, C3H/HeN and CBA/2N of $H\text{-}2^k$ haplotype, and C57BL/6N of $H\text{-}2^b$ haplotype were each divided into three groups consisting of five to six mice in one cage and s.c. infected with three grades of doses, i.e., 1.3×10^6, 1.3×10^4, or 1.3×10^2 Yreka organisms. The results in the form of a death/survival curve against the days after challenge are depicted in the left half of Fig. III.27.

At the conclusion of these experiments, it was revealed that the genetic control of natural resistance does not correlate with the H-2 haplotype; another serendipitous finding was that two of five DBA/2N mice survived against challenge with 1.3×10^2 Yreka organisms, i.e., a certainly lethal dose for all of the other inbred-mouse strains. Significantly prolonged survival times of DBA/2N mice against the challenge with 1.3×10^4 and 1.3×10^6 organisms were also recorded.

In order to clarify the genetic mechanisms of the latter serendipitous finding, ten female DBA/2N mice were s.c. challenged with 1.2×10^2 Yreka organisms. All mice were dead by the ninth day of infection, suggesting a sex-associated difference in natural resistance.

Starting from these and the previous findings, F1-hybrid and back-cross mice of various combinations between DBA/2N and BALB/CJ mice were produced and 12 each of male and female mice were s.c. challenged with 2.0×10^2, 2.0×10^4, or 2.0×10^6 Yreka organisms in order to analyze the genetic control of the natural resistance gene(s) unique to the male DBA/2N mice. The results obtained on back-cross mice are shown in the left half of Fig. III.28. When the presence of survival mice against the challenge with the certainly lethal dose of Yreka organisms was defined as the indicator of this natural resistance gene(s), this was found in [female (BALB/CJ × DBA/2N) F1 × male BALB/CJ] mice as shown in this figure. To draw a conclusion from these results, this natural resistance gene(s) is associated with the X-chromosome of DBA/2N, because the Y-chromosome of male DBA/2N should have been lost during the procedure of breeding for the production of these back-cross mice. But the expression of this gene(s) is repressed in female DBA/2N, i.e., in the form of an XX allele. Results from similar experiments using F1-hybrid mice between DBA/2N and BALB/CJ did not contradict this interpretation (data omitted).

Exchanging the host animals to the guinea pigs, Wake (1980) s.c. injected various inbred strains of guinea pig. The daily survivals after challenge with 4.4×10^3 and 4.4×10^4 organisms of *Y. pestis* Yreka organisms were recorded and significant strain differences in natural resistance were recognized between JY2 (GPLA, i.e., guinea pig MHC type of Ia.2b, Ia.13b, P.3, P.4, P.6), and JY1

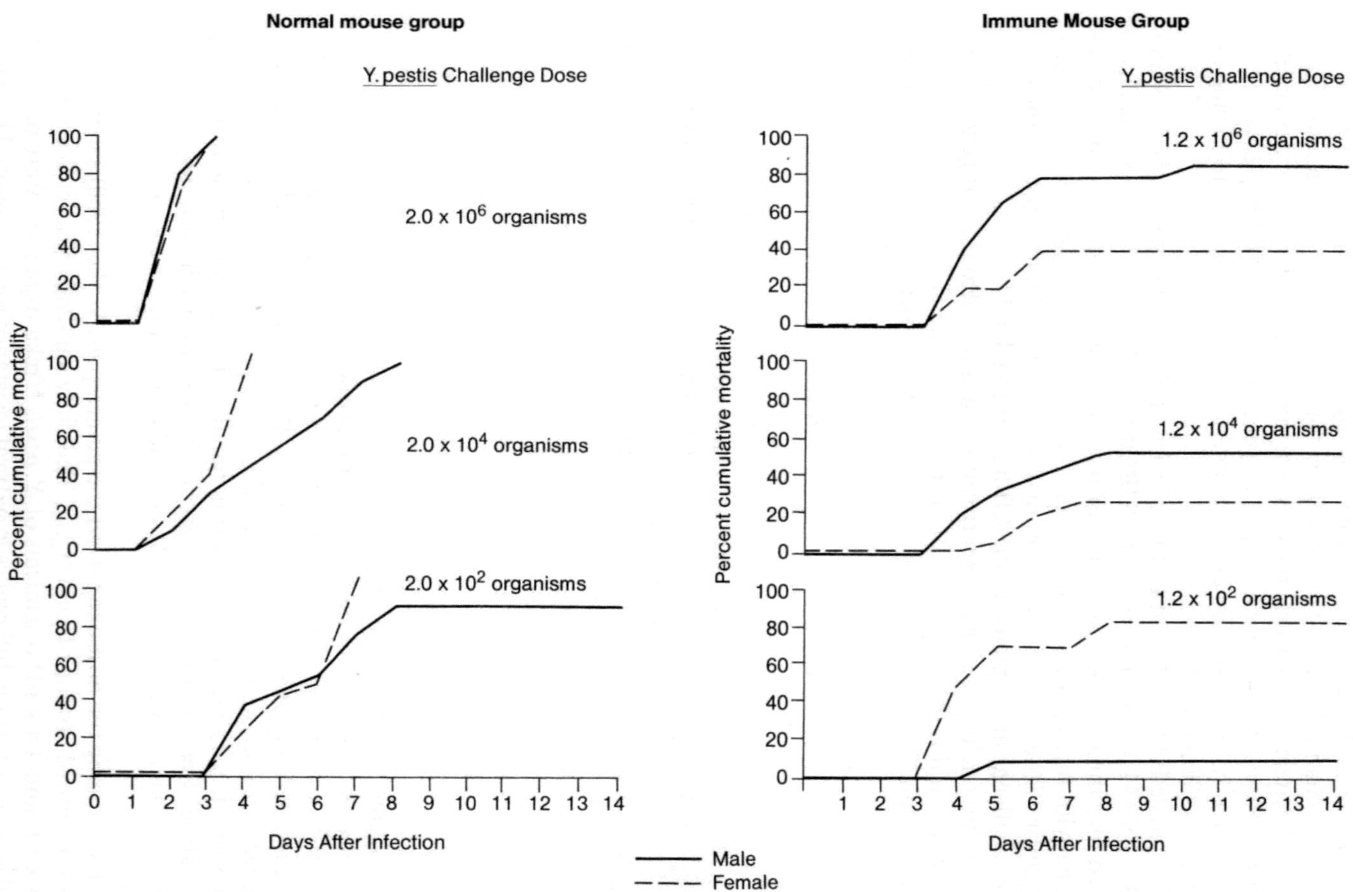

Fig. III.28. Sexual differences of [(BALB/CJ × DBA/2N) × BALB/CJ] back-cross mouse groups (*left*) and [BALB/CJ × (BALB/CJ × DAB/2N)] back-cross ones (*right*) with regard to the natural (innate) and cell-mediated immune resistances against *Y. pestis* infection. Each of 12 female and male back-cross mice were infected s.c. with indicated doses of *Y. pestis* Yreka organisms

(GPLA of Ia.13a, Ia.13b, P.5), NIHNo.2 (GPLA of Ia.2a, Ia.2b, P.1, P.2, P.3, P.4) and NIHNo.13 ($P<0.01$, according to the parallel-line assay). Mean survival times in these strains were 3.6, 6.8, 6.4, and 6.8 days, respectively.

In summary, natural resistance to bacterial infection which shows integrated effects of multifaceted host-defense mechanisms is feasible to study independently from the influence of acquired (specific) resistance mechanisms, i.e., immunity in the *Y. pestis*-host experimental system. However, the experimental results obtained by the use of artificially isolated host components require careful interpretation to exclude the possible artifactual influences from unnatural environmental conditions. This section demonstrated that immunogenetic analysis of antibacterial natural resistance is feasible by the use of the *Y. pestis*-host experimental system. The conclusions drawn from these studies indicated promising targets for future analytical studies using artificially-isolated host components. For instance, researchers will be able to determine which facet of mouse defense mechanisms is critical in providing natural resistance against *Y. pestis* infection by comparing the activities of selectively isolated host-defensive components, e.g., bactericidal activity of polymorphonuclear cells (PMNs) taken from male DBA/2N mice and those from female or other inbred-mouse strains.

Humoral Aspects of Anti-*Y. pestis* Acquired Immunity

Preceding chapters delineated that the chemical constituents of *Y. pestis* cell outer membrane and envelope, i.e., capsular substances before infection to host animal, are determined qualitatively as well as quantitatively by genetic control and by their interaction with outer-environmental, physical, and chemical conditions, e.g., those inherent to flea digestive organs.

In addition, the changes occurring after infection were also delineated. After the deaths of bacilli at any stage of this changing process, various cytoplasmic components will be released by the degradation of bacilli due to the actions of host-destructive mechanisms. Similar releases of various substances should occur after the in vivo degradation of killed bacilli which were administered in the form of killed-antiplague vaccine. Most of these substances thus produced in vivo are equipped with respectively unique antigenic epitopes (determinants) and induce immune responses via the mechanisms delineated in Chaps. 1, 7, and 8, Part II.

Of these immune responses, this section deals with the production of immunoglobulins of selected specificities with regard to the significance of (1) serological diagnosis of *Y. pestis* and (2) protection against the infection with virulent *Y. pestis* organisms.

One of the established virulence factors, i.e., F1 (Part III, Chap. 2, p. 189), has been admitted as the antigen specific to *Y. pestis,* in spite of the F1 – mutant strains in which its production is repressed to trace amount. Therefore, the use of antigenic specificity of F1 in serological surveys, i.e., passive hemagglutination (PHA) tests (Chen 1952), radioimmunoassay (RIA), enzyme-linked immunoassay (EIA), and fluorescent-antibody (FA) techniques, have been applied widely in epidemiological studies of plague.

With regard to humoral-immune responses providing protection against *Y. pestis* infection, Schütze (1939) described the divergence between mice and guinea pigs. Although F1 had not been defined at that time, the immunization with live-attenuated *Y. pestis* organisms equipped with appreciable amounts of envelope conferred mice good protection. However, those organisms without visible envelopes provided guinea pigs better protection than the former type in particular. Accordingly, Jawetz and Meyer (1943) discriminated "rat or mouse antigen" from the "guinea pig antigen". Baker et al. (1952) noticed that F1 purified by them conferred good protection to mice, but the effective fraction for guinea pigs resided in the residual bacterial bodies after extracting an F1-containing fraction with 2.5% NaCl.

This apparent controversy on the antigen(s) effective to protect either mice or guinea pigs was explained by Spivack et al. (1958) who demonstrated that F1 preparations of less than 50 μg with adjuvant stimulate the production of antibodies detectable by specific-complement fixation (CF) and passive hemagglutination (PHA) tests. Guinea pigs were protected against *Y. pestis* s.c. challenge in parallel with their anti-F1 antibody titers. However, immunization of guinea pigs with larger doses of F1 did not confer protection. Therefore, the trace amount of F1, remaining unextracted in the residual bacterial bodies, did not suppress, but induced, the anti-F1 immunoglobulin production. Because an antibody is defined as an effector molecule with specific recognition capabilities, the expression of which is inducible or *suppressible* (Cohn 1980, Part II, Chap. 1, p. 82), it is not surprising that the immune mechanisms of guinea pigs, but not of mice, responded to the F1-antigenic epitopes of more than 50 μg in quantity by suppressed-antibody production. A similar immune reaction has been known to occur by the administration of excessive quantities of pneumococcal capsular polysaccharide and this phenomenon is called "immunoparalysis" (Spivack and Karler 1958). Along with the progress of modern immunology, the term "high-dose immune tolerance" is preferred and its mechanisms have been studied at the cellular as well as at the molecular level (refer to Part II, Chap. 8, p. 142). In this context, the protective potency of anti-F1 immunoglobulin against *Y. pestis* infection was established.

For the assay of protective potency of anti-*Y. pestis* humoral antibodies, Meyer and Foster (1948) defined a mouse-protection index (MPI) as the percentage of mice that die within 14 days after s.c. infection with 100 MLD of a fully virulent *Y. pestis* organisms just after i.v. injection of the serum (0.5 ml) to be tested into at least ten mice divided by the average survival time in days. With plague-immune-human sera, the index is from 0 to 10; a value of 0 (no deaths) indicates the strongest possible protection. With normal sera, the index is usually 15 or more. When this index was used to relate the protective activity of convalescent sera with their anti-F1 hemagglutination (PHA) titers, guinea pig and human sera with PHA titers less than 320 correlated with the MPI index of 3.90 to 3.20 (Chen 1952). However, with regard to the PHA anti-F1 titers of the guinea pig sera from animals that resisted lethal challenge with a fully virulent *Y. pestis* Yreka, the titers range from 16 to 256 and the frequency of PHA anti-F1 titers as low as 32 to 64 was high among these animals (Wake, unpublished data). Therefore, anti-F1 immunoglobulin solely does not seem to play a protective role against *Y. pestis* infection and the involvement of anti-V immunoglobulin (Lawton et al. 1963, Part III, Chap. 2, p. 195) and cell-mediated immunity (refer to next section) should be considered to explain the high resistance of plague-convalescent animals or humans.

Cell-Mediated Immunity as Anti-*Y. pestis* Bactericidal Mechanisms

The preceding section delineated that immunization of host animals with killed *Y. pestis* vaccine or its purified components, i.e., F1 or V, elicits the production of immunoglobulin specific to the protective antigen(s), providing a prophylactic measure against plague epidemics. However, immunization of host animals or humans with attenuated-live vaccine also has been employed throughout the world as was previously mentioned (introduction to this chapter). In spite of these experiences, the guinea pigs thus immunized with live vaccine and which survived reinfection with virulent *Y. pestis* organisms, had sera with low passive hemagglutination (PHA) titers against the protective antigen F1.

In this connection, the involvement of cell-mediated immunity, the mechanism of which was delineated in Chaps. 8–11, Part II, was suspected in as much as *Y. pestis* is a facultatively intracellular bacterium.

In fact, Korobkova (1956) prepared a reagent designated "pestin" in analogy to "tuberculin" for the use of the detection of delayed-type hypersensitivity reaction and demonstrated that pestin-positive animals are resistant to the infection with virulent *Y. pestis* organisms. Wake et al. (1972) prepared pestin containing 26 μg protein ml^{-1} as a millipore-filtered, dialyzed 60% ammonium sulfate-precipitated fraction of *Y. pestis* Yreka culture supernatant. This kind of reagent was prepared from *Y. pseudotuberculosis* cultures of serovars I, II, III, IV, and V, in addition. All these reagents and fraction I (F1) preparations, induced delayed-type hypersensitivity skin reactions in the guinea pigs immunized with the attenuated *Y. pestis* EV organisms. Furthermore, pestin could be used for the detection of the positive, delayed-type hypersensitivity reaction of mice immunized with live vwa^{+} cal^{+} *Y. pseudotuberculosis* in the foot pad test (Wake et al. 1972). This is evidence for the involvement of cell-mediated immunity via the recognition by T_{DTH} cells of widely ranged antigenic epitopes that are common among *Yersinia* organisms (refer to Chaps. 1 and 8, Part III).

Both of the above indicators of delayed-type hypersensitivity reaction are characterized by the calibrations of lengths, i.e., diameters or thickness of the localized inflammation sites to where the leukocytes, above all monocytes, are

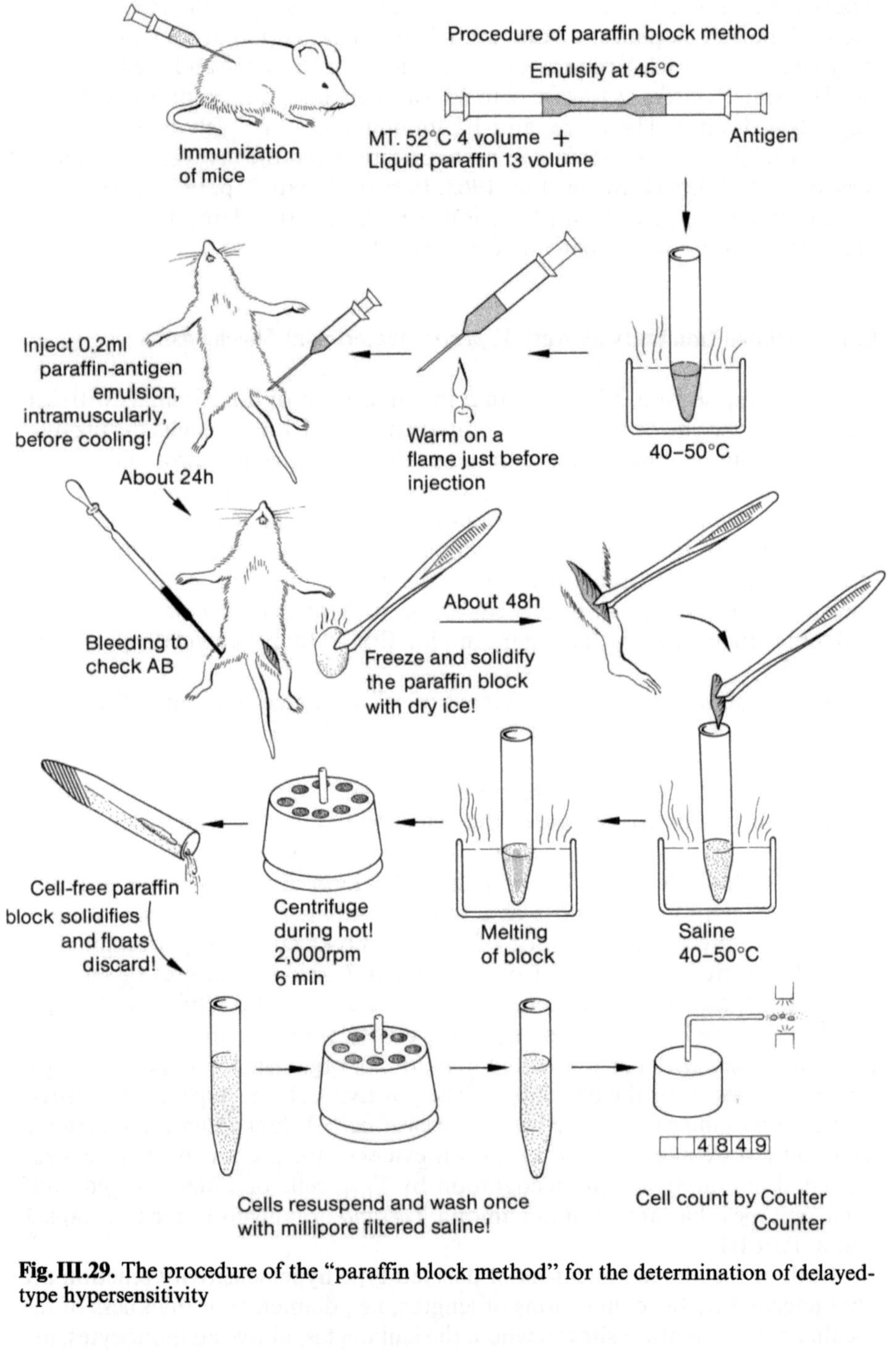

Fig. III.29. The procedure of the "paraffin block method" for the determination of delayed-type hypersensitivity

Table III.22. Measurement of delayed hypersensitivity by the paraffin block method

Time after antigen injection	20–26 h			44–48 h		
Immunizing antigen	SCY	222/+	None	SCY	222/+	None
No. of recruited leukocytes[a] (×10⁴)	9.6±4.7	64.2±34.1	11.8±7.2	14.4±8.3	75.8±63.5	20.5±17.4

[a] Values given the mean ± SE of numbers of leukocytes recovered from antigen-paraffin as determined using a Coulter counter. (From Wake et al. 1978).

recruited. In order to replace these qualitative calibrations with quantitative enumerations of the recruited leukocytes, Wake et al. (1978) devised a method diagrammatically shown in Fig. III.29. In short, the test antigen, in this case formalin-killed *Y. pestis* organisms, is emulsified into a particular paraffin, the melting point of which is 36 °C to 38 °C. Two-tenths ml of this antigen-paraffin emulsion is injected intramuscularly (i.m.) into the femoral muscle of a mouse. After 24 h and 48 h, respectively, the mice are bled and the site of injection containing the paraffin is cooled by gentle rubbing with dry ice. When the paraffin has solidified and become visible as a white mass through the skin, an incision is made and the paraffin block is removed and melted in 10 ml saline at 45 °C. This is then centrifuged at 800 × g for 6 min. During centrifugation, most cells mixed with, or adhering to, the paraffin block are released and precipitated and the paraffin solidifies and floats to the top. The paraffin is discarded and the packed cells are resuspended in the saline. The cells are counted in an automatic cell counter. This method was applied to show the cell-mediated immunity and was termed "paraffin block method" (Wake et al. 1978, Table III.22).

For exploration of the effective mechanisms of cell-mediated immunity in the acquired resistance against *Y. pestis* infection, important was the prevalent belief that in vivo multiplication of the immunizing bacteria is the prerequisite for the production of cell-mediated immunity exerting a protective effect (refer to Part III, Chap. 2 and Table III.5, where the survival of mice against the second challenge with virulent *Y. pestis* organisms was used as evidence of prior in vivo bacterial multiplication, for an example. Furthermore, previous chapters delineated that ferric iron (Fe^{3+}) stimulates the in vivo multiplication of P− *Yersinia* organisms. However, subcutaneous injection of Fl+VW+Cal+Pl+P-Pu+ *Y. pestis* EV organisms suspended in chondroitin sulfate ferric iron (Blutal) kill some mice and the cause is attributable to the effect of murine toxin that reached the lethal-dose level due to the in vivo multiplication of EV organisms. In contrast, the subcutaneous injection with a 10^2 to 10^3

dose of murine-toxin-nonproducing Fl−VW+Cal+Pl−P−Pu+ *Y. pseudotuberculosis* 222/+ organisms suspended in the same iron drug has been found to be nonlethal in mice, although their in vivo multiplication causes the formation of localized abscesses that heal within 14 days, although with remaining scars. Histopathological studies using the indirect fluorescent-antibody technique demonstrated the invasion of 222/+ organisms to the regional lymph nodes, but not to the spleen (Wake et al. 1973).

In spite of the complete deficiency of Fl production by *Y. pseudotuberculosis,* the immunization of mice with 222/+ organism using this particular method confer on the animals protective resistance against *Y. pestis* infection as will be delineated throughout this und the next sections.

Experiments showing the protective efficacy of Fl-nonproducing organisms were also reported by Alonso et al. (1978). The mice infected with *Y. enterocolitica* by the intragastric (i.g.), s.c., or i.v. route became resistant to a *Y. pestis* lethal challenge, although no bacteria from the initial infection could be isolated from their tissues. The passive transfer of serum from *Y. enterocolitica* infected-convalescent mice failed to confer resistance against subsequent challenge (Alonso et al. 1980), suggesting the involvement of cell-mediated immunity in this resistance mechanism. In this context, the so-called lesser *Yersinia* (Brubaker 1984), are equipped with certain antigenic epitope(s) responsible for the induction of cell-mediated immunity conferring protective resistance to major *Yersinia,* i.e., *Y. pestis.* On the other hand, Wake has confirmed that the immunization with the mutant strains of *Y. pseudotuberculosis* 222/+ cured from *vwa−cal* plasmid fail to confer on the mice protective resistance against *Y. pestis* infection in repeated trials, enabling the postulation that the lesser *Yersinia* harboring *vwa−cal* plasmid confer on mice protective resistance to *Y. pestis* infection via the function of this plasmid.

In order to examine the validity of this postulation, an isogenic derivative *vwa*$^-$ *cal*$^-$ mutant WA5-1, i.e., cured of this plasmid by spontaneous segregation from WA2-1, was used to immunize the inbred-mouse strain C57BL/6J. Mice primed with WA2-1 organisms developed protective immunity against *Y. pestis* Yreka infection as measured by the survival rate (4/9) and the prolongation of mean survival time; on the other hand, the priming with WA5-1 organisms shortened the mean survival time and only one mouse survived the infection (Fig. III.30). The novel survival-time-shortening effect of immunization with *vwa*$^-$*cal*$^-$ WA5-1 belonged to the phenomenon termed "bacterial subversion of immunity" (Wake et al. 1983), resulting from the lack of bactericidal mechanisms (refer to Part II, Chap. 9, p. 149). Presumably, the enhanced phagocytic activities of macrophages involved in the cell-mediated immune responses were triggered by the bacterial antigens irrelevant to VW+ Cal+ phenotype, but these macrophages permitted the intracellular multiplication of the *Y. pestis* organisms resulting in septicemia.

In this regard, to confirm the production of the VW antigens by the effective WA2-1 organisms, Ouchterlony agar-gel diffusion tests were performed using bacterial suspensions and anti-VW specific-rabbit serum. The results,

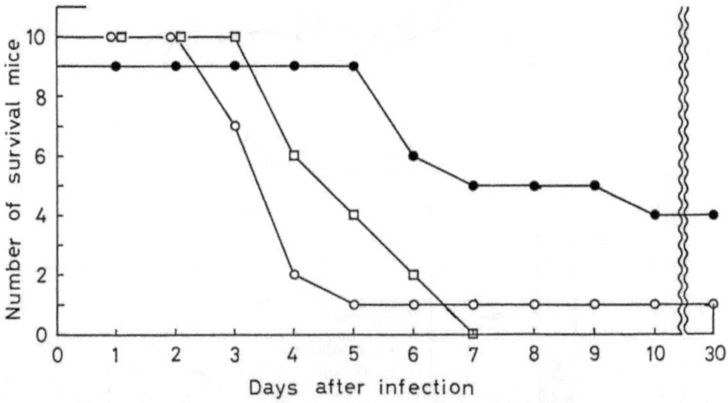

Fig. III.30. Effects of immunization with VW+Cal+ *Y. enterocolitica* WA2-1 and its VW−Cal− derivative WA5-1 on the resistance of inbred C57BL/6J mice to *Y. pestis* infection. Each of ten animals in three groups were injected s.c. with either 3×10^3 organisms of WA2-1, 4×10^3 organisms of WA5-1 suspended in 0.1 ml iron drug (Blutal) containing 0.4 mg Fe^{2+}, or the same dose of iron drug alone. After 12 days, one of the WA2-1 immunized mice died. The 1×10^2 organisms of a fully virulent *Y. pestis* strain Yreka, in 0.2 ml of saline, were injected s.c. 15 days after immunization. Each *point* is expressed as the number of surviving mice at the time of observation. (●): Mice immunized with WA2-1; (○): mice immunized with WA5-1; (□): mice injected with 0.1 ml Blutal

Fig. III.31. Agar-gel diffusion tests showing that all VW+CAL+ *Yersinia* species share common V and W antigens, but VW−Cal− *Y. enterocolitica* do not. Well *1* contains rabbit antiserum prepared by repeated immunization with VW+Cal+. *Y. enterocolitica* WA2-1; wells *2* contain bacillary saline suspension of VW+Cal+ *Y. enterocolitica* WA2-1; well *3* contains bacillary saline suspension of VW−Cal− *Y. enterocolitica* WA5-1 (spontaneous mutant of WA2-1); wells *4* contain bacillary suspension of VW+Cal+ *Y. pseudotuberculosis* 222/+; well *5* contains bacillary suspension of VW+Cal+ *Y. pestis* EV. All bacterial strains were grown at 37 /C

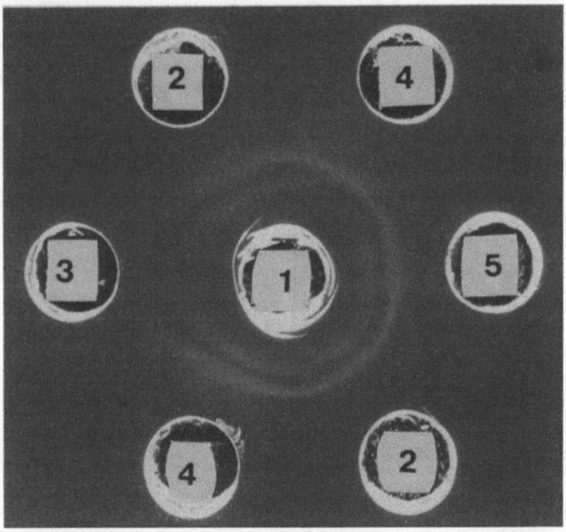

Fig. III.31, indicated the production of one antigen, presumably to be designated V, by both WA2-1 and *Y. pestis* organisms; and another, to be designated W. However, remarkable is that each precipitin line between these two different species of organisms and the serum are forming spurs mutually with regard to V, whereas forming one fused line with regard to W. Because WA2-1

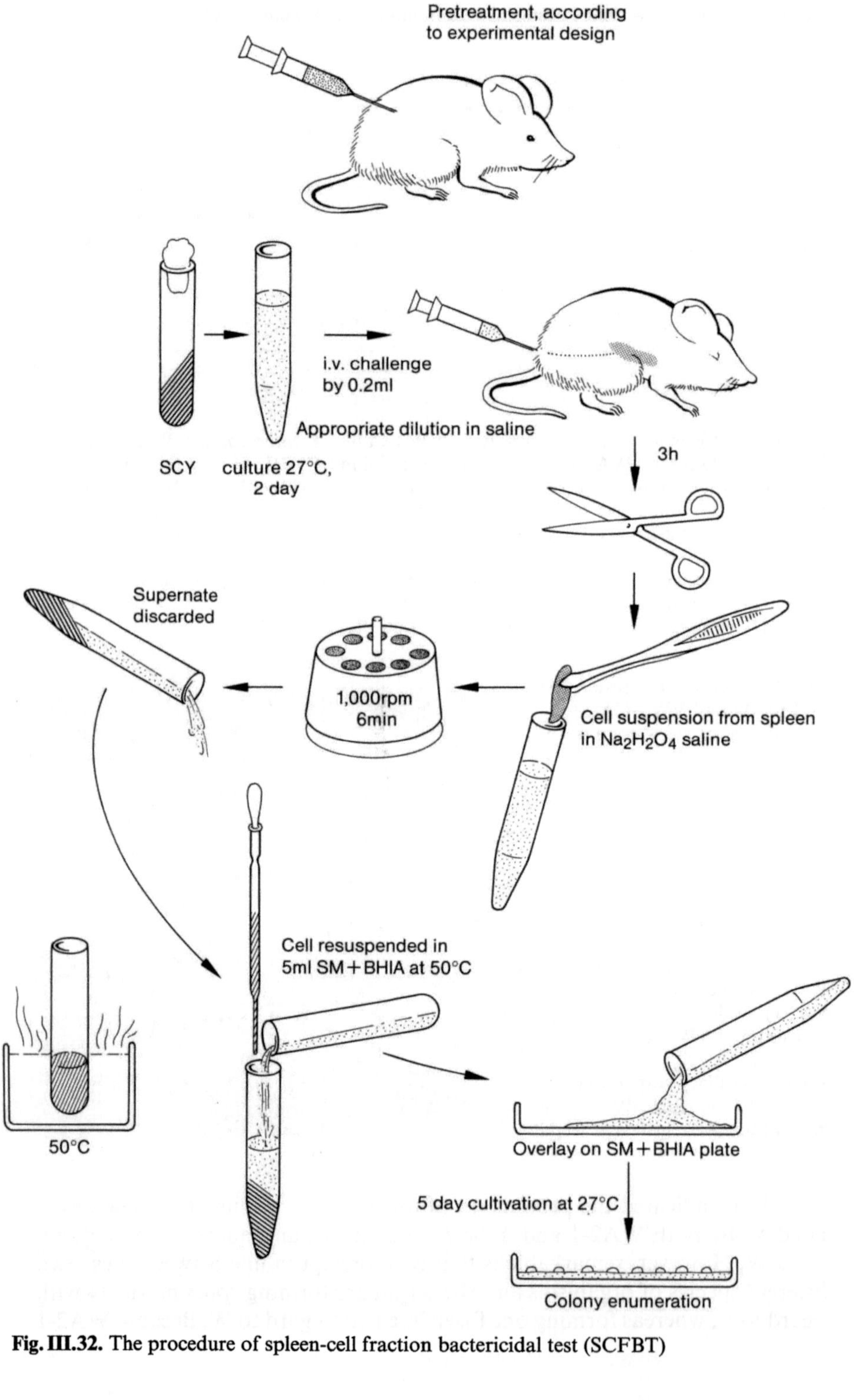

Fig. III.32. The procedure of spleen-cell fraction bactericidal test (SCFBT)

Table III.23. Measurement of bactericidal activity of macrophages (SCFBT)

Immunizing antigen	No. of colony-forming units per mouse spleen		
	SCY	222/+	None
Cell fraction[a]	168.70±62.52	93.40±49.27	193.50±98.00[c]
Supernatant fraction	45.00±25.30	28.80±17.96	63.30±31.16[d]
SCFBT index[b]	1.15	2.1	1.0
Index applied to supernatant	1.4	2.2	1.0

[a] Values given are mean ± SE of colony-forming units (CFU) of SCY, recovered from mouse-spleen cell suspensions.
[b] See text.
[c] Significant difference among three groups within $P<0.05$.
[d] Significant difference among three groups within $P<0.01$. (From Wake et al. 1978).

and Yreka were calcium-dependent (Cal+), whereas WA5-1, calcium-independent (Cal−), the former two organisms should have harbored a 40–48 Mdal plasmid family according to Portnoy et al. (1981).

Accordingly, the *vwa − cal* plasmid of 40–48 Mdal size is concluded to control the VW antigen production that is responsible for the induction of immunity conferring mice protective resistance in cooperation with the function of cal^+ gene(s) locating on the same plasmid.

Clearly, the aforementioned discussions postulated that the mechanisms of protective resistance against *Y. pestis* organism include the bactericidal mechanisms. Chapter 9 of Part II has delineated that the bactericidal mechanisms function intracellularly as well as extracellularly, presumably both in vivo.

In the case of the *Y. pestis*-mouse experimental system, the effectiveness of bactericidal mechanisms can be assayed by the use of a formulated experimental procedure designated "the spleen-cell fraction bactericidal test" (SCFBT) developed by Wake et al. (1978) as follows.

A particular avirulent antibiotic-resistant *Y. pestis* strain SCY (refer to Chap. 2, Part III, p. 232) is i.v. injected into ten mice to be assayed for bactericidal activity with a dose of about 10^3 organisms suspended in 0.1 ml saline containing antibiotics. At predetermined intervals, ten spleens were minced to suspend isolated cells into 10 ml saline containing antibiotics and anticoagulant. After three washings by the saline containing antibiotics, the collected spleen cells were resuspended in 4 ml brain-heart infusion agar (BHIA) containing antibiotics melted by heating at 50 °C. These suspensions were quickly overlaid onto BHIA plates containing antibiotics. After incubating these plates at 27 °C for 5 days, the numbers of colonies of the indicator strain are averaged. Spreading of the spleen cells is recommended to be done between 3 to 6 h after the i.v. injection of the indicator *Y. pestis* organisms, because the preliminary experiments concluded that the mean value thus obtained represents the specific bactericidal activity of the mice assayed. Furthermore, the mean value obtained by the use of control mice is divided by the mean value obtained from the assayed mice, to be expressed as the SCBFT index. For example, the experiments in which the bactericidal activities of the mice s.c. immunized with *Y. pseudotuberculosis* 222/+ suspended in the iron drug Blutal were calculated for this SCBFT index every day after the immunizing injection, gave index values of less than 2.0 up to the eighth day, but it increased to greater than 2.0 after the ninth day. These results were compatible with such an interpretation that 8 days were required to perfect the immunity mechanism capable of expressing bactericidal activity. As another example, ^{60}Co irradiation (400 rad) at the time of 222/+ immunization lowered this index to 1.0, indicating that immune mechanisms were destroyed by the effect of radioactivity. When the immunized mice were irradiated by a 600 rad dose of ^{60}Co at the 14th day of the immunizing injection, the SCBFT index was 2.0, indicating that this dose of radioactive irradiation did not destroy the already perfected bactericidal immunity. To summarize, the procedure of SCFBT is diagrammatically shown in Fig. III.32 and the respresentative results of one such experiment using this method are presented in Table III.23.

To summarize all the results of these serial experiments intended to clarify the nature of immunity particularly induced by the s.c. injection of *vwa*$^+$ *cal*$^+$ *Y. pseudotuberculosis* 222/+ suspended in the iron drug Blutal, it was concluded that this immunity confers on mice bactericidal activity to *Y. pestis* organisms specifically by the use of some antigenic determinant(s) [epitope(s)] shared in common by these two bacteria: the VW phenotype controlled by a 40 to 48 Mdal plasmid family that is shared by the major and lesser *Yersinia* and that is responsible for its induction. However, a definitive conclusion for the involvement of cell-mediated immunity is demonstrable only by the transfer of anti-*Y. pestis* protective potency via the transfer of immune cells free from humoral antibodies from donor mice to the recipient. In addition, these immune cells must be recognized as "self" by the recipient mice in order to express their specific function. On the basis of these prerequisites, Alonso et al. (1980) transferred i.v. the dissociated and washed, viable Peyer's patches-mononuclear cells from the inbred mice infected i.g. twice with *Y. enterocolitica* at 27-day-intervals, into the inbred mice of the same haplotype, i.e., using C57BL/6 or C3H/HeJ mice. When these mice adoptively receiving immune cells were challenged with a lethal dose of *Y. pestis* 6/69M organisms, the C3H/HeJ mice receiving the Peyer's patches cells collected on the ninth day after the second *Y. enterocolitica* (strain No. 4052 of serovar 0:3) infection of the same inbred-mouse strain, the number of the *Y. pestis* cells recoverable from the homogenated spleen decreased significantly, indicating that the immunity was transferred adoptively via the mononuclear cells of Peyer's patches.

However, similar evidence has been obtained by the use of another, lesser *Yersinia,* i.e., *Y. pseudotuberculosis* strain 222/+, as the immunizing bacteria as follows.

The donor-inbred C57BL/6J mice were immunized s.c. with 10^2–10^3 222/+ organisms suspended in 0.1 ml iron drug (Blutal) then again injected

Table III.24. Transfer of antiplague protection to C57BL/6J recipients of syngeneic spleen cells from Blutal-suspended live 222/+ vaccine donors administered or not administered a 222/+ (VW+Cal+ *Y. pseudotuberculosis*) booster

Immunizing 222/+ dose (No. of organisms)	Booster[a] 222/+ dose ($\times 10^6$ organisms)	Post-booster (day)	Donor cells transferred: Spleen cells ($\times 10^8$)	Donor cells transferred: Viability	Yreka challenge dose ($\times 10^3$)	Recipient survival Yreka challenge[b] (%)	Mean death time[c] (day)
450	1.4	12	44–53	93–99	28	40	5.0±0
0	0		0			0	3.8±0.3
400	0		14–75	89–98	28	0	4.0±0.3
400	0		54	0[d]	510	0	4.1±0.3
0	0		45	90	510	0	3.8±0.4
0	0		0		510	0	3.2±0.1
370	1.4	12	16	90	55	10	4.9±0.4
370	1.4	12	6.2	0[d]	55	0	2.8±0.4
0	0		0		55	0	2.9±0.3

[a] The immunized mice received intraperitoneal injections of 1.4×10^6 organisms of VW+Cal+ *Y. pseudotuberculosis* 222/+ 20–21 days after the primary immunization.
[b] Percentages of the survival mice 30 days after challenge infection with indicated doses of *Y. pestis* Yreka organisms.
[c] Calculated as the arithmetical means of the days from the infection to death of only the killed mice in each group.
[d] The transferred donor cells were incubated at 56°C for 30 min in a water bath before transfer.

i.p. with 10^6 222/+ organisms after 20–21 days. Spleen cells from these mice were suspended into RPMI1640 tissue-culture medium with antibiotics and washed twice with antibiotic-free medium. The syngeneic recipient mice received cells from two donors each (i.p.) and were then s.c. infected with a lethal dose of a fully virulent *Y. pestis* Yreka organisms 20 h after the cell transfer and observed for 30 days.

The results, summarized in Table III.24 indicate that 10^9 viable immune spleen cells could transfer protective immunity against s.c. challenge with *Y. pestis* organisms when the donor had received a booster immunization 12 days before cell transfer.

However, the number of mice protected was relatively small, presumably because of the large challenging dose (larger than 460 MLD) and because the strains used had perhaps only one antigenic determinant (*vwa – cal*). The short, mean death time among the killed mice of all groups suggests that no humoral antibodies were present to opsonize the challenge organisms.

Efforts to detect antibodies against *Y. pestis* in the serum of mice immunized s.c. with $10^{2\text{-}3}$ organisms of 222/+ suspended in 0.1 ml Blutal by the use of agar-gel diffusion assay (Wake et al. 1978) or radioimmunoassay were unsuccessful, suggesting that this method of immunization directed the induction

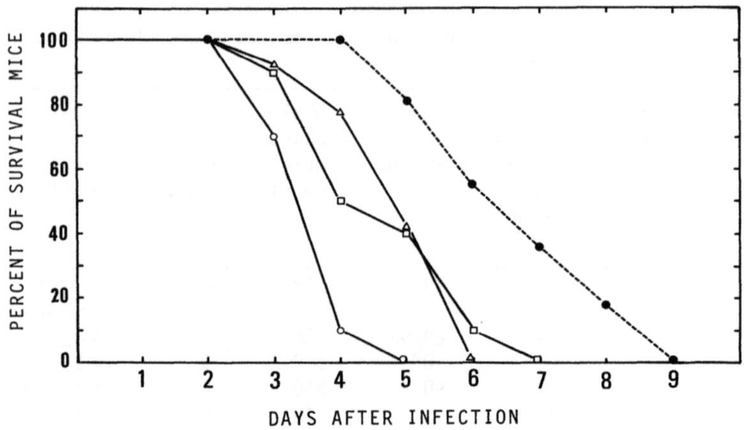

Fig. III.33. Effect of priming athymic mice with BCG, killed *C. parvum* or live VW+Cal+ *Y. pseudotuberculosis* 222/+. Without priming (○——○); primed i.v. with 0.5 mg BCG 14 days before (△——△); primed i.p. with 0.7 mg killed *C. parvum* 7 days before (□——□); and primed s.c. with 320 organisms of *Y. pseudotuberculosis* 222/+ with 0.1 ml chondroitin sulfate colloidal iron (Blutal) 14 days before (●——●). All mice were challenged s.c. with 850 organisms of *Y. pestis* Yreka

of cell-mediated immune (CMI) response, but not the production of humoral antibodies.

The next series of experiments (Wake and Sutoh 1983) intended to examine whether the immune cells transferred from the donor mice of a certain major histocompatibility haplotype (H-2) interact with the macrophages of another haplotype to form the cell-mediated immunity with the anti-*Y. pestis* bactericidal activity, i.e., the involvement or noninvolvement of H-2 restriction (refer to Part II, Chap. 11, p. 166).

Clearly, only athymic mice can serve as the recipients because these are devoid of mature T-cells and do not reject the implanted foreign cells, but are equipped with macrophages of the same H-2 as the normal inbred mice from which they derived.

First of all, the possibility that nonspecific stimulation to these mice might confer the bactericidal activity was excluded from the following experiment.

Priming of athymic mice originated from the BALB/C-inbred strain with *Mycobacterium bovis* (Japanese BCG strain) or killed vaccine of *Corynebacterium parvum* (CN6134, Batch. PX383, 7 mg dry wt. ml^{-1}, Wellcome Research Laboratories) did not significantly prolong their survival time following subsequent s.c. challenge with *Y. pestis*. Priming s.c. with VW+Cal+ *Y. pseudotuberculosis* 222/+ suspended in 0.1 ml iron drug (Blutal) caused rather severe local inflammation around the injection site, which lasted up to 2 weeks and significantly ($P<0.05$) prolonged survival time compared to that of the control group upon subsequent challenge with *Y. pestis*.

However, no mouse survived longer than 9 days in spite of priming with living or killed vaccines (Fig. III.33). The defense mechanisms responsible for

Table III.25. Antibody titers[a] to *Y. pestis* of athymic mice and their littermates primed with *Y. pseudotuberculosis* 222/+

Pretreatment of serum-donor mice	Thymus	Number of serum samples with antibody titer of					
		IgM			IgG		
		$<10^3$	10^3	32×10^6	$<10^3$	10^3	32×10^6
Nonprimed (individual sample)	+	11			11		
	−	10			10		
7 Days after priming with 222/+ (individual sample)	+	10			8	2	
	−	9	1		10		
14 Days after priming with 222/+ (individual sample)	+	10	1		10		
	−	12			11		
4 Days after priming with EV (pooled sample as control for antiplague IgM antibody)	+			1			
Hyperimmunized[b] (pooled sample as control for antiplague IgG antibody)	+						1

[a] Titrated by radioimmunoassay against *Y. pestis* strain EV. Antibody titers are expressed in the reciprocal of the highest dilution binding 50% or more of the total background counts.
[b] Sera taken from 19 convalescent mice 7 days after the infection with Yreka to be pooled. These mice resisted the virulent infection because they had received injections with 222/+, Yreka, and the three-priming with EV.

such survival-time-prolonging effects in these athymic mice were not defined, but no increased bactericidal activity against *Y. pestis* was observed. The possibility that humoral antibodies were effective in prolonging the survival time was excluded by the results of antibody measurement using radioimmunoassay (Table III.25).

Priming with 222/+ organisms conferred effective mechanisms for bactericidal killing on the recipient when later the spleen cells from BALB/C thymus-bearing (euthymic) littermate primed with 222/+ organisms were introduced i.p. 24 h before the *Y. pestis* challenge (Fig. III.34). However, all of the recipient athymic mice that were not primed with 222/+ organisms and that were primed with *C. parvum* vaccine before receiving immune-cell transfer did not acquire the protective resistance, even after the mice received the immune spleen cells adoptively (Fig. III.34).

The effectiveness of the adoptive transfer of normal spleen cells to potentiate the induction of cell-mediated immunity with bactericidal activity in mice was also examined using two inbred-mouse strains of different haplotypes, H-2^b and H-2^d. The latter, but not former, haplotype was shared by the recipient athymic mice.

One group of ten athymic mice received intraperitoneally (i.p.) 6.5×10^8 spleen cells from their littermate (BALB/C nu/+, H-2^d), a second group re-

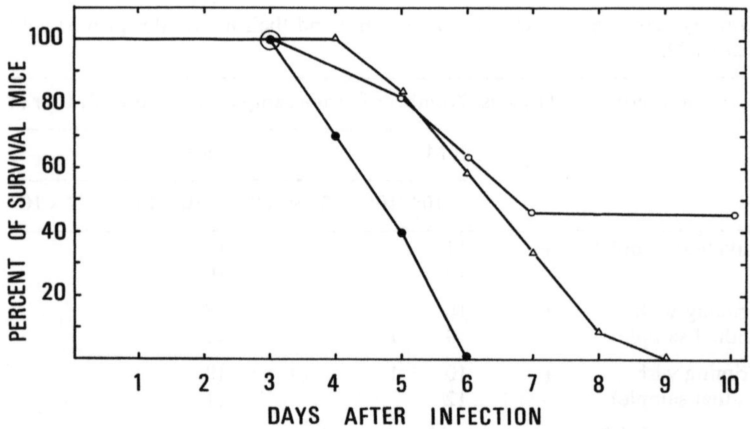

Fig. III.34. Effect of transfer of spleen cells from immunized thymus-bearing littermates to athymic mice primed with VW+Cal+ *Y. pseudotuberculosis* 222/+. Sixty BALB/C (nu/+) mice were primed s.c. with 240 organisms of 222/+ suspended in 0.1 ml Blutal and were boostered i.p. with 9.7×10^5 of the same organisms suspended in 0.1 ml saline 20 days later; these mice were used as immunized spleen-cell donors. Eleven athymic mice used as recipients of the immune-spleen cells were primed with 340 organisms of 222/+ suspended in 0.1 ml Blutal 14 days earlier (○——○). Another group of 12 animals was primed i.p. with 0.7 mg killed *C. parvum* vaccine 7 days earlier (△——△). Twelve days after the booster-priming of donor mice, the first group received 1.0×10^9 immunized-spleen cells i.p., with 0–130 spleen-cell-associated "passenger" 222/+ organisms; the second group of mice received 1.4×10^9 immune-spleen cells with 20–30 222/+ passenger organisms. The other nonprimed group of ten animals received 1.6×10^9 immune-spleen cells with 3–117 222/+ passenger organisms (●——●). On the next day of cell transfer, all mice were challenged s.c. with 620 organisms of Yreka strain

ceived 6.4×10^8 spleen cells from C57BL/6J mice (H-2^b), and these two groups were primed s.c. with 222/+ organisms on the next day. When these mice were challenged s.c. with 290 organisms of *Y. pestis* Yreka strain 14 days later, 7 of 10 animals in the group receiving BALB/C spleen cells died within 7 days and the remaining 3 animals survived the infection. All animals in the groups receiving C57BL/6J spleen cells died within 9 days. In the control group all athymic mice receiving no spleen cells were killed within 5 days following challenge.

These results suggested that T-cells in the adoptively transferred BALB/C spleen cells reconstituted the potency of the cell-mediated immunity in mice to enable the donor T_{DTH} cells to interact with the recipient macrophages activated with VW+Cal+ *Y. pseudotuberculosis* 222/+ organisms via the mutual recognition of the same H-2 complex expressed on donor T_{DTH} cells and recipient macrophages. Through this self-recognition, the recipient macrophages apparently can present the antigenic epitopes to the donor T_{DTH} cells to stimulate release of lymphokines or stimulate direct-contact killings. In other words, the associated recognition (between both anti-X and X and anti-R and R delineated in Part II, Chap. 1) was accomplished.

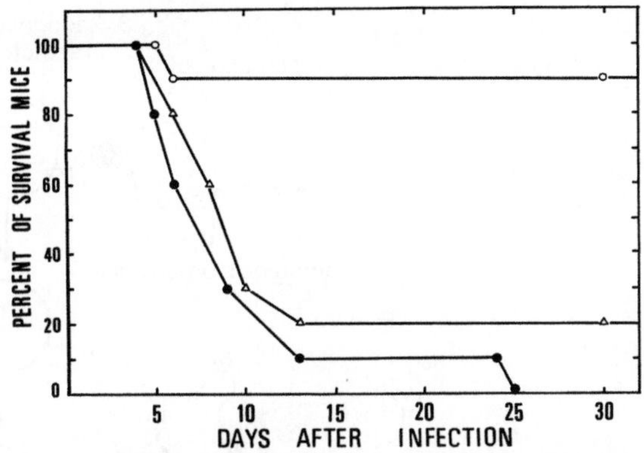

Fig. III.35. H-2 restriction to spleen cells of immunized thymus-bearing donors and athymic recipients. Groups of BALB/C (nu/+), C57BL/6J, and (C57BL/6J × BALB/C) Fl-hybrid mice of 20, 20, and 23 animals, respectively, were each primed s.c. with 860 organisms of 222/+ suspended in 0.1 ml Blutal to serve as immune spleen-cell donors. Twelve days after the booster i.p. injection with 1.7×10^6 222/+ organisms, 1.2×10^9, 1.1×10^9, and 1.2×10^9 spleen cells with no spleen-cell-associated 222/+ were transferred i.p. from each mouse group to three groups, each comprised of ten athymic mice were primed s.c. with 530 organisms of 222/+ 14 days before. The next day, all recipient mice were s.c. injected with 170 organisms of Yreka strain and observed for 30 days. Only one (10%) of the BALB/C (nu/+) cell recipients died in 8 days (○——○). All C57BL/6J recipients died within 25 days (●——●). Two animals of Fl-cell recipients survived for 30 days (△——△)

On the other hand, these results suggest that the donor-spleen cells of C57BL/6J haplotype (H-2^b) might have activated, via an allogenic effect, the recipient macrophages of BALB/C haplotype (H-2^d) in cooperation with the priming by 222/+ organisms to produce prolongation of survival time following challenge with *Y. pestis*, but that T_{DTH} cells of H-2^d haplotype cooperated with the recipient macrophages of the same haplotype and exerted the bactericidal activity resulting in the survival of mice against *Y. pestis* infection. In immunology, this is called H-2 restriction.

The H-2 restriction was confirmed by another experiment examining the effect of the transfer of spleen cells from immunized littermates to mice primed with VW+Cal+ *Y. pseudotuberculosis* 222/+. The results and the experimental procedures are given in Fig. III.35.

The transfer of spleen cells from (C57BL/6J × BALB/C) F1-hybrid mice protected 20% of the BALB/C-derived athymic recipient mice (H-2^d) against the *Y. pestis* challenge. In other words, semiallogenic, i.e., H-$2^{b/d}$ donors were partially effective in protecting the H-2^d recipient athymic mice. Therefore, these results were in parallel with the studies on cell-mediated immunity using the *Listeria monocytogenes*-mouse experimental system (Farr et al. 1979; Zinkernagel 1974; Zinkernagel et al. 1977), delineated in Part II, Chap. 11.

In this context, definitive evidence for cell-mediated immunity mechanisms with bactericidal activity was obtained from the experimental system in which

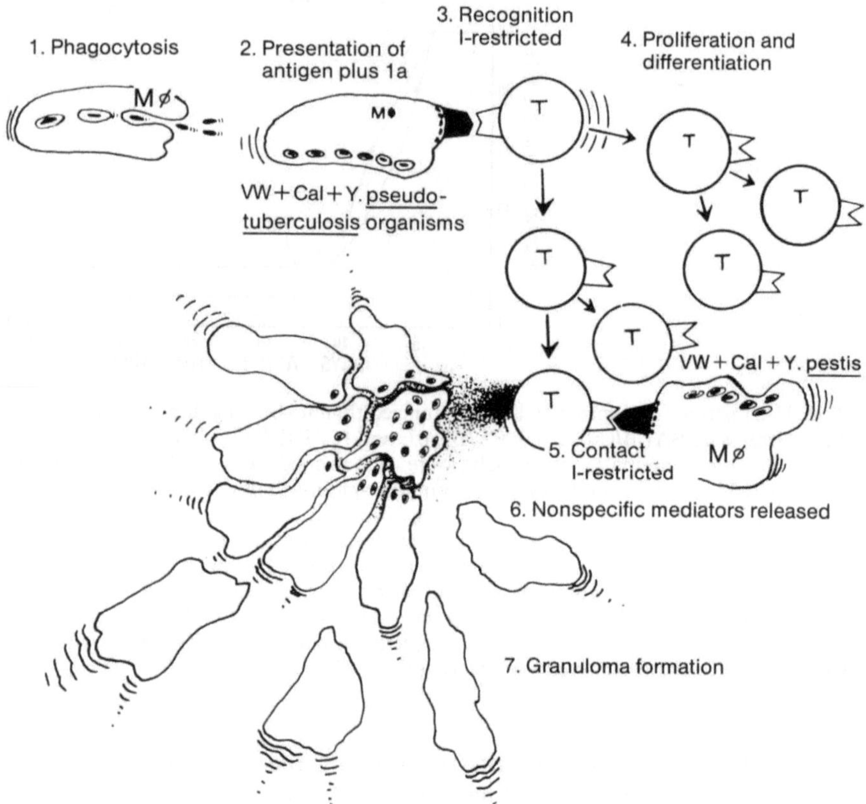

Fig. III.36. The single steps of cell-mediated immunity as analyzed in the mouse system. After infection macrophages phagocytize, but do not eliminate intracellular bacteria (*1*). Instead, some bacteria are processed and antigenic products presented by macrophages in association with H-2I-encoded surface structures (*2*). Recognition of antigen in association with H-2I products (*3*) results in proliferation and differentiation of specific T-cells (*4*) that, after a second contact with antigen in association with H-2I products (*5*), now are triggered to release lymphokines (*6*). In a nonspecific way lymphokines attract and activate further macrophages to form granulomas and eliminate intracellular bacteria (*7*). Only steps *5–7* have been characterized, whereas steps *2–4* have been reproduced in vitro, but are inferred by analogy to step *5* and to other in vitro systems

mice were s.c. immunized with $vwa^+ cal^+$ *Y. pseudotuberculosis* suspended in an iron drug, Blutal, i.e., H-2 restriction.

In connection with the significance of injecting the immunizing bacteria *suspended in an iron drug* into the recipient athymic mice, it is postulated that the macrophages without expression of Ia molecules (refer to Part II, Chap. 1), i.e., Ia-negative macrophages, are converted into those with expression of them, i.e., Ia-positive ones, by the interaction with these bacteria at the site of inflammation. To support this postulation, Snyder et al. (1982) reported that macrophages isolated from *Listeria*-infected mice had 40 to 80% Ia-positive

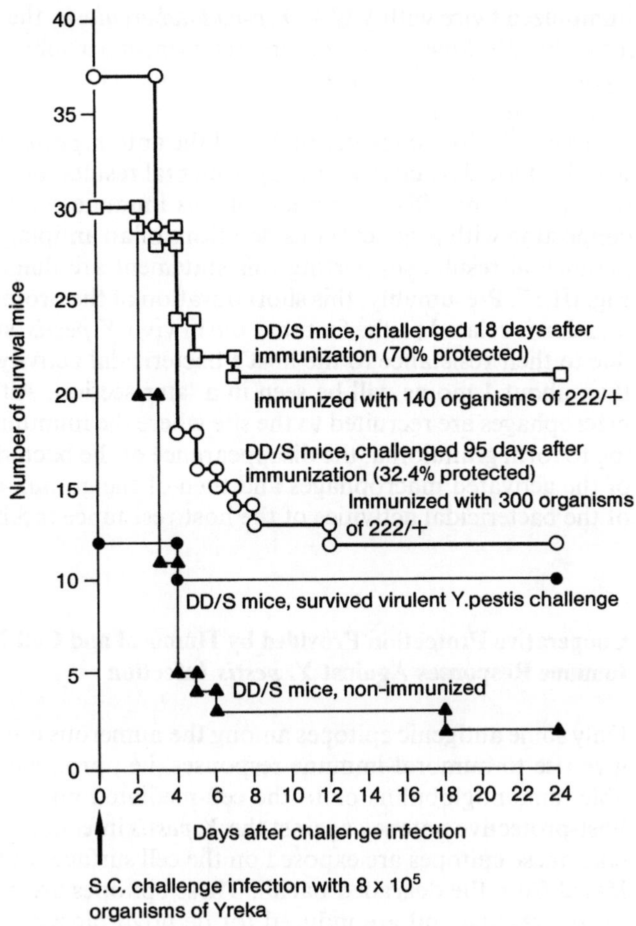

Fig. III.37. Duration of 222/+ Blutal-induced protective cell-mediated immunity against *Y. pestis* infection

cells versus resident macrophages from normal mice with 5 to 10% Ia-positive cells. This report stated furthermore that the macrophages activated by the infected *L. monocytogenes* organisms produce thromboxane (TX)A_2 and its stable metabolite TXB_2 surprisingly potent antagonist to the effects of prostaglandin E_2 (refer to Part II, Chap. 6, Fig. II.18). Because prostaglandin E_2 inhibitis Ia expression on the surface of macrophages, TXA_2 and TXB_2 enable the expression of Ia molecules. If this scheme is applicable to the macrophages activated by the infection with *Y. pseudotuberculosis,* those macrophages of athymic mice that are parasitized with VW+ *Y. pseudotuberculosis,* also, express Ia molecules on the cell surface. After dual recognition of these antigenic epitopes, i.e., VW phenotype and Ia molecule by the specific T_{DTH} cells directed to VW epitope after the spleen cell transfer from the donor, which had been immunized twice with VW+ *Y. pseudotuberculosis*, the activated T_{DTH} cells release lymphokine that confers recipient-macrophage bactericidal activity against the challenged *Y. pestis* organisms. Thus, the conclusion is diagrammatically illustrated in Fig. III.36.

In conclusion to the discussion of the anti-*Y. pestis* cell-mediated immunity with bactericidal activity, the experimental results showing the relatively short duration of the effective period of this immune state deserve mentioning in connection with practical considerations of an antiplague live vaccine. The experimental results supporting this statement are diagrammatically shown in Fig. III.37. Presumably, this short duration of the protective effect after immunization is related to the fact that the in vivo *Y. pestis*, surviving intracellularly due to their resistance to the host's bactericidal activity, cannot persist longer than about 1 mo as will be seen in a later section. Although the monocytes-macrophages are recruited to the site where the immunizing bacteria are locating to form granuloma, the disappearance of the bacteria causes the limitation of the activated macrophages and then of the granulomas and, thus, the loss of the bactericidal activities of the host-resistance mechanisms.

Cooperative Protection Provided by Humoral and Cell-Mediated Immune Responses Against *Y. pestis* Infection

Only some antigenic epitopes among the numerous ones of *Yersinia* organisms give rise to humoral-immune responses, i.e., production of diversely classifiable immunoglobulins or to the cell-mediated immune responses conferring host-protective potency against the *Y. pestis* infection. In the case of killed vaccine, these epitopes are exposed on the cell surfaces of the dead bacteria or released from the degraded bacilli. These epitopes are also involved in the case of live vaccine and are induced for de novo biosynthesis via derepression of the bacterial genes along with the time course of few hours after vaccine administration as a consequence of bacterial interaction with host-microenvironmental factors, e.g., those of extra- or intraphagocytic sites of residence.

All of these epitopes are the components of the molecules responsible for any virulence-factor functions, because the immune responses thus induced interfere with the expression of virulence and show protective efficacy for the host by keeping the particular facet of host-defense mechanisms, which is the target of that virulence factor, unconquered. In fact, the two preceding sections delineated the anti-F1 immunoglobulins that interfere with the antiphagocytic function of the F1-envelope substance and the cell-mediated immunity that neutralizes the bacterial resistance to intraphagocytic bactericidal mechanisms via enhancement of the latter by the specific response of the host to the VW=Cal phenotype. Furthermore, these observations also indicate that F1 epitope is prone to induce humoral responses, whereas the epitope unique to VW is prone to induce cell-mediated immunity with the aid of the Cal phenotype. However, this is not always the case, because extensive immunization with *vwa*$^+$ *cal*$^+$ *Yersinia* organisms enables the production of the anti-V and anti-W immunoglobulins (Burrows et al. 1956; Wake et al. 1983a), and the F1 substance is effective in inducing delayed-type hypersensitivity skin reaction in the guinea pig immunized with an attenuated F1+VW+Cal+P1+P−Pu+ *Y. pestis* strain EV (refer to previous section).

In spite of such complex situations, many experimental studies reveal the host's choice of responding to any antigenic epitope by producing immunoglobulin, i.e., humoral response, or by inducing a cell-mediated immune response, to be directed depending on the following items: (1) the chemical nature of the antigenic epitopes as was exemplified above by F1 or VW−Cal; (2) the site and mode of interactions being determined by the administration route of epitopes by the state of the epitopes, e.g., killed organisms whose epitopes are already preformed or live organisms whose epitopes are synthesized as the consequence of interactions with in vivo microenvironmental conditions; (3) whether the immune response is inducible or suppressible, although changeable depending on the quantities of epitopes contained in the vaccine; and (4) immunogenetic constitution of the host, e.g., mouse or guinea pig, and the strain differences within a single animal species (see the next section).

Combining the above knowledge with other information that the established virulence factors of *Y. pestis* are F1, VW−Cal, P1, P, and Pu and the virulences of lesser *Yersinia*, i.e., *Y. pseudotuberculosis* and *Y. enterocolitica*, are related to the presence of VW−Cal, the whole set of immune responses (being induced within a single host by various epitopes in parallel) is feasible to study in an analytical way. Then the cooperative protection provided by humoral and cell-mediated immune responses can be concluded from the consideration of their integrated effects. For this purpose, comparisons of protective effects conferred to the mice by the mutant-attenuated *Y. pestis* organisms lacking some established virulence factor(s) were performed. These mutants were derived from a fully virulent F1+VW+Cal+P1+P+Pu+ *Y. pestis* strain Yreka using the method described in Chap. 2, Part III. Briefly, a drug-resistance plasmid was transferred from *E. coli* to Yreka organisms by conjugation followed by plasmid-curing treatments and each substrain obtained was assayed for its established virulence factors in vitro. The mutants with diversely established virulence-factor designations, were injected as the live vaccine either intraveneously (i.v.), which is the preferable route to induce humoral immunity, or subcutaneously (s.c.), the preferable route to induce cell-mediated immunity. However, another directing item, i.e., the chemical nature of the antigenic epitopes, was expected to be responsible for the determination of whether humoral or cell-mediated immunity is induced by the particular epitope present on the virulence-factor molecule.

The immunizing-mutant strains used were: (1) F1−VW+Cal+P1−P−Pu+ strain No. 51; (2) F1−VW−Cal−P1−P+Pu+ strain No. 4; (3) F1−VW−Cal−P1−P−Pu+ strain No. 45; and (4) F1+VW+Cal+P1+P+Pu strain 37-6e9. The last strain was tetracycline-resistant and aviru-

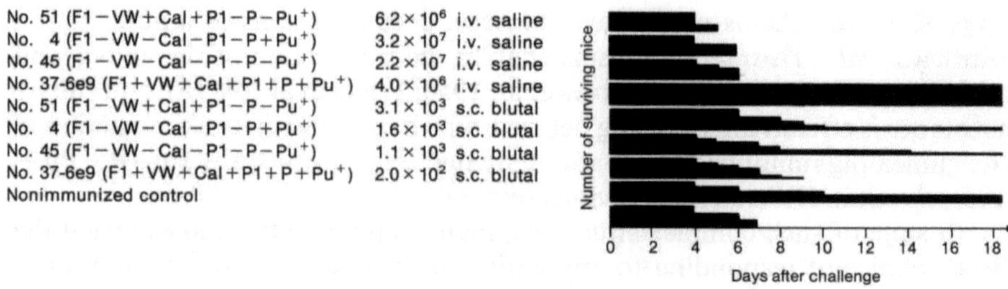

Fig. III.38. Protection provided by *Y. pestis* mutants lacking one or more virulence factors to challenge with *Y. pestis* Yreka

lent in spite of the full positive expressions of all established virulence factors of *Y. pestis* as was delineated in Chap. 2, Part III. Those strains suspended in saline were injected i.v. with the approximate doses of 10^6–10^7 organisms into ten C57BL/6J mice each, whereas the same set of strains suspended in an iron drug (Blutal) were injected s.c. with the approximate doses of 10^2–10^3 organisms into ten C57BL/6J mice each. Fourteen days after immunization, all mouse groups together with ten nonimmunized C57BL/6J control mice were s.c. challenged with 1.2×10^4 *Y. pestis* Yreka organisms (ca. 1000 LD_{50}) and the death/survivals were daily recorded for 3 weeks as is diagrammatically shown in Fig. III.38. In this experimental design, the immunizing organisms were suspended in Blutal before s.c. injection on the basis of previously acquired knowledge that this method induces cell-mediated immunity without participation of antibody-immunoglobulin (refer to the preceding two sections).

To summarize the results shown in Fig. III.38 succinctly; (1) all mutant *Y. pestis* strains lacking Fl did not induce protective immunity when administered by the method expecting humoral-immune response; (2) the strain 376e9 provided the mice 100% protection by the same immunizing method. This strain presumably produces anti-Fl immunoglobulin and some immunoglobulin specific to the virulence factors VW, Pl, and P. The involvement of anti-VW, cell-mediated immunity was not proven in reference to the nonprotection of mice i.v. immunized by Fl−VW+Cal+Pl−P−Pu+ strain No. 51. (3) The strain 376e9 provided mice of the s.c. immunized group the protective potency that prolonged the survival times in six mice, but four mice resisted the lethal infection, thus, suggesting the effective function of bactericidal activity due to the cell-mediated immunity. However, two mice each also resisted lethal infection in two mouse groups s.c. immunized with Fl−VW+Cal+Pl+Pl−P−Pu+ strain No. 51 or with Fl−VW−Cal−Pl−P+Pu+ strain No. 4, suggesting that the VW−Cal phenotype or P+ phenotype induced cell-mediated immunity with bactericidal activity. The last result, suggesting the protective potency of P+ phenotype, was unexpected and deserves further investigation. However, the most remarkable result in this experiment was the protection provided by i.v. injected Fl+VW+Cal+Pl+P+Pu+ strain 376e9 and

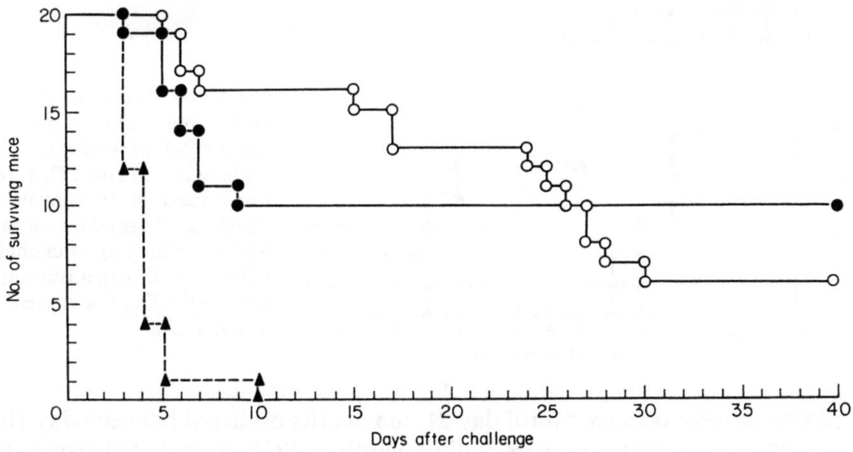

Fig. III.39. Protection resulting from active immunization of mice, showing survival of SCY (○), 222/+ (●), and nonimmunized (▲) mice, after s.c. challenge with 5.2×10^5 organisms (approx 10^5 LD_{50}) of a virulent *Y. pestis* strain Yreka

the nonprotection by Fl – VW – Cal – Pl – P – Pu + strain 45, irrespective of the immunization route. There is no evidence supporting the involvement of cell-mediated immunity in the former mouse group, but the cooperation of humoral and cell-mediated immunity cannot be excluded.

Before such cooperation of humoral and cell-mediated immunity is unequivocally demonstrated, another approach based on the construction of an experimental system, in which the humoral-immune response solely or the cell-mediated immune response solely is the mechanism that confers mice protective potency against *Y. pestis* infection (Wake et al. 1978), is delineated as follows:

For the purpose of inducing cell-mediated immunity solely, the s.c. immunization with F1 – VW + Cal + P1 – P – Pu + *Y. pseudotuberculosis* 222/+ organisms suspended in the iron drug (Blutal) was used as delineated in the preceding section. For inducing humoral immune response solely the i.v. immunization with F1 + VW + Cal – P1 + P + Pu + $Sm^{r+}Cp^{r}Tc^{r}$ *Y. pestis* strain SCY (NR1) whose deriving process from Yreka has been delineated in Chap. 2, Part III (p. 232) was used. Fourteen days after these immunizations using an outbred-mouse strain ddY/S, all mice and control mice were s.c. challenged with 5.2×10^5 Yreka organisms (ca. 10^5 LD_{50}). The daily death/survival curves are diagrammatically depicted in Fig. III.39. Half of the 222/+ immunized group died within 10 days, but the other half resisted the lethal challenge with Yreka organisms. This protection shown by the latter group of mice was termed as long-term immunity in reference to the survival-time-prolonging effect of SCY immunization, which is termed as short-term immunity on the basis of the unique pattern of the daily death/survival record (Wake et al. 1978). After challenging SCY-immunized mice with Yreka organisms, few

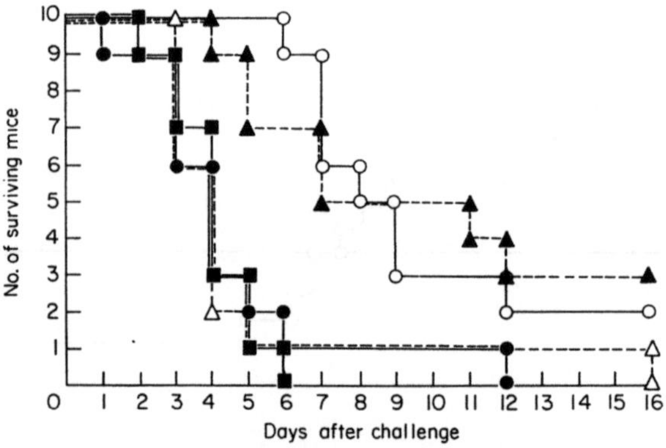

Fig. III.40. Protection from passive immunization of mice showing survival of mice injected i.v. with: SCY (○) and 222/+ (●) immunized serum; convalescent serum (▲); normal serum (■); and no serum (△), after challenge with 1.3×10^6 organisms (approx. 10^5 LD_{50}) of virulent strain Yreka

deaths of mice occurred until day 21, ten deaths occurred between day 10 and day 30, but no deaths occurred subsequently in SCY-immunized group. Thus, active immunization with VW+Cal+ *Y. pseudotuberculosis* 222/+ and F1+VW+Cal− *Y. pestis* SCY induced protective immunity with clearly contrasting characteristics. The participation of protective antibodies in 222/+ immunized mice was suspected because of the early deaths of ten mice within this group. Therefore, groups of SCY or 222/+ immunized mice and nonimmunized mice were bled after immunization. The separated serum of each group was pooled and 0.5 ml injected i.v. into ten mice as immune serum. Convalescent serum (0.25 ml), which was separated from the pooled blood of mice which survived the preceding active immunization followed by Yreka challenge, was also tested and injected i.v. into ten mice. One hour after injection these four groups (receiving anti-SCY serum, anti-222/+ serum, normal serum and convalescent serum, respectively) of mice and the ten nonimmune mice were challenged s.c. with 1.3×10^6 Yreka organisms. The results are shown in Fig. III.40. Animals passively immunized with normal serum or anti-222/+ immune serum and normal animals gave similar daily death rates. On the other hand, the two groups receiving anti-SCY immune serum or convalescent serum survived longer, but with only two to three survivors at the termination of the experiment on the 16th day. It was concluded that serum from SCY-immunized and convalescent mice contained protective humoral antibodies, i.e. immunoglobulins, but the serum from 222/+ immunized mice did not.

These two distinct types of protective immunity designated as "long-term" and "short-term" immunity could cooperate to form the best protective immunity in mice if the immunizing organisms expressed the F1 antigen, thus stimulating antibody production so that the toxic deaths of the mice during immunogenesis could be avoided. However, testing this hypothesis has been impossible because no F1+VW+Cal+ *Y. pestis* strain lacking the murine toxin has been isolated.

In contrast, such cooperation of humoral immune response with cell-mediated immune response conferring host effective resistance has been found to

Table III.26. Humoral antibodies of EV-Blutal immunized guinea pigs. The animals were bleeded 14 days after intracutaneous immunization by 2.7×10^3 CFU of EV strain suspended in 0.1 ml Blutal from their foot veins without sacrificing the animals

Guinea pig designation	MPI[a]	HA titer[b]
a	13.9	< 2
b	12.0	4
c	10.0	2
d	8.2	< 2
e	7.0	2
f	6.4	16
g	6.4	4
h	6.4	2
i	6.4	2
j	6.3	2
k	6.0	2
l	6.0	4
m	5.7	8
n	5.5	< 2
	non-protective level	
	protective level	
o	4.4	8
p	3.8	4
q	3.1	8
r	1.7	8
s	1.4	8
t	1.0	8
	rabbit hyperimmune serum[c]	>20480
	saline	< 2
non-immune control		
u	37.0	< 2
v	34.5	< 2

[a] MPIs were calculated 14 days after i.p. injection of each 0.5 ml of the one guinea pig serum into 10 mice and immediate challenging with 2.4×10^5 CFU of Yreka strain.
[b] HA titers were microtitrically determined by Fraction I sensitized tanned sheep red blood cell agglutination according to Chen (1952).
[c] Rabbit hyperimmune serum is one of our stock lots which have been prepared by repeated intravenous immunizations of rabbits with an attenuated *Y. pestis* strain A1122.

be inducible in guinea pigs that are resistant to murine toxin by the use of $fra^+ vwa^+ cal^+ pst^+ pgm^- pur^+$ *Y. pestis* organisms EV suspended in 0.1 ml of the iron drug (Blutal) for s.c. immunization. In fact, the Hartley guinea pigs immunized in this way was showed 100% protection against challenge with 10^8 organisms ($> 10^7$ LD_{50}) of a fully virulent *Y. pestis* Yreka (Wake, unpublished data). However, the humoral antibody titers of these immunized guinea pigs before challenge had been found to be extremely low as shown in Table III.26.

Sera from only 6 of 20 animals could protect mice significantly by passive immunization. Therefore the strong protective resistance of the immunized guinea pigs could be explained only by the cooperative protection provided by humoral and cell-mediated immune responses against *Y. pestis* infection.

To summarize this section, the *Yersinia*-host parasite relationship experimental system has enabled the dissection of the mechanisms of anti-*Y. pestis* protective immunity on the basis of the knowledge of each facet of the whole set of immune responses, the theoretical scope for developing an effective attenuated live vaccine i.e. F1+VW+Cal+P1+P−Pu+ *Y. pestis* EV suspended in an iron drug was suggested. However, this vaccine should be checked for safety in practical use. Furthermore, the duration of the effectiveness of the acquired immunity for antiplague prophylaxis may not be long with reference to the preceding section.

The Strain Differences of Anti-*Y. pestis* Acquired Immunity in Mice and Guinea Pigs

The dissociability of natural- and acquired-immune resistances to bacterial infection is unique to the animal experimental systems using *Y. pestis* organisms as the challenging bacteria. Therefore, the genetic control of natural resistance in mice and guinea pigs was already delineated in the first section of this chapter before discussing the acquired-immune resistance against *Y. pestis* infection. For the purpose of reconfirming the aforementioned dissociability, however, all experiments discussed in this section were simultaneously performed in parallel with those described in the section dealing with natural resistance. Therefore, all details of challenging animals with the virulent *Y. pestis* Yreka organisms are the same between the preceding section and this section, and all figures presented in the former section are also available for use in this section.

For the purpose of elucidating strain differences of anti-*Y. pestis* acquired immunity the mechanisms of which are cell-mediated immune responses with bactericidal activity, each of 10–11 mice of C57BL/6J, C3H/He, BALB/C, athymic nu/nu (derived from BALB/C), and an outbred ddY/S strain were s.c. immunized with 200–300 organisms of VW+Cal+ *Y. pseudotuberculosis* 222/+ suspended in the iron drug (Blutal). Subcutaneous challenges with 4.2×10^5 organisms of *Y. pestis* Yreka organisms were performed 14 days after immunization. Remarkable strain differences were shown, because 80% of C57BL/6J; 50% of C3H/He; 0% of athymic nu/nu; 10% of BALB/C; and 25% of ddY/S resisted the lethal challenge dose of Yreka organisms until the observation was finished 30 days after challenge. To confirm the reproducibility of these strain differences, similar experiments were repeated twice. In one experiment, 91% of C57BL/6J, 75% of C3H/He, and 30% of ddY/S resisted the s.c. challenge with 8.0×10^5 Yreka organisms. In another, 91% of C57BL/6J, 10%

of BALB/C, and 40% of (C57BL/6J × BALB/C) F1-hybrid mice resisted the s.c. challenge with 6.6×10^4 Yreka organisms.

These results indicate that C57BL/6J is a high, BALB/C a low, and C3H/He an intermediate responder to *Y. pestis* after priming with a cross-reactive organism, VW+Cal+ *Y. pseudotuberculosis*. It must be considered, however, that genetic characteristics of inbred mice may be more or less divergent depending on the sources. All mice employed in this study were supplied from the Department of Veterinary Medicine, National Institute of Health, Tokyo. Therefore, Wake and Akiyama (unpublished) repeated similar experiments by the use of various inbred-mouse strains, supplied from another source (Ohmura Zikkendoubutsu Co., Japan). The origins of inbred-mouse strains were, e.g., C57BL/6J and BALB/CJ, the former supplier; and C57BL/6N and BALB/CN, the latter supplier.

Another purpose of this study was to examine whether or not the H-2 haplotype is correlated with the strain differences of inbred mice. BALB/CN and DBA/2N were H-2^d haplotype; C3H/HeN and CBA/2N, H-2^k; and C57BL/6N were H-2^b haplotype. Using 5–6 mice per group maintained in one cage, all immunized mouse groups (s.c. immunizing dose 1.9×10^3, 222/+ organisms suspended in 0.1 ml Blutal) were s.c. infected with 1.3×10^6, 1.3×10^4, and 1.3×10^2 organisms of a fully virulent *Y. pestis* Yreka on the 14th day after immunization. The results are depicted in the right half of Fig. III.27.

From these results and results obtained by the use of mice supplied from another source, general conclusions were as follows:

1. The genetic control of cell-mediated immune resistance of a given inbred-mouse strain can be altered during maintenance at the supplier. The investigators of immunogenetics *must* keep this problem in mind.
2. In spite of this, the C57BL/6 strain can be regarded as a higher responder which induces cell-mediated immunity against *Y. pestis* infection as long as the standard procedures, including challenge doses of 10^5–10^6 Yreka organisms, are strictly followed.

On the other hand, the BALB/C strain can be regarded as a low responder. A result suggesting that this strain difference might be controlled by a single gene, is depicted in Fig. III.41.

3. The genetic control of cell-mediated immune resistance to *Y. pestis* infection *does not* correlate with the H-2 haplotype of the inbred mice.
4. An inbred-mouse strain with exceptionally high natural resistance (refer to the preceding section of this chapter) as well as cell-mediated immune resistance to *Y. pestis* was found. It is the DBA/2N strain employed in this study.

Among these general conclusions, the hypothetical one-gene control of cell-mediated immune-resistance regulation should be noted, because Skamene et al. (1982) proposed that genetic regulation of resistance to intracellular pathogens, e.g., *Mycobacterium bovis, M. tuberculosis, M. leprae, Salmonella typhimurium,* and *Leishmania donovanii,* is controlled by the same chromosome 1 locus of the inbred mice. Therefore, the same chromosome 1 locus might control the cell-mediated immune-resistance regulation in *Y. pestis* in-

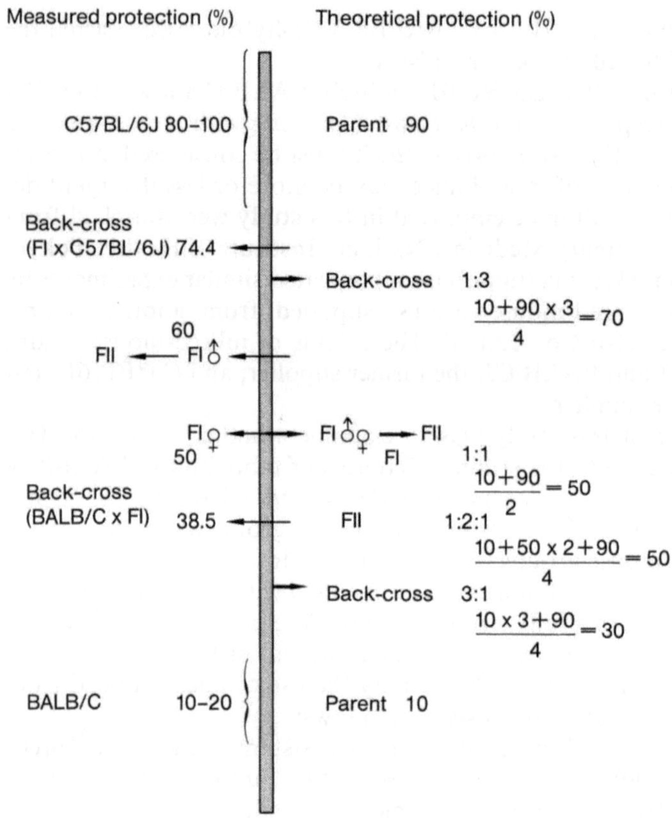

Fig. III.41. Suggested single-gene control of cell-mediated immune resistance of inbred mice and their FI, FII, back-cross mice against *Y. pestis* infection. Ten C57BL/6J, 10 BALB/C, 79 (C57BL/6J × BALB/C) FI, 66 FII, 17 (BALB/C × FI) back-cross, and 41 (FI × C57BL/6J) back-cross mice were assessed for their cell-mediated immune resistance to *Y. pestis* Yreka infection, according to the standard procedure (see p. 283 of this book). Similar assessments using each of the ten C57BL/6J and BALB/C mice were repeated several times, and the results were pooled in this diagram. All parent mice were supplied from the Department of Veterinary Medicine. The mice derived from these parent mice were raised in the laboratory of Wake, First Department of Bacteriology of the National Institute of Health, Tokyo, Japan

fection. Here, again, natural (innate) resistance and cell-mediated immune resistance in the *Y. pestis* experimental system could be dissociated because all inbred-mouse strains, except DBA/2N, are killed within 1 week, before the effective level of cell-mediated immune resistance is attained on the nineth day after s.c. immunization. On the other hand, intracellular pathogens with the sole possible exception of *Sal. typhimurium* enumerated by Skamene et al. (1982) would kill the inbred mice after the cell-mediated resistance attains the effective level. Therefore, the natural (innate) resistance against these pathogens might include cell-mediated immune resistance.

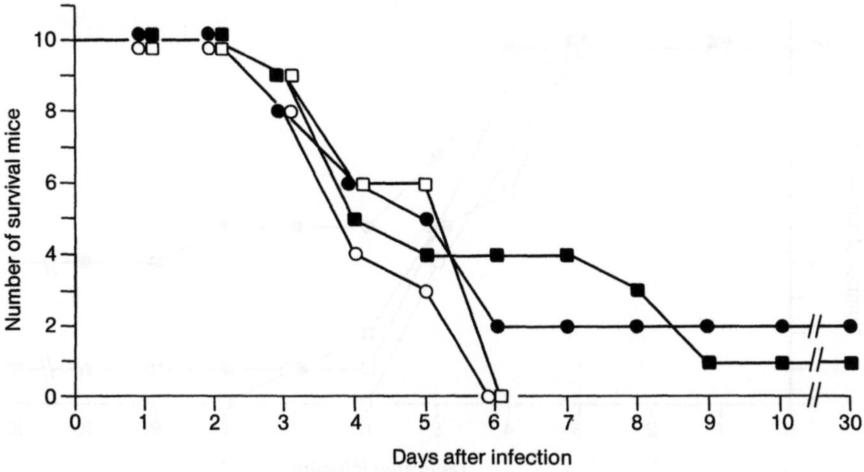

Fig. III.42. Effects of immunization with VW+Cal+ *Y. enterocolitica* Te1001 (0:3) on the resistance of inbred high- or low-responder mice against *Y. pestis* infection. Ten animals of each mouse strain were immunized s.c. with 2×10^4 Te1001 organisms suspended in 0.1 ml Blutal. The 7×10^2 organisms of a fully virulent strain Yreka, in 0.2 ml saline were injected s.c. 14 days after immunization. Each *point* is expressed as the number of surviving mice at the time of observation: (●——●) immunized C57BL/6J mice; (○——○) nonimmunized C57BL/6J mice; (■——■) immunized BALB/C mice; (□——□) nonimmunized BALB/C mice

The exceptionally high cell-mediated immune resistance of DBA/2N mice to *Y. pestis* might have been derived from an additional effect in this inbred strain. If so, the resistance of inbred mice may be controlled not by single gene locus, but rather two.

These experiments also suggested that the natural (innate) resistance gene expresses its function only against challenge with small doses (10^2 organisms) of *Y. pestis* and could be hidden by the more effective function of the cell-mediated immune resistance gene when larger doses (10^4 or 10^6 organisms) of *Y. pestis* organisms were used. The mouse-strain difference between high responder C57BL/6N mice and low responder BALB/CN mice with regard to immune resistance against *Y. pestis* infection was also seen when VW+Cal+ *Y. enterocolitica* was used as the immunizing bacteria. Male, 4–5 weeks old, inbred C57BL/6J or BALB/CJ mice (supplied by the Department of Veterinary Medicine, National Institute of Health, Tokyo) were immunized s.c. with 2×10^3–1×10^4 live *Y. enterocolitica* cells suspended in 0.1 ml Blutal. The immunized and control mouse groups were challenged s.c. with certainly lethal doses of *Y. pestis* Yreka 14–15 days after immunization and observed daily for 30 days.

Figures II.42 and III.43 show that both VW+Cal+ *Y. enterocolitica* strains Te1001 (serotype 0:3) and WA2-1 (serotype 0:8), immunized C57BL/6J and BALB/C mice against lethal challenge with *Y. pestis*. High and low responses were seen, respectively, consistent with the studies in which VW+

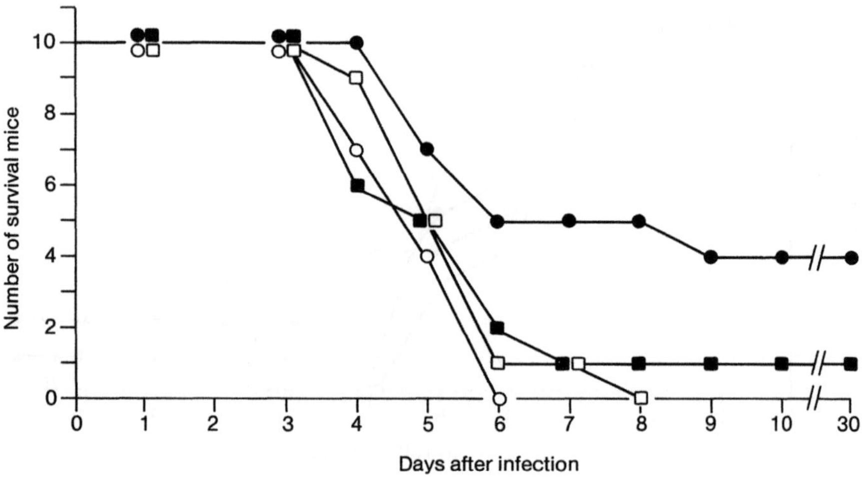

Fig. III.43. Effects of immunization with VW+Cal+ *Y. enterocolitica* WA2-1 (0:8) on the resistance of inbred high- or low-responder mice to *Y. pestis* challenge. Ten animals of each mouse strain were immunized s.c. with 2×10^3 WA2-1 organisms suspended in 0.1 ml Blutal. The 5×10^2 organisms of a fully virulent strain Yreka, in 0.2 ml saline, were injected s.c. 14 days after immunization. Each *point* is expressed as the number of surviving mice at the time of observation: (●——●) immunized C57BL/6J mice; (○——○) nonimmunized C57BL/6J mice; (■——■) immunized BALB/mice; (□——□) nonimmunized BALB/C mice

Cal+ *Y. pseudotuberculosis* was used as immunogen. According to the one-way lay-out method of statistical analysis of the average survival days before the tenth day of infection, significant differences ($P<0.01$) were found between the C57BL/6J mice and that of BALB/C mice in Fig. III.43, in which *Y. enterocolitica* WA2-1 (0:8) was used as immunogen. On the other hand, the strain difference with regard to immune resistance induced by *Y. enterocolitica* Te1001 (0:3) against *Y. pestis* was statistically insignificant ($P<0.05$), but two of ten C57BL/6J mice and one of ten BALB/C mice resisted challenge with *Y. pestis* Yreka. These results support the interpretation that not the host bacteria themselves parasitized by the *vwa – cal* plasmid, but the parasitizing plasmid itself, is responsible for the induction of cell-mediated immunity with bactericidal activity: triple biology.

In connection with the animal hosts, then, a wide range of mammals are sensitive to *Y. pestis* infection. These plague-sensitive mammalian species can be classified into plague-murine toxin-sensitive and toxin-resistant species. The mouse belongs to the former; therefore, the immunogen serving the immunogenetic studies described in the preceding section had to be living, but murine-toxin-nonproducing *Y. pseudotuberculosis* or *Y. enterocolitica*. Guinea pigs are representative of murine-toxin-resistant animals; therefore, these animals can be immunized with live F1+VW+Cal+P1+P−Pu+ *Y. pestis* organisms, e.g., strain EV. By the assessment of acquired-immune resistance of this guinea pig model, one can see that the protective immunity is the cooperative effect of humoral-antibody production stimulated by F1 and cell-mediated immune resistance conferred by VW – Cal (p. 280). This kind of study is of interest from the practical point of view, if humans (*Homo sapiens*) belong to the murine-toxin-resistant species similar to guinea pigs.

For the assessment of genetically-controlled acquired-immune resistance, inbred guinea-pig groups of six each JY1, JY2, NIH No. 2, and NIH No. 13 were immunized intracutaneously (i.c.) with 9.7×10^3 organisms of F1 + VW + Cal + P1 + P – Pu + *Y. pestis* EV suspended in 0.1 ml Blutal. All animals were challenged s.c. with 1.2×10^4 organisms of *Y. pestis* Yreka and observed for 30 days. One of six JY2 was dead 13 days after infection. All other animals survived for 30 days. The surviving animals were challenged s.c. with 1.3×10^7 (10^6 LD_{50}s) organisms of Yreka again, and observed daily for 30 days. All animals resisted this challenge. Apparently JY1, JY2, NIH No. 2, and NIH No. 13 guinea pigs are all capable of acquiring adequate levels of immune resistance against *Y. pestis* infection. All of these inbred guinea-pig strains were partially bled before immunization, i.e., the 14th day after immunization and 30 days after a second challenge infection. The IgM and IgG titers specific for *Y. pestis* EV were determined using solid-phase radioimmunoassay (RIA), according to Takeda's modification of the method of Smith and Gehle (1977, p. 275). All inbred guinea-pig strains responded well with regard to IgM and IgG production and no strain differences were recognized in the ability to produce humoral antibodies specific for *Y. pestis*, even though strain JY2 has shown significantly lower natural resistance to *Y. pestis* infection.

These results suggest that high levels of acquired-immune resistance have developed also in guinea pigs with low natural resistance. The high levels of acquired-immune resistance detected in these guinea pig groups, including JY2 that showed a lower level of natural resistance, presumably resulted from the cooperation of humoral-immune response and cell-mediated immune response, because the immunizing bacteria were F1 + VW + Cal + P1 + P – *Y. pestis* EV.

Chapter 5 The Phylogenetic Influence of *Yersinia* Infection on Human Major Histocompatibility Complex (HLA)

One of the principal thoughts of this book is that host-parasite relationships have evolved as the result of intimate mutual interactions between the host and microorganisms (Chap. 12, Summary of Part II). In this regard, the involvement of *Y. pestis* as the dreadful pathogen in the genetic alteration of the European population during the Middle Ages (Vogel et al. 1960) was noted in Chap. 1, Part III (p. 181). At the present time, however, it is difficult to trace the influence of *Y. pestis* on the human major histocompatibility complex (HLA). Instead, some chronic diseases that have been recognized as autoimmune diseases are suspected as being related to HLA, and *Y. enterocolitica* infection is also included among them from human statistical immunogenetic data using the term "relative risk" in relation to a particular HLA type. According to Svejgaard et al. (1980), a positive association exists between disease and the particular HLA type if the relative-risk index is above one, and if the index is below one, a negative association exists between them. Based on this definition of "relative risk", the indirect association of *Y. enterocolitica* infection with a certain HLA type attracted attention, because Catteral (1976) reported a Reiter's disease syndrome that particularly occurred after infection with *Y. enterocolitica* and the possession of HLA-D/DR3 has a relative risk of 3.7 to Reiter's syndrome. However, Svejgaard et al. (1980) showed a relative risk of 37.0 of HLA-B27 type (Table III.27) to Reiter's syndrome. With regard to Hashimoto's thyroiditis, the relative risk of which is 3.7, associated with HLA-D/DR5 (Table III.27), and was studied on Caucasians revealing a cross-reactivity between typhoid epithelial cells and some antigen of serovar 0:3 *Y. enterocolitica* organisms. On the other hand, Zen-Yoji (1980) wrote that antibodies specific for three other serovars, i.e., 0:5, 0:6, and 0:9 of *Y. enterocolitica,* were increased in the Japanese patients with Hashimoto's thyroiditis. In another report of Larsen (1976), significant relationships between Reiter's syndrome, arthropathies, and yersiniosis were concluded based on clinical diagnosis and maximal serum titers against *Y. enterocolitica* serovar 0:3 in 188 patients with arthritis as shown in Table III.28. Leirisalo et al. (1980) pointed out that these arthropathies seem to have a more severe prognosis in HLA-B27-positive patients, although it remains to be established how often Reiter's disease and reactive arthropathies develop into ankylosing spondylitis. These reports have called to the attention of immunogeneticists the fact that secondary complications of yersiniosis are associated with HLA.

Table III.27. HLA-associated disorders[a]. (Modified from Svejgaard et al. 1980)

Condition	HLA	Frequency (%) Patients	Frequency (%) Controls	Relative risk
Hodgkin's disease	A1	40	32.0	1.4
Idiopathic hemochromatosis	A3	76	28.2	8.2
	B14	16	3.8	4.7
Behcet's disease	B5	41	10.1	6.3
Congenital adrenal hyperplasia	B47	9	.6	15.4
Ankylosing spondylitis	B27	90	9.4	87.4
Reiter's disease	B27	79	9.4	37.0
Acute anterior uveitis	B27	52	9.4	10.4
Subacute thyroiditis	B35	70	14.6	13.7
Psoriasis vulgaris	Cw6	87	33.1	13.3
Dermatitis herpetiformis	D/DR3	85	26.3	15.4
Coeliac disease	D/DR3	79	26.3	10.8
	D/DR7	Also increased		
Sicca syndrome	D/DR3	78	26.3	9.7
Idiopathic Addison's disease	D/DR3	69	26.3	6.3
Graves' disease	D/DR3	57	26.3	3.7
Insulin-dependent diabetes	D/DR3	56	28.2	3.3
	D/DR4	75	32.2	6.4
	D/DR2	10	30.5	.2
Myasthenia gravis	D/DR3	50	28.2	2.5
	B8	47	24.6	2.7
S.L.E.	D/DR3	70	28.2	5.8
Idiopathic membranous nephropathy	D/DR3	75	20.0	12.0
Multiple sclerosis	D/DR2	59	25.8	4.1
Optic neuritis	D/DR2	46	25.8	2.4
C2 deficiency	D/DR2			
	B18			
Goodpasture's syndrome	D/DR2	88	32.0	15.9
Rheumatoid arthritis	D/DR4	50	19.4	4.2
Pemphigus (Jews)	D/DR4	87	32.1	14.4
IgA nephropathy	D/DR4	49	19.5	4.0
Hydralazine-induced SLE	D/DR4	73	32.7	5.6
Hashimoto's thyroiditis	D/DR5	19	6.9	3.2
Pernicious anaemia	D/DR5	25	5.8	5.4
Juvenile rheumatoid arthritis:				
pauciart	D/DR5	50	16.2	5.2
all cases	D/DRw8	23	7.5	3.6

[a] This information is a quotation from Svejgaard et al. (1980) who extracted from the HLA and Disease Registry in Copenhagen (Ryder et al. 1979; Svejgaard et al. 1980). Antigen frequencies refer mainly to a Danish population, but when only one or a few studies are available, the control frequencies published by these authors are given. HLA-D and DR antigens have not been distinguished from each other for simplicity, although there are occasional differences, for example, DR4 shows a stronger association with insulin-dependent diabetes than Dw4.

Table III.28. Clinical diagnosis and maximal serum titer against *Yersinia enterocolitica* serotype 0:3, in 188 patients with acute arthritis[a]

Diagnosis	Number of patients	Agglutinin titers					% With titer 80
		40	80	160	320	640	
Arthralgia	59	28	18	8	3	2	52.5
Monoarthritis	40	18	7	4	6	5	55.0
Acute polyarthritis	47	12	9	11	5	10	74.5
Acute juvenile polyarthritis	8	1		1	3	3	87.5
Acute seronegative RA	13	4	3	4	1	1	69.2
Reiter's syndrome	6	2	2	1	1		66.7
Schönlein-Henoch's syndrome	3	1				2	66.7
Rheumatic fever?	12	1	1		2	2	75.0

(From Larsen 1976)

[a] All had anti-serotype 0:9 titers of 8; 20% of patients tested by only one serum sample.

In this regard, the fact that both HLA-ABC and D/DR antigens are involved in T-cell-mediated immunity (Svejgaard et al. 1980) should be considered. The D/DR antigens control cooperation between T-helper lymphocytes, and B-lymphocytes; the HLA-D/DR antigens may thus be identical with the *Ir* determinants that regulate the immune responses, and if they are not, the genes controlling HLA-D/DR antigens are at least believed to map very close to the *Ir* determinants of human chromosome 6 (Fig. II.3) and probably have a pronounced linkage disequilibrium with them. Therefore, it is not surprising that some *Yersinia* antigen(s) being inducible or suppressible for immune responses could trigger the sequential reactions in the human bodies of a particular HLA type leading to some chronic type of disease as a secondary complication of *Yersinia* infection.

With regard to the various reactive arthropathies, *Yersinia* infections are occasionally followed by acute arthritis, particularly in B27-positive individuals. According to Svejgaard et al. (1980), it is possible that these reactive arthropathies are due to circulating immune complexes trapped in the capillaries of the joints. If this is true, some variant of complement could be involved. This variant could be responsible for the poor phagocytosis of certain immune complexes and/or could allow them to reach a size that causes them to be trapped in the capillaries. However, this is not the sole hypothesis to explain the association of *Yersinia* infection with a particular HLA type. Presumably, the whole integrated immune system immunogenetically determined in a unique way to that HLA type is related to the prognosis of *Yersinia* infection. The reason why the particular HLA type was selected for this association with *Yersinia* antigens is attributable to the phylogenetic influence of *Yersinia* infection to HLA, i.e., the evolutional history of *Yersinia*-human relationships.

If the secondary complications of yersiniosis are associated with HLA, ethnic variations with regard to the HLA-type distributions are naturally involved. Table III.27 contains the results obtained among Caucasians. The secondary complications of yersiniosis are less frequent among Mongolians, particularly in Japan. Whether frequent gastrointestinal infections due to oral infections with *Y. enterocolitica* and *Y. pseudotuberculosis* (Chap. 1, Part III) are, in contrast, unique to Japan, is suspected, but not proven. Epidemiological surveys in other parts of the world are needed for comparison. Thus, the *Yersinia*-host relationships are the target of scientific analysis from two directions of genetic approach; one involving microbial genetics, the second involving mammalian genetics, particularly human genetics via modern concepts of immunology.

Chapter 6 Summary

Part III is constructed around *Yersinia pestis,* the causative agent of bubonic plague, and the less virulent species *Y. pseudotuberculosis* and *Y. enterocolitica.* This is based on the fact that *Yersinia* have provided a major experimental system enabling medical microbiologists to resolve a variety of important mechanisms of bacterial virulence at the molecular level (Brubaker 1984).

Previously, the Summary of Part I proposed to revisit the classical, but yet novel, concept, "host-parasite relationships" by assimilating all the more progressing knowledge of molecular genetics and modern immunology. Part III is a particular example to show this concept as feasible to experimental study.

The particular choice of *Yersinia* in Part III can not be unrelated to the fact that these organisms cause zoonoses. Thus, the first obstacle in studying microbial virulence experimentally was eliminated, because we have a very natural experimental model system mimicking human disease at hand.

Being aided by the above merits of the use of *Yersinia,* Burrows (1962) opened the novel road of approach, i.e., methodology in the genetic studies of bacterial virulence (p. 185). This methodology is still applicable principally, although modern techniques of molecular genetics, e.g., Southern blotting for identifying genetic codes and Western blotting for the identification of crucial peptide amino acids, are available for further progress in this research field. Consequently, this methodology has led to the recognition of the *"established virulence factors of Y. pestis,* namely, F1, VW, Cal, P1, P, and Pu" which were described individually in Part III as their phenotypes in reference to the determinations of their corresponding genotypes, i.e., *fra, vwa, cal, pst, pgm,* and *pur,* respectively.

In parallel with the success achieved by microbial geneticists who used animal, i.e., in vivo experiments, biochemists who used in vitro techniques, e.g., purifications, isolations of target chemical substances, and fundamental knowledge of metabolic pathways, started their studies from the hypothesis of close taxonomic relationships of *Yersinia* with *Escherichia coli.* For these biochemists, it was easy to understand how the growth of *Yersinia* is limited in natural environments, since it is normally blocked in amino acid catabolism and the hexose monophosphate pathway owing to the lack of detectable aspartase and glucose-6-phosphate dehydrogenase activity, respectively. Furthermore, *Y. pestis* evidently possesses an additional block in the tricarboxylic acid cycle, as judged by the existence of a constitutive glyoxylate bypass, and is unable to regulate a number of catabolic enzymes, especially serine deaminase. None of these lesions exist in *Y. pseudotuberculosis* or *Y. enterocolitica,* which undergo regulatory changes as efficiently as any other commensal or saprophytic enteric bacterium.

This direction of biochemical approach has been effective in exploring the enigmas, e.g., how each phenotype of the established virulence factors is functioning in the expression of virulence and how their genotypes are expressed to phenotypes.

However, Part III intended to determine the localizations of the genetic codes regulating the individual phenotypes, thus, avoiding the elaborate discussions on the biochemical metabolic pathways, because the former problem, i.e., the localizations of the virulence-determinant genetic codes, is the more fundamental than the phenotypic biochemical expressions which are the *results* of integrated expressions of the former, i.e. virulence-determinant genetic codes.

Concluding from the accumulated evidence in Part III, the virulence determinant *pur* was determined to occupy a region of the bacterial chromosome DNA. There is some evidence indicating that the less active homologues of virulence determinants, i.e., *pgm* regulating siderophore production and *fra* regulating F1 production, may be also integrated into the chromosome, but the essential *pgm* and *fra* determinants which regulate the phenotypic expressions of P and F1 are located in plasmids. These phenotypes are recognizable by the activities of the latter, essential plasmid determinants. Besides, the virulence determinant *pst* associated with fibrinolysin and coagulase production is on the same species of 6–7 Mdal plasmids, the function of which can be detected by the activity of a bacteriocin-designated pesticin, P1. In a similar way, the genetic codes for the common virulence factors of all virulent *Yersinia* VW and Cal, occupy regions on the 40 to 48 Mdal plasmid family.

These above-mentioned, modern techniques of molecular genetics were efficient in exploring these intricate mechanisms, when these techniques were applied both before and after the transfer of another species of plasmids, e.g., R- or F-lac from *E. coli* donors as well as after curing the various virulence plasmids using appropriate drugs or high temperatures on the basis of knowledge of molecular genetics. The use of these foreign plasmids meant that transposon (Tn) and insertion units (IS) were used as the effective tools for the modifications of DNA sequences. Furthermore, a novel IS, i.e., IS100, that is inherent to *Yersinia* was also found, and virulence determinants of *Yersinia* were found to be movable among different sites of chromosome, plasmids, and phage replicons.

Along with these experiments, *Yersinia*-mouse models were used to assay the virulences of these various mutants, and these enabled researchers to resolve important mechanisms of bacterial virulence at the molecular level.

Returning to the main theme of this book, "host-parasite relationships", researchers should refer to Part II, where the background knowledge of multifaceted, defense mechanisms is elucidated. Another reason for the choice of

Y. pestis in Part III is that the process of the defense mechanisms, discussed in individual sections of Part II, can be conquered by some virulence factor of *Y. pestis,* which belongs to the disease category of pathogens as an invasive (penetrating) infection.

Prior to Part III, Part I clarified that the other disease category of pathogens utilizes a fewer number of virulence factors, e.g., toxins, adherence or colonizing factors, invasins that permit only restricted invasions into epithelial cells, thus utilizing at most, antiphagocytic factors. Furthermore, Part I also clarified that other intracellular microorganisms, which can resist the intracellular-killing mechanisms, utilize an additional number of virulence factors, e.g., a host-self-mimicking or not-stimulating factor of *M. tuberculosis.* However, there is divergency among the intracellular viabilities enabling these microorganisms to persist and resist the intracellular killing mechanisms of the host, and *Y. pestis* is the one with the most appropriate length of time feasible for the experimental studies. In addition, the fact that *Y. pseudotuberculosis* and *Y. enterocolitica* are less virulent, is determined by the characteristic lack of the established virulence factor P1 and the presence of more or less hydrophilic surface structures of lipopolysaccharide chains, enabling these organisms to survive in contaminating foods and host stomach.

Thus, the next point of view which was to be elucidated in Part III, is that the multifacets of host-defense mechanisms are dissectable by the use of the *Yersinia* experimental system.

The *Y. pestis* organisms surmount the superficial host-defense mechanisms, primarily the mammalian skin via flea bites, although direct inhalation of airborne bacteria into lung alveoli is the route of infection in the case of pneumonic plague. Thereafter, all other facets of host-defense mechanisms delineated in Part II are conquered in the time course of host-parasite relationships as the result of effective displays of the established virulence factors, F1, VW, Cal, P, P1, and Pu until the host animals are killed by endotoxemia in most cases of sensitive animals.

However, even these dreadful *Y. pestis* organisms can not kill the insensitive, i.e., resistant animal species; or can show only low virulence to those animal or human individuals in which some facet(s) of the defense mechanisms are potent enough to restrict the in vivo multiplication or invasion of *Y. pestis* organisms. These capabilities of natural resistance, the mechanisms of which can be variable, are genetically determined and constitute themselves as the targets of immunogenetics. Furthermore, even the detailed mechanisms of some particular facet contributing to resistance can become the target of future research at the biochemical and genetic levels.

In parallel with such studies on natural resistance to *Y. pestis* infection, Part III has delineated the mechanisms of acquired, specific immunity to *Y. pestis* infection.

In addition to the aspect of humoral immunity against *Y. pestis* infection, the mechanisms of cell-mediated immunity and the cooperative effects of both (humoral and cell-mediated) immunities are also delineated in Part III. In this regard, the identifications of the responsible inducing antigens are remarkable. Thus, it was demonstrated that *Yersinia* organisms are as potent microorganisms as *Listeria monocytogenes* for future studies on fundamental immunology.

Finally, the *Yersinia*-host relationship is regarded as the product of phylogenetic developments, i.e., evolutions of both mutually interacting components, i.e., ancestral forms of *Yersinia* and the mammalian host.

The evolutionary process of the mammalian host and its defense mechanisms was delineated in the summary of Part II. The suggestion that the surviving European population during the Middle Ages was altered genetically by the epidemics of plague (Vogel et al. 1960, p. 181) means that the sensitivity or resistance to plague may be related to the human histocompatibility complex type. In this context, an important implication is that the invasive actions of intracellular parasites may have played the role of a selective pressure by eliminating sensitive parts of population and may have helped the diversification of the T-cell receptor repertoire. The presence of MHC (H-2, in particular) restriction in the protective, cell-mediated immunity against *Y. pestis* infection (p. 277) provides evidence supporting T-cell involvement in the *Yersinia*-host relationship. In more detail, the antigen receptors of these T-cells are specifically directed to a particular epitope, which constitutes a part of the major virulence factor of *Yersinia*, presumably the V-protein itself. It should be noted that this V-protein is coded by a 40 to 48 Mdal plasmid family, which might be regarded as the extrachromosomal replicon of exogenous origin.

Of course, all mammalian animals that are insensitive or less sensitive to *Y. pestis* infection can be characterized by their unique, high-potent defense mechanisms, e.g., the high body temperature of birds. As a result, even this dreadful pathogen *Y. pestis* has its own unique host range.

In this context, the main mechanisms for the divergent evolution of primordial *Yersinia* organisms into *Y. pestis* and less virulent *Y. pseudotuberculosis* and *Y. enterocolitica,* seem to be attributable to their preferences in the routes of infection through which they have succeeded to infect in the host animals and to maintain the identities of their species or subspecies.

With special regard to this problem of the divergent evolutions of *Yersinia,* the following quotation from the review of Brubaker (1984) deserves consideration in concluding the summary of Part III:

"It is significant that *Y. pestis* carries a number of cryptic chromosomal genes that are normally expressed in the lesser yersiniae. At present, there is no adequate genetic explanation for this phenomenon; since CRIMs (cryptic intermediates: A.W.) have not been reported, the meiotrophic even may serve to open blocked promoters or otherwise restore superrepressed operons. There is no obvious physiological relationship between these cryptic gene products and no reason to suppose that their loss through mutation promotes some subtle advantage in vivo. *Y. pseudotuberculosis* and *Y. enterocolitica* are primarily transmitted orally and thus must be able to maintain themselves in natural environments. Accordingly, the selective pressures generated by the need to compete with the normal flora of soil and water are evidently sufficient to ensure the retention of maximal biochemical capabilities. *Y. pestis* is normally contained within the rigidly fixed, nutritionally enriched environments comprising its flea vector and mammalian host... Indeed, the species may now be undergoing a process of *degenerative* evolution analogous to that assumed to have occurred in certain obligate intracellular parasites, such as rickettsiae and chlamydiae."

References

Adler HI, Fisher WD, Cohen A, Hardigree AA (1967) Miniature *Escherichia coli* cells deficient in DNA. Proc Natl Acad Sci USA 57:321–326

Ajl SJ, Reedal JS, Durrum EL, Warren J (1955) Studies on plague. I. Purification and properties of the toxin of *Pasteurella pestis*. J Bacteriol 70:158–169

Albizo JM, Surgalla MJ (1970) Isolation and biological characterization of *Pasteurella pestis* endotoxin. Infect Immun 2:229–236

Alonso JM, Joseph-Francois A, Mazigh D, Bercovier H, Mollaret HH (1978) Résistance à la peste de souris expérimentalement infectées par *Yersinia enterocolitica* (1978). Ann Microbiol (Inst Pasteur) 192B:203–207

Alonso JM, Vilmer E, Mazigh D, Mollaret HH (1980) Mechanisms of acquired resistance to plague in mice infected by *Yersinia enterocolitica* 03. Curr Microbiol 4:117–122

Amano T, Morioka T, Seki Y, Kashiba S, Fujikawa K, Ichikawa S (1955) Studies on the immune bacteriolysis VIII. The mechanisms of the immune bacteriolysis. Med J Osaka Univ 6:709–716

Amano T, Morioka T, Seki Y, Kashiba S, Fujikawa K, Ichikawa S (1956) Studies on the immune bacteriolysis IX. The effect of immune bacteriolytic system and lysozyme on the cell wall preparation. Med J Osaka Univ 6:1022–1026

Anonymous (1976) Wkly Epidem Rec No. 33 (13. Aug. 1976) 262

Anonymous (1977) MMWR (CDC) 26:53–54

Arnow LE (1937) Colorimetric determination of the components of 3,4-dihydroxyphenylalamine-tyrosine mixtures. J Biol Chem 118:531–537

Arthus M, Breton M (1903) Lésions cutanées produites par les injections de sérum de cheval chez le lapin anaphylactisé par et pour ce sérum. C R Soc Biol 55:817–820

Asakawa Y, Asakawa S, Kagata N, Noguchi M (1973) Two community outbreaks of human infection with *Yersinia enterocolitica*. J Hyg Camb 71:715–723

Askenase PW (1980) Effector cells in late and delayed hypersensitivity reactions that are dependent on antibodies or T cells. In: Fougereau M, Dausset J (eds) Immunology 80. Academic Press, London, pp 829–845

Atkins E, Bodel P (1972) Fever. N Engl J Med 286:27–34

Bacon GA, Burrows TW, Yates M (1951) The effects of biochemical mutation on the virulence of *Bacterium typhosum:* the loss of virulence of certain mutants. Br J Exp Pathol 32:85–96

Baker EE, Sommer H, Foster LE, Meyer E, Meyer KF (1952) Studies on immunization against plague. I. The isolation and characterization of the soluble antigen of *Pasteurella pestis*. J Immunol 68:131–145

Bardana EJ Jr, McClatchy JK, Farr RS, Minden P (1973) Universal occurence of antibodies to tubercle bacilli in sera from non-tuberculous and tuberculous individuals. Clin Exp Immunol 13:65–77

Barksdale L, Garmise L, Horibata K (1960) Virulence, toxinogeny and lysogeny in *Corynebacterium diphtheriae*. Ann NY Acad Sci 88:1093–1108

Baseman JB, Hages NS, Spitznagel JK, Hayes EC (1979) Virulence determinants among the spirochetes. In: Schlessinger D (ed) Microbiology – 1979, American Society of Microbiology, Washington DC, pp 203–208

Bast RC Jr, Cleveland RP, Littman BH, Zbar B, Rapp HJ (1974) Acquired cellular immunity: extracellular killing of *Listeria monocytogenes* by a product of immunologically activated macrophages. Cell Immunol 10:248–259

Baugh CL, Andrews AW, Surgalla MJ (1964) Effects of bicarbonate on growth of *Pasteurella pestis*. III. Replacement of bicarbonate by pyrimidines. J Bacteriol 88:1394–1398

Beesley ED, Brubaker RR, Janssen WA, Surgalla MJ (1967) Pesticins III. Expression of coagulase and mechanism of fibrinolysis. J Bacteriol 94:19–26

Behring E von, Kitasato S (1890) Über das Zustandekommen der Diphtherie-Immunität bei Tieren. Dtsch Med Wochenschr 16:1113–1398

Bekierkunst A, Levij IS, Yarkoni E, Vilkas E, Lederer E (1969) Granuloma formation induced in mice by chemically defined mycobacterial fractions. J Bacteriol 100:95–102

Benacerraf B, McDevitt HO (1972) Histocompatibility-linked immune response genes. Science 175:273–279

Ben-Gurion R, Hertman I (1958) Bacteriocin-like material produced by *Pasteurella pestis*. J Gen Microbiol 19:289–297

Ben-Gurion R, Shafferman A (1981) Essential virulence determinants of different *Yersinia* species are carried on a common plasmid. Plasmid 5:183–187

Bennet HS, Luft JH, Hampton JC (1956) Morphological classification of vertebral blood capillaries. Am J Physiol 196:381

Bennett IL Jr, Nicastri A (1960) Fever as a mechanism of resistance. Bacteriol Rev 24:16

Bennett LG, Tornabene TG (1974) Characterization of the antigenic subunits of the envelope protein of *Yersinia pestis*. J Bacteriol 117:48–55

Bercovier H, Mollaret HH (1984) Genus XIV. *Yersinia* Van Loghem 1944, 15^{AL}. In: Krieg NR, Holt JG (eds) Bergey's manual of systematic bacteriology, vol 1. Williams and Wilkins, Baltimore, pp 498–506

Beresford CH, Neale RJ, Brooks OG (1971) Iron absorption and pyrexia. Lancet March 20:568

Berg CM, Berg DE (1981) Bacterial transposons. Microbiology – 1981. American Society of Microbiology, Washington DC, pp 107–116

Bernheimer AW, Avigad LS, Kim KS (1974) Staphylococcal sphingomylinase (beta-hemolysin). Ann NY Acad Sci 236:292–306

Bernard C (1857) Leçons de physiologie générale. Balliere, Paris

Bhaskaran K (1974) Vibrio genetics. In: Barua D, Burrows W (eds) Cholera. Saunders, Philadelphia, pp 41–59

Biozzi G, Siqueira M, Stiffel C, Ibanez OM, Mouton D, Ferreira VCA (1980) Genetic selections for relevant immunological functions. In: Fougereau M, Dausset J (eds) Immunology 80. Academic Press, London, pp 432–457 (Progress in Immunology IV)

Blanden RV, Langman RE (1972) Cell-mediated immunity to bacterial infection in the mouse. Thymus-derived cells as effectors of acquired resistance to *Listeria monocytogenes*. Scand J Immunol 1:379–391

Bloch H (1950) Studies on the virulence of tubercle bacilli. J Exp Med 91:95–102

Bloom W, Fawcett DW (1968) A text book of histology, 9th edn. Saunders, Philadelphia

Broom JD (1961) Evidence that the 1-asparaginase activity of guinea pig serum is responsible for its antilymphoma effects. Nature 191:1114

Brown KA, Ratledge C (1975) Iron transport in *Mycobacterium smegmatis:* ferrimycobactin reductase [NAD(P)H: ferrimycobactin oxidoreductase], the enzyme releasing iron from its carrier. FEBS Lett 53:262–266

Brown SD, Montie TC (1977) Beta-adrenergic blocking activity of *Yersinia pestis* murine toxin. Infect Immun 18:85–93

Brubaker RR (1972) The genus yersinia: biochemistry and genetics of virulence. Curr Top Microbiol 57:111–158

Brubaker RR (1979) Expression of virulence in *Yersiniae*. Microbiology – 1979. American Society of Microbiology, Washington DC

Brubaker RR (1984) Molecular biology of the dread black death. AMS News 50:240–245

Brubaker RR, Surgalla MJ (1961) Pesticins I. Pesticin-bacterium interrelationships and environmental factors influencing activity. J Bacteriol 82:940–949

Brubaker RR, Surgalla MJ (1962) Genotypic alterations associated with avirulence in streptomycin-resistant *Pasteurella pestis*. J Bacteriol 84:615–624

Brubaker RR, Surgalla MJ, Beesley ED (1965) Pesticinogeny and bacterial virulence. Zentralbl Bakteriol Orig 196:302–315

Brubaker RR, Beesley ED, Surgalla MJ (1965a) *Pasteurella pestis;* role of pesticin I iron in experimental plague. Science 149:442–424

Bullen JJ, Rogers HJ (1969) Bacterial iron metabolism and immunity to *Pasteurella septia* and *Escherichia coli*. Nature 224:380–382

Bullen JJ, Wallis SN (1977) Reversal of the bactericidal effects of polymorphs by a ferritin-antibody complex. FEMS Lett 1:117–120

Bullen JJ, Rogers HJ, Griffiths E (1978) Role of iron in bacterial infection. Curr Top Microbiol Immunol 80:1–35

Burrows TW (1962) Genetics of virulence in bacteria. Br Med Bull 18:69–73

Burrows TW (1965) A possible role for pesticin in virulence of *Pasteurella pestis*. Zentralbl Bakteriol I Abt Orig 196:315–317

Burrows TW (1973) Observations on the pigmentation of *Yersinia pseudotuberculosis*. Contrib Microbiol Immunol 2:184–189

Burrows TW, Bacon GA (1956a) The basis of virulence in *Pasteurella pestis:* the development of resistance to phagocytosis in vitro. Br J Exp Pathol 37:286–299

Burrows TW, Bacon GA (1956b) The basis of virulence in *Pasteurella pestis:* an antigen determining virulence. Br J Exp Pathol 37:481–491

Burrows TW, Bacon GA (1958) The effects of loss of different virulence determinants on the virulence and immunogenicity of strains of *Pasteurella pestis*. Br J Exp Pathol 39:278–291

Burrows TW, Bacon GA (1960) V and W antigens in strains of *Pasteurella pestis*. Br J Exp Pathol 41:38–44

Burrows TW, Gillet WA (1971) Host specificity of Brazilian strains of *Pasteurella pestis*. Nature 229:51–52

Burrows TW, Farrell JMF, Gillet WA (1964) The catalase activities of *Pasteurella pestis* and other bacteria. Br J Exp Pathol 45:579–588

Butler T, Fu Y-S, Furman L, Almeida C, Almeida A (1982) Experimental *Yersinia pestis* infection in rodents after intragastric inoculation and ingestion of bacteria. Infect Immun 36:1160–1167

Butler T (1983) Plague and other *Yersinia* infections. Plenum, New York

Camien MN, Dunn MS (1957) Studies of the composition of adjuvants which selectively enhance delayed-type hypersensitivity to lipid conjugated protein antigens. J Immunol 114:76–80

Cannon WB (1929) Organization for physiological homeostasis. Phys Rev 9:399–431

Capron ARG, Capron M, Dessaint J-P (1980) ADCC as primary mechanisms of defence against metazoan parasites. In: Fougereau M, Dausset J (eds) Immunology 80, Academic Press, London, pp 782–793 (Progress in Immunology IV)

Carter PB, Zahorchak RJ, Brubaker RR (1980) Plague virulence antigens from *Yersinia enterocolitica*. Infect Immun 28:638–640

Catterall RD (1976) The role of microbial infection in Reiter's syndrome. In: Dumonde DC (ed) Infection and immunology in the rheumatic disease. Blackwell, London, pp 147–150

Cavanaugh DC, Randall R (1959) The role of multiplication of *Pasteurella pestis* in mononuclear phagocytes in the pathogenesis of fleaborne plague. J Immunol 83:348–363

Charnetzky WT, Brubaker RR (1982) RNA synthesis in *Yersinia pestis* during growth restriction in calcium-deficient medium. J Bacteriol 149:1089–1095

Charnetzky WT, Shuford WW (1985) Survival and growth of *Yersinia pestis* within macrophages and an effect of the loss of the 47-Megadalton plasmid on growth in macrophages. Infect Immun 47:234–241

Chen TH (1952) Studies on immunization against plague IV. The method of the hemagglutination test and some observations on the antigen. J Immunol 69:587–596

Chen TH, Elberg SS, Boyles J, Velez MA (1975) *Yersinia pestis:* Correlation of ultrastructure and immunological status. Infect Immun 11:1382–1390

Christie AB, Chen TH, Elberg S (1980) Plague in camels and goats: their role in human epidemics. J Infect Dis 141:724–926

Clements JD, Finkelstein (1979) Isolation and characterization of homogenous heat-labile enterotoxins with high specific activity from *Escherichia coli* cultures. Infect Immun 24:760–769

Cocking EC, Keppie J, Witt K, Smith H (1960) The chemical basis if the virulence of *Pasteurella pestis* II. The toxicity for guinea pigs and mice of products of *Pasteurella pestis*. Br J Exp Pathol 41:460–471

Cohn ZA (1970) Lysosomes in mononuclear phagocytes. In: van Furth R (ed) Mononuclear phagocytes. Blackwell, Oxford, p 50

Cohn ZA, Scott WA (1982) New role for the mononuclear phagocyte. In: Mizuno D, Cohn ZA, Takeya K, Ishida N (eds) Self-defense mechanisms: role of macrophages. University of Tokyo Press, Tokyo

Cohn M, Langman R, Geckeler W (1980) Diversity 80. In: Fougreau M, Dausset J (eds) Immunology 80. Academic Press, London

Cronkite EP, Vincent PC (1970) Granulocytopoisis. In: Stohlman F (ed) Hemopoietic cellular proliferation. Grune and Stratton, New York, p 211

Csáky TZ (1948) On the estimation of bound hydroxylamine in biological materials. Acta Chem Scand 2:450–454

Daires AJS (1980) Theme 5; Summary: Ontogeny and differentiation of T and B lymphocytes. In: Fougreau M, Dausset J (eds) Immunology 80. Academic Press, London, pp 351–356

Dajani AS, Wannamaker LW (1969) Demonstration of a bactericidal substance against betahemolytic streptococci in supernatant fluids of staphylococcal cultures. J Bacteriol 97:985–991

Dallas WS, Falkow S (1980) Amino acid sequence homology between cholera toxin and *Escherichia coli* heat-labile toxin. Nature 4 Dec 288:499–501

D'Arcy Hart P, Armstrong JA (1974) Strain virulence and the lysosomal response in macrophages infected with *Mycobacterium tuberculosis*. Infect Immun 10:742–746

Darveau RP, Charnetzky WT, Hurlbert RE (1980) Outer membrane protein composition of *Yersinia pestis* at different growth stages and incubation temperatures. J Bacteriol 143:942–949

Darveau RP, Charnetzky WT, Hurlbert RE, Hancock REW (1983) Effects of growth temperature, 47-megadalton plasmid, and calcium deficiency on the outer membrane protein porin and lipopolysaccharide composition of *Yersinia pestis* EV76. Infect Immun 42:1092–1101

Davies DAL (1961) Dideoxysugars of *Pasteurella pseudotuberculosis* – specific polysaccharides, and the occurrence of ascarylose. Nature 191:43–44

Dean AG, Ching YC, Williams RG, Handen LB (1972) Test for *Escherichia coli* enterotoxin using infant mice: application in a study of diarrhea in children in Honolulu. J Infect Dis 125:407–411

DeChatelet LR, Shirley PS, McPhail LC, Huntley CC, Muss HB, Bass DA (1977) Oxidative metabolism of the human eosinophil. Blood 50:525–535

Delwiche EA, Fukui GM, Andrews AW, Surgalla MJ (1959) Environmental conditions affecting the population dynamics and the retention of virulence of *Pasteurella pestis:* the role of carbon dioxide. J Bacteriol 77:355–360

Desser RK, Himmerlhoch SR, Evans WH, Januska M, Mage, Shelton E (1972) Guinea pig heterophil and eosinophil peroxidase. Arch Biochem Biophys 148:452–465

De Vries RRP, Van Rood JJ (1977) HLA-linked control of immunity to infectious diseases. In: Bodmer WF, Batchelor JR, Bodmer JG, Festenstein H, Morris PJ (eds) Histocompatibility testing – 1977. Munksgaard, Copenhagen, p 409

Dienes L, Mallory TB (1937) The influence of allergy on the development of early tuberculous lesions. Am J Pathol 13:897–910
Dizon JJ, Fukumi H, Barua D, Valera J, Jayme F, Gomez F, Yamamoto S, Wake A, Gomez CZ, Takahira Y, Parran A, Rolda L, Alvero M, Aboa-Gareeb AH, Kobari K, Azurin JC (1967) Studies on cholera carriers. Bull WHO 37:737–743
Domaradskii IV, Yarmyuk GA, Vastukhina LV, Korotayeva AV (1963) Coagulation of blood plasma by plague and pseudotuberculosis bacilli. Bull Exp Biol Med (USSR) (Engl Transl) 56:79–82
Donta ST, Moon HW, Whipp SC (1974) Detection of heat-labile *Escherichia coli* enterotoxin with the use of adrenal cells in tissue culture. Science 183:334–336
Dreyfus LA, Brubaker RR (1978) Consequences of aspartase deficiency in *Yersinia pestis*. J Bacteriol 136:757–764
Dubach R, Callender STE, Moore CV (1948) Studies in iron transportation and metabolism, VI. Absorption of radioactive iron in patients with anemias of varied etiology. Blood 3:526
Dubos RJ (1947) The effect of lipids and serum albumin on bacterial growth. J Exp Med 85:9–22
Dubos RJ, Middlebrook G (1947) Media for tubercle bacilli. Am Rev Tuberc 56:334–345
Dupont HL, Formal SB, Hornick RB, Snyder MJ, Libouati JP, Sheahan DG, LaBrec EH, Kalas JP (1971) Pathogenesis of *Escherichia coli* diarrhea. N Engl J Med 285:1–9
Dye ES, Kapral FA (1980) Partial characterization of a bactericidal system in staphylococcal abscesses. Infect Immun 30:198–203
Eisler DM, Von Metz E (1963) Anti-*Pasteurella pestis* factor in the organ of normal mice and guinea pigs. I. Biologic characteristics. J Immunol 91:287–294
Elgat M, Ben-Gurion R (1969) Mode of action of pesticin. J Bacteriol 98:359–367
Elsbach P, Weiss J, Franson RC, Beckerdite-Quagliata S, Scheider A, Harris L (1979) Separation and purification of a potent bactericidal permeability-increasing protein and a closely associated phospholipase A2 from rabbit polymorphonuclear leukocytes. J Biol Chem 254:11000–11009
Elwell LP, Shipley PL (1980) Plasmid-mediated factors associated with virulence of bacteria to animals. Annu Rev Microbiol 34:465–496
Engley FB Jr (1952) The neurotoxin of *Shigella dysenteriae* (Shiga). Bacteriol Rev 16:153–178
Evans DC, Silver RP, Evans DJ Jr, Chase DG, Gorbach (1975) Plasmid controlled colonization factor associated with virulence in *Escherichia coli* enterotoxigenic for human. Infect Immun 12:656–667
Farr AG, Wechter WJ, Kiely JM, Unanue ER (1979) Induction of cytocidal macrophages after in vitro interactions between *Listeria*-immune T cells and macrophages – Role of H-2. J Immunol 122:2405–2412
Feingold DS, Youngner JS, Chen J (1968) Interferon production in mice by cell wall mutants of *Salmonella typhimurium*. II. Effect of purified glycolipid from S and Re chemotypes. Biochem Biophys Res Commun 32:554–557
Feldman WH (1938) Avian tuberculosis infections. Williams and Wilkins, Baltimore
Felsenfeld O, Decker WJ, Wohlhieter JA, Rafyi A (1965) Studies in borreliae II. Some immunologic biochemical and physical properties of the antigenic components of *Borrelia turicatae*. J Immunol 94:805–817
Ferber DM, Brubaker RR (1979) Mode of action of pesticin: N-acetylglucosaminidase activity. J Bacteriol 139:495–501
Ferber DM, Brubaker RR (1981) Plasmid in *Yersinia pestis*. Infect Immun 31:839–841
Fessenden-Raden JM (1972) Effect of fatty acids on the movement and staining of membrane proteins in polyacrylamide gel electrophoresis. Biochem Biophys Res Commun 46:1347–1353
Finkelstein RA, Hanne LF (1982) Hemagglutinin (Colonizing factors?) produced by *Vibrio cholerae*. Microbiology – 1982. American Society of Microbiology, Washington DC, pp 324–326

Finkelstein RA, LoSpalluto JJ (1969) Pathogenesis of experimental cholera: preparation and isolation of choleragen and choleragenoid. J Exp Med 130:185–202

Finkelstein RA, Solocinski PZ, Atthasampunna PM, Cherunmethee P (1966) Pathogenesis of experimental cholera: identification of choleragen (Procholeragen A) by disc immunoelectrophoresis and its differentiation from cholera mucinase. J Immunol 97:25–33

Formal SB, LaBrec EH, Kent TH, Falkow S (1965) Abortive intestinal infection with an *Escherichia coli-Shigella flexneri* hybrid strain. J Bacteriol 89:1374–1382

Formal SB, Gemski P, Baron LS, LaBrec EH (1971) A chromosomal locus which controls the ability of *Shigella flexneri* to evoke keratoconjunctivitis. Infect Immun 3:73–79

Fredericq P (1954) Transduction génetiques des proprieteés colicinogenes chez *Escherichia coli* et *Shigella sonnei.* C R Soc Biol 148:399–402

Freedman SO, Turcotte R, Fish AJ, Sehon AH (1963) The in vitro detection of "cell-fixed" hemagglutinating antibodies to tuberculin purified protein derivative (PPD) in humans. J Immunol 90:52–59

Freeman VJ (1951) Studies on the virulence of bacteriophage infected strains of *Corynebacterium diphtheriae.* J Bacteriol 61:675–688

Freter R (1982) Bacterial association with the mucus gel system of the gut. Microbiology – 1982. American Society of Microbiology, Washington DC, pp 278–281

Fruton JS, Simmons S (1958) General biochemistry. Wiley & Sons, New York

Fukui GM, Ogg JE, Wessman GE, Surgalla MG (1957) Studies on the relation of cultural conditions and virulence of *Pasteurella pestis.* J Bacteriol 74:714–717

Fukui GM, Lawton WD, Ham DA, Janssen WA, Surgalla MJ (1960) The effect of temperature on the synthesis of virulence factors by *Pasteurella pestis.* Ann NY Acad Sci 88:1146–1151

Garber ED (1956) A nutrition-inhibition hypothesis of pathogenicity. Am Nat 90:183

Garibaldi JA (1972) Influence of temperature on the biosynthesis of iron transport compounds by *Salmonella typhimurium.* J Bacteriol 110:262–265

Gemski P Jr, Takeuchi A, Washington O, Formals SB (1972) Shigellosis due to *Shigella dysenteriae* 1: Relative importance of mucosal invasion versus toxin production in pathogenesis. I. Infect Dis 126:523–530

Gemski P, Lazere JR, Casey T (1980 a) A plasmid associated with pathogenicity and calcium dependency of *Yersinia enterocolitica.* Infect Immun 27:682–685

Gemski P, Lazere JR, Casey T, Wohlhieter JA (1980 b) Presence of a virulence-associated plasmid in *Yersinia pseudotuberculosis.* Infect Immun 28:1044–1047

Giannella RA (1975) Pathogenesis of Salmonella enteritis and diarrhea. Microbiology – 1975. American Society of Microbiology, Washington DC, pp 170–173

Giannella RA, Washington O, Gemski P, Formal S (1973) Invasion of HeLa cells by *Salmonella typhimurium:* a model for study of invasiveness of salmonella. J Infect Dis 128:69–75

Gill DM, Meren R (1978) ADP-ribosylation of membrane proteins catalyzed by cholera toxin: Basis of the activation of adenylate cyclase. Proc Natl Acad Sci USA 75:3050–3054

Ginoza HS, Matney TS (1963) Transmission of a resistance transfer factor from *Escherichia coli* to two species of *Pasteurella.* J Bacteriol 85:1177

Gladstone GP, Walton E (1971) The effect of iron and haematin on the killing of staphylococci by rabbit polymorphs. Br J Exp Pathol 52:452–464

Głośnicka R, Gruszkiewicz E (1980) Chemical composition and biological activity of the *Yersinia pestis* envelope substance. Infect Immun 30:506–512

Gohda A, Kazuno Y, Sasaki S (1973) The ligated ileal loop method. In: Sasaki S (ed) Designs of experimental models for the studies of infections (in Japanese). Kindaishuppan, Tokyo, pp 75–85

Goldschmidt MC, Koffler H (1950) Effect of surface-active agents on penicillin yields. Ind Eng Chem, Anal Ed 42:1819–1823

Gorbach SL, Banwell JG, Chatterjee BD, Jacobs B, Sack RB (1971) Acute undifferentiated human diarrhea in the tropic 1. Alteration in intestinal microflora. J Clin Invest 50:881–889

Gorbach SL, Khurana CM (1972) Toxigenic *Escherichia coli*. A cause of infentile diarrhea in Chicago. N Engl J Med 287:791–795
Goren MB (1975) Cord factor revisited: a tribute to the late Dr. Hubert Bloch. Tubercle 56:65–71
Goren MB (1977) Phagocyte lysosomes: interactions with infectious agents, phagosomes, and experimental perturbations in function. Annu Rev Microbiol 31:507–533
Goren MB, Brokl O, Schaefer WB (1974) Lipids of putative relevance to virulence with elaboration of sulfatides and strongly acidic lipids. Infect Immun 9:142–149
Goren MB, Brokl O, Roller P, Fales HM, Das C (1976 a) Sulfatides of *Mycobacterium tuberculosis:* the structure of the principal sulfatide (SL-1). J Biochem 15:2728–2735
Goren MB, D'Arcy Hart P, Young MR, Armstrong JA (1976 b) Prevention of phagosome-lysosome fusion in cultured macrophages by sulfatides of *Mycobacterium tuberculosis*. Proc Natl Acad Sci USA 73:2510–2514
Grignachi VI, Sperperato AM, Etcheverry MJ, Macario AJL (1963) Un nuevo cuadro citoquimico: Negatividad espontanea de las reacciones de peroxidasas, oxidasas y lipido en la progenie neutrofila y en los monocitos de dos hermanos. Rev Assoc Med Argent 77:218–221
Guentzel MN, Berry LT (1975) Motility as a virulence factor for *Vibrio cholerae*. Infect Immun 11:890–897
Gyles CL, Barnum DA (1969) A heat-labile enterotoxin from strains of *Escherichia coli* enteropathogenic for pigs. J Infect Dis 120:419–426
Habig W, Hudson BW, Marshall JD Jr, Cavanaugh DC, Rust JH Jr (1971) Evidence for molecular heterogeneity of the specific antigen (fraction-1) of *Pasteurella pestis*. Infect Immun 3:498–499
Hahn H (1974) Fieber. Immun Infekt 2:69–72
Hall PJ, Yang GCH, Little RV, Brubaker RR (1974) Effect of Ca^{2+} on morphology and division of *Yersinia pestis*. Infect Immun 9:1105–1113
Hardesty BA, Michell HK (1963) The interaction of fatty acids with mammalian cytochrome c. Arch Biochem Biophys 100:1–8
Hayashi H, Koono M, Yoshinaga M, Muto M (1969) The role of a SH-dependent protease and its inhibitor in the Arthus type hypersensitivity reaction, with special reference to chemical mediation of increased vascular permeability and leucocyte emigration. In: Bertellia A, Houck JC (eds) Inflammation biochemistry and drug interaction. Excerpta Med Found, Amsterdam, pp 34–52
Hayashi H, Yoshinaga M, Yamamoto S (1974) The nature of a mediator of leucocyte chemotaxis in inflammation. In: Sorkin E (ed) Chemostaxis its biology and biochemistry. Karger, Basel, pp 296–332
Hayes NS, Muse KE, Collier AM, Baseman JB (1977) Parasitism by virulent *Treponema pallidum* on host cell surface. Infect Immun 17:174–186
Henley DA, Nungester WJ (1955) Comparative study of host resistance of guinea pig and rat. III. Pneumocidal extracts from the liver and spleen of the susceptible rat and the resistant guinea pig. J Infect Dis 97:8–21
Higuchi K, Smith JL (1961) Studies on the nutrition and physiology of *Pasteurella pestis* VI. A differential plating medium for the estimation of the mutation rate to avirulence. J Bacteriol 81:605–608
Hildebrand GJ, Ng J, von Metz EK, Eisler DM (1966) Studies on the mechanism of circulatory failure induced in rat by *Pasteurella pestis* murine toxin. J Infect Dis 116:615–629
Hirata F, Corcoran BA, Venkatasubramanian K, Schiffmann E, Axelrod J (1979) Chemattractants stimulate degradation of methylated phospholipids and release of arachidonic acid in rabbit leukocytes. Proc Natl Acad Sci USA 76:2640–2643
Hirsch JG (1972) The phagocytic defense system. In: Smith H, Pearce JH (eds) Microbial pathogenicity in man and animals. Cambridge Univ Press, Cambridge, p 59
Honda T, Takeda Y, Miwatani T (1981 a) Isolation of special antibodies which react only with homologous enterotoxins from *Vibrio cholerae* and enterotoxigenic *Escherichia coli*. Infect Immun 34:333–336

Honda T, Takeda Y, Miwatani T (1981 b) Immunological nonidentity of heat-labile enterotoxins from human and porcine enterotoxigenic *Escherichia coli.* Infect Immun 34:337–340

Honjo S, Takasaka M, Fujiwara T, Nakagawa M, Ando K, Ogawa H, Takahashi R, Imaizumi K (1964) Shigellosis in cynomolgus monkeys (Macaca irus) II. Experimental infection with *Shigella flexneri* 2a with special references to clinical and bacteriological finding. Jpn J Med Sci Biol 17:307–319

Honjo T, Nishizuka Y, Hayaishi O, Kato I (1968) Diphtheria toxin-dependent adenosine diphosphate ribosylation of aminoacyl transferase II and inhibition of protein synthesis. J Biol Chem 243:3533–3555

Honjo T, Nishizuka Y, Kato I, Hayaishi O (1971) Adenosine diphosphate ribosylation of aminoacyl transferase II and inhibition of protein synthesis by diphtheria toxin. J Biol Chem 246:4251–4260

Hood LE, Weissman JL, Wood WB (1978) Immunology. Benjamin/Cummings, Menlo Park, Cal

Hu PC, Brubaker RR (1974) Characterization of pesticin. Separation of antibacterial activities. J Biol Chem 249:4749–4753

Hu PC, Yang GCH, Brubaker RR (1972) Specificity, induction and absorption of pesticin. J Bacteriol 112:212–219

Hubert WT (1972) Yeresiniosis in mammals and birds in the United States. Case reports and review. Am J Trop Med Hyg 21:458–463

Inoue M, Nakashima H, Ueba O, Ishida T, Date H, Kobashi S, Takagi K (1984) Community outbreak of *Yersinia pseudotuberculosis*. Microbiol Immunol 28:883–891

Isaacs A, Lindenman J (1957) Virus interference. I. The interferon. Proc R Soc Lond Ser B 147:258–267

Jackson S, Burrows TW (1956 a) The pigmentation of *Pasteurella pestis* on a defined medium containing haemin. Br J Exp Pathol 30:570–576

Jackson S, Burrows TW (1956 b) The virulence-enhancing effect of iron on non-pigmented mutants of virulent strains of *Pasteurella pestis*. Br J Exp Pathol 30:577–583

Jacob F, Wallman EL (1961) Sexuality and the genetics of bacteria. Academic Press, New York

Janssen WA, Fukui GM, Surgalla MJ (1958) A study of the fate of *Pasteurella pestis* following intracardiac injection into guinea pigs. J Infect Dis 103:183–187

Janssen WA, Lawton WD, Fukui GM, Surgalla MJ (1963) The pathogenesis of plague. 1. A study of the phagocytosis resistance of some strains of *Pasteurella pestis*. J Infect Dis 113:139–143

Jawetz E, Meyer KF (1943) Avirulent strains of *Pasteurella pestis*. J Infect Dis 73:124–144

Jawetz E, Meyer KF (1944) Studies on plague immunity in experimental animals. J Immunol 49:15–30

Jerne NK (1973) The immune system. Sci Am 229:52–60

Jones GW, Richardson LA (1981) The attachment to an invasion of HeLa cells by *Salmonella typhimurium:* the contribution of mannose-sensitive and mannose-resistant haemagglutinating activities. J Gen Microbiol 127:361–370

Jones GW, Abrams GD, Freter R (1976) Adhesive properties of *Vibrio cholerae* adhesion to isolated rabbit brush border membranes and hemagglutinating activity. Infect Immun 14:232–239

Jones GW, Rabert DK, Svinarich DM, Whitfield HJ (1982) Association of adhesive, invasive, and virulent phenotypes of *Salmonella typhimurium* with autonomous 60-megadalton plasmids. Infect Immun 38:476–486

Julianella LA (1941) Biological and immunological studies of Listerella. J Bacteriol 42:367–383

Jungermann K, Möhler H (1980) Biochemie. Springer, Berlin Heidelberg New York, pp 733

Kadis S, Ajl SJ (1970) Site and mode of action of murine toxin of *Pasteurella pestis*. Microbiol. Toxins 3:39–67

Kadis S, Cohen M, Ajl SL (1965) The effect of plague murine toxin on the electron transport system. Biochem Biophys Acta 96:179–185
Kamal AM (1971) Outbreak of gastroenteritis by non-agglutinable [NAG] vibrios in the Republic of the Sudan. October-November, 1968. J Egypt Public Health Assoc 46:125–152
Kamio Y, Nikaido H (1976) Outer membrane of *Salmonella typhimurium:* Accessibility of phospholipid head groups to phospholipase C and cyanogen bromide activated dextran in the external medium. Biochemistry 15:2561–2570
Kanai K (1967 a) Affinity of lysosomal acid phosphatase to the surface of the tubercle bacillus. Jpn J Med Sci Biol 20:401–411
Kanai K (1967 b) Detection of host-originated acid phosphatase on the surface of "in vivo grown tubercle bacilli". Jpn J Med Sci Biol 20:73–90
Kanai K (1967 c) Resistance to sodium hydroxide treatment of "in vivo grown tubercle bacilli". Jpn J Med Sci Biol 20:91–96
Kanai K, Kondo E (1970) A suggested role of the lysosomal membrane as a part of the defense mechanism against tuberculous infection. Jpn J Med Sci Biol 23:295–302
Kanai K, Kondo E (1979) Antibacterial and cytotoxic aspects of long-chain fatty acids as cell surface events. Selected topics. Jpn J Med Sci Biol 32:135–174
Kanai K, Wiegeshaus E, Smith DW (1970) Demonstration of mycolic acid and phthiocerol dimycocerosate in "in vivo grown tubercle bacilli". Jpn J Med Sci Biol 23:327–333
Kaneko K, Hashimoto K (1983) Fecal excretion associated with Ca^{2+} dependency of *Yersinia enterocolitica* 03 and 09 and pseudotuberculosis in mice. Microbiol Immunol 27:199–202
Kaplan SS, Basford RE (1979) Exogenous iron and impairment of intraleukocytic bacterial killing, or the leukocyte tells her story. Microbiology – 1979, American Society of Microbiology, Washington DC, pp 92–95
Kapral FA, Miller MM (1971) Product of *Staphylococcus aureus* responsible for the scalded-skin syndrome. Infect Immun 4:541–545
Kasai GJ, Burrows W (1966) The titration of cholera toxin and antitoxin in the rabbit ileal loop. J Infect Dis 116:606–614
Kasuga T (1977) The history of plague epidemics in Japan. (In Japanese) Kagaku 47:687–700
Kato M (1968) Studies of a biochemical lesion in experimental tuberculosis in mice. VIII. Effect of derivatives and chemical analogues of cord factor on structure and function of mouse liver mitochondria. Am Rev Respir Dis 98:668–676
Kato M (1971) The role of cord factor in pathogenesis of tuberculosis. Jpn J Med Sci Biol 24:47–51
Kato M, Kusunose M, Matsunaga K, Yamamura Y (1959) The mechanism of the toxicity of cord factor. Am Rev Tuberc 80:240–248
Kaufmann SHE, Hahn H (1982) Biological functions of T cell lines with specificity for the intracellular bacterium *Listeria monocytogenes* in vitro and in vivo. J Exp Med 155:1754–1765
Kearns RJ, De Freitas EC (1983) In vitro propagation of antigen-specific T lymphocytes that adoptively transfer resistance to *Listeria monocytogenes*. Infect Immun 40:713–719
Keyhani M, Wiley BB, Glasgow LA (1975) Chromosomal synthesis of staphylococcal exfoliative toxin. Infect Immun 12:193–197
Kitamura Y, Go S, Shimada M, Matsuda H, Hatanaka K, Seki M (1979) Distribution of mast-cell precursors in hematopoietic and lymphopoietic tissues of mice. J Exp Med 150:482–490
Kitos PA, Campbell JJR, Tomlinson N (1953) Influence of temperature on the trace element requirements for citric acid production by *Aspergillus niger*. Appl Microbiol 1:156–159
Klebanoff ST (1980) Cytocidal mechanisms of phagocytic cells. In: Fougereau M, Dausset J (eds) Immunology 80. Academic Press, London, pp 720–736
Klecker N (1981) Transposable elements in prokaryotes. Annu Rev Genet 15:341–404

Klecker N, Roth J, Botstein D (1977) Genetic engineering *in vivo* using translocatable drug-resistance elements. J Mol Biol 116:125–159

Klein J (1975) Biology of the mouse histocompatibility 2 complex. Springer, Berlin Heidelberg New York

Kluger MJ, Rothenburg BA (1979) Fever and reduced iron: their interaction as a host defense response to bacterial infection. Science 203:374–376

Kluger MJ, Vaughn LK (1978) Fever and survival in rabbits infected with *Pasteurella multocida*. J Physiol (Lond) 282:243–251

Knapp W (1963) Klinisch bakteriologische und epidemiologische Befunde bei der Pseudotuberkulose des Menschen. Arch Hyg (Berlin) 147:369–380

Kobayashi R (1940) On the causative bacteria and the genesis of sepsis (in Japanese) Nihonijishinpoh (Jpn Med News) 940:4–10

Kobayashi R (1942) On the primary and secondary sepsis caused by typhoid-paratyphoid bacteria and immunity against them (in Japanese) Nihonigakuoyobikenkohhoken (Jpn Med and Health Ins) 3266:5–13

Kobayashi R (1944) On the concept of infection, disease-onset and immunity in bacterial disease (in Japanese). Nihonigaku (Jpn Med) 3361:3–11, ibid 3362:61–69

Kobayashi R, Ushiba D (1952) Studies on immunity of experimental typhoid I. A review of experiments carried out in our laboratory during the past twenty years. Keio J Med 1:35–47

Kochan I (1973) The role of iron in bacterial infections, with special consideration of host-tubercle bacillus interaction. Curr Top Microbiol Immunol 60:1–30

Kochan I (1977) Roles of siderophores in nutritional immunity and bacterial parasitism. In: Weinberg ED (ed) Microorganisms and minerals. Dekker. New York

Kochan I, Pellis NR, Pfohl DG (1972) Effects of normal and activated cell fractions on the growth of tubercle bacilli. Infect Immun 6:142–148

Kol'tsova EG, Suchkov YG, Lebedeva SA (1973) Transmission of a bacteriocinogenic factor in *Pasteurella pestis*. Sov Genet 7:507–510

Kondo E, Kanai K (1981 a) Host lipids in tuberculous infection II. Cholesterol Kekkaku 56:41–47

Kondo E, Kanai K (1981 b) Host lipids in tuberculous infection III. Long-chain fatty acids against tubercle bacilli. Kekkaku 56:109–262

Kondo E, Kanai K (1981 c) Modification fo the fatty acid pattern of phagocytic cell phospholipids and the course of tuberculous infection in mice fed with diets deficient in essential fatty acid. Kekkaku 56:423–517

Kondo I, Sakurai S, Sarai Y (1973) Purification of exfoliatin produced by *Staphylococcus aureus* of bacteriophage group II and its physicochemical properties. Infect Immun 8:156–164

Kondo I, Sakurai S, Sarai Y, Futaki S (1975) Two serotypes of exfoliatin and their distribution in staphylococcal strains isolated from patients with scalded skin syndrome. J Clin Microbiol 1:397–400

Kopecko DJ, Washington O, Formal SB (1979) Molecular characterization of plasmids from virulent and spontaneously occurring avirulent colonial variants of *Shigella flexneri*. Infect Immun 24:580–582

Kopecko DJ, Washington O, Formal SB (1980) Genetic and physical evidence for plasmid control of *Shigella sonnei* form I cell surface antigen. Infect Immun 29:207–214

Kopecko DJ, Sansonetti PJ, Formal SB, Baron LS (1981) Use of transposons to identify and manipulate *Shigella* virulence plasmid. Proc Third Tokyo Symp Microb Drug Resist pp 26–27

Korobkova EI (1956) Living antiplague vaccines (in Russian). Medgiz, Moscow

Kudoh Y, Zen-Yoji H, Matsushita S, Sakai S, Maruyama T (1977) Outbreaks of acute enteritis due to heat-stable enterotoxin producing strains of *Escherichia coli*. Microbiol Immunol 21:175–178

Kuehl FA Jr, Egan RW (1980) Prostaglandines, arachidonic acid, and inflammation. Science 210:978–984

Kugelmass NI (1959) Biochemistry of blood in health and disease. Charles Thomas, Springfield, Ill
Kurauchi K, Kasuga T, Homma H, Ki R (1937) The results of human prophylactic vaccination with Kurauchi's „Pestimmunogen" (Studies on S.I.F. of plague bacilli No. 3) (in Japanese). Saikingakuzasshi 493:48–73
Kuroda M, Endo A (1977) Inhibition of in vitro cholesterol synthesis by fatty acids. Biochim Biophys Acta 486:70–81
Kurokawa M, Murata R (1961) On the toxicity of the "toxoid" preparation responsible for the Kyoto catastrophe in 1948. Jpn J Med Sci Biol 14:249–256
LaBrec EH, Schneider H, Magnani TJ, Formal SB (1964) Epithelial cell penetration as an essential step in the pathogenesis of bacillary dysentery. J Bacteriol 88:1503–1518
Lane FC, Unanue ER (1972) Requirement of thymus (T) lymphocyte for resistance to listeriosis. J Exp Med 135:1104–1112
Langer WL (1964) The black death. Sci Am 210:114–121
Langford EV (1972) *Pasteurella pseudotuberculosis* infections in Western Canada. Can Vet J 13:85–87
Larsen JH (1976) Yersinia enterocolitica infections and arthritis. In: Dumonde DC (ed) Infection and immunology in the rheumatic disease. Blackwell, London, pp 133–141
Lawton WD, Fukui GM, Surgalla MJ (1960) Studies on the antigens of *Pasteurella pestis* and *Pasteurella pseudotuberculosis*. J Immunol 84:475–479
Lawton WD, Erdman RL, Surgalla MJ (1963) Biosynthesis and purification of V and W antigen in *Pasteurella pestis*. J Immunol 91:179–184
Lehninger AL (1975) Biochemistry, 2nd edn. Worth, New York
Lein J, Lein P (1949) Studies on a Neurospora mutant requiring unsaturated fatty acids for growth. J Bacteriol 58:595–599
Leirisalo M, Skylv G, Kausa M, Voipio-Pulkki L-M, Hvidmen L, Darm-Neilsen E, Nissisla M, Laitinen O (1980) Data submitted to the HLA and disease registry
Levine MM, Dupont HL, Formal SB, Hornick RB, Takeuchi A, Gangarosa EJ, Snyder MJ, Libonati JP (1973) Pathogenesis of *Shigella dysenteriae* 1 (Shiga) dysentery. J Infect Dis 127:261–270
Lindberg AA (1973) Bacteriophage receptors. Annu Rev Microbiol 27:205–241
Losert W (1975) Prostaglandine: Biologische Bedeutung. Umsch Wiss Tech 75:197–202
Lwoff A, Lwoff M (1958) L'inhibition du développment, du virus poliomyélitique a 39 °C et le problem du rôle de l'hyperthermie dans l'evolution des infections virales. C R Acad Sci 246:190–192
Lyell A, Dick HM, Alexander JOD (1969) Outbreak of toxic epidermal necrolysis associated with staphylococci. Lancet 1:787–789
Macham LP, Ratledge C, Nocton JC (1975) Extracellular iron acquisition by mycobacteria: role of the exochelins and evidence against the participation of mycobactin. Infect Immun 12:1242–1251
Mackaness GB (1962) Cellular resistance to infection. J Exp Med 116:381–406
Mackaness GB (1964) The immunological basis of acquired cellular resistance. J Exp Med 120:105–120
Mackaness GB (1969) The influence of immunologically committed lymphoid cells on macrophage activity in vitro. J Exp Med 129:973–992
Madison RR (1936) Fibrinolytic specificity of *Bacillus pestis*. Proc Soc Exp Biol Med 34:301–302
Mair NS (1968) *Pasteurella pseudotuberculosis* infections in Great Britain 1959–1966. Symp Ser Immunobiol Stand 9:121–128
Mair NS (1973) Yersiniosis in wildlife and its public health. J Wildl Dis 9:64–71
Mair NS (1975) Yersiniosis. In: Hubert WT, McCulloch WF, Schnurrenberger PR (eds) Diseases transmitted from animals to man, 6th edn. Charles C. Thomas, Springfield, Ill, pp 174–185
Marchalonis JJ (1974) Phylogenetic origins of antibodies and immune recognition. In: Brent L, Holborow J (eds) International Congress of Immunology, Brighton. North-Holland, Amsterdam (Progress in Immunology II)

Martinevskii IL (1967) Taxonomie du genre Yersinia et nomenclature de ses representants. Symp Ser Immunobiol Stand 9:361–368

Marx JL (1975) Antibody structure: Now in three dimensions. Science 189:1075–1076

Mashoff W Von (1953) Eine neuartige Form der mesenterialen Lymphadenitis. Dtsch Med Wochenschr 78:532–535

Matsuyama T (1982) Surface parasites – on the host cell surface as a microenvironment. Seibutsu Kagaku (Biol Sci, Tokyo) 34:57–67

McClintock B (1951) Chromosome organization and genetic expression. Cold Spring Harbor Symp Quant Biol 16:13–47

McDevitt HO, Deak BD, Schreffler DC, Klein J, Stimpfing JH, Snell GD (1972) Genetic control of the immune response. Mapping of the Ir-1 locus. J Exp Med 135:1259–1278

Melching L, Vas SI (1971) Effects of serum components on gramnegative bacteria during bactericidal reactions. Infect Immun 3:107–115

Melish ME, Glasgow LA (1970) The staphylococcal scalded-skin syndrome: development of an experimental model. N Engl J Med 282:1114–1119

Melish ME, Glasgow LA, Turner MD (1972) The staphylococcal scalded-skin syndrome: isolation and partial characterization of the exfoliative toxin. J Infect Dis 125:129–140

Meyer KF (1953) Recent studies on the immunity response to administration of different plague vaccines. Bull WHO 9:616–636

Meyer KF (1958) Pasteurella. In: Dubois RJ (ed) Bacterial and mycotic infections of man, 3rd ed. Lippencott, Philadelphia, Pa, pp 405–420

Meyer KF, Foster LE (1948) Measurement of protective serum antibodies in human volunteers inoculated with plague prophylactics. Stanford Med Bull 6:75–79

Meyers JA, Sanchez D, Elwell LP, Falkow S (1976) Simple agarose gel electrophoretic method for the identification and characterization of plasmid deoxyribonucleic acid. J Bacteriol 127:1529–1537

Middlebrook G, Dubois RJ (1948) Specific serum agglutinations of erythrocytes sensitisized with extracts of tubercle bacilli. J Exp Med 88:521–528

Migler R, DeChatelet LR, Bass DA (1978) Human eosinophilic peroxidase: role in bactericidal activity. Blood 51:445–456

Miki K, Mackaness GB (1964) The passive transfer of acquired resistance to *Listeria monocytogenes*. J Exp Med 120:93–103

Miles AA, Khimiji PL (1975) Enterobacterial chelator of iron: their occurrence, detection, and relation to pathogenicity. J Med Microbiol 8:477–490

Miller JFAP (1962) Immunological significance of the thymus of the adult mouse. Nature 195:1318–1319

Mitsuoka T (1978) A story of intestinal bacterial flora. Iwanamishoten, Tokyo

Mitsuoka T, Hayakawa K (1972) Die Fäkalflora bei Menschen I. Mitteilung: Die Zusammensetzung der Fäkalflora der verschiedenen Altersgruppen. Zentralbl Bakteriol Hyg Abt Orig A223:333–342

Mitsuoka T (1982) A color atlas of anaerobic bacteria. Sobunsha, Tokyo, p 341

Mollaret HH (1968) L'infection humaine et animale à bacille de Malassez et Vignal, en France, de 1959a 1967. Symp Ser Immunobiol Stand 9:45–58

Moncla BJ, Hillier SL, Charnetzky WT (1983) Constitutive uptake and degradation of fatty acids by *Yersinia pestis*. J Bacteriol 153:304–344

Montie TC, Montie DB (1971) Protein toxins of *Pasteurella pestis* subunit composition and acid finding. Biochemistry 70:2094–2100

Moscona AA (1975) Embryonic cell surfaces: mechanisms of cell recognition and morphogenetic cell adhesion. In: McMahon D, Fox CF (eds) Developmental biology. W. A. Benjamin, Menlo Park, Cal

Moulder JW (1962) The biochemistry of intracellular parasitism. Univ Chicago Press, Chicago

Müller-Eberhard HJ (1980) Complement reaction pathway. In: Fougereau M, Dausset J (eds) Immunology 80. Academic Press, London, pp 1001–1024

Muramatsu S, Inaba K (1982) Functions of Ia^+ nonlymphoid-nonmacrophage cells in the initiation of immune response. In: Mizuno D, Cohn ZA, Takeya K, Ishida N (eds) Self-defense mechanisms, role of macrophages. Tokyo Press, Tokyo

Muschel LH, Larsen LJ (1970) Effect of hypertonic sucrose upon the immune bactericidal reaction. Infect Immun 1:51–55

Nagano Y, Kojima Y (1954) Pouvoir immunisant du virus vaccinal inactivé par des rayons ultraviolets. C R Soc Biol 148:1700–1702

Nakamura A (1967) Virulence of shigella isolated from healthy carriers. Jpn J Med Sci Biol 20:213–223

Nakamura A (1973) Pathogenesis of bacillary dysentery. Mod Media 19:373–384

Nakano M (1961) An electron-microscopic study of the liver of mice infected with *Salmonella enteritidis*. J Electron Microsc 10:26–36

Nakano M, Saito K (1969) Chemical components in the cell wall of *Salmonella typhimurium* affecting its virulence and immunogenicity in mice. Nature 222:1085–1086

Nathan CF, Nakagawa A (1982) Role of reactive oxygen intermediates in macrophage killing of intracellular pathogens: a review. In: Mizuno D, Cohn ZA, Takeya K, Ishida N (eds) Self-defense mechanisms: role of macrophages. Univ Tokyo Press, Tokyo, pp 279–294

Nathan CF, Brukner LH, Silverstein SC, Cohn ZA (1979a) Extracellular cytolysis by activated macrophages and granulocytes I. Pharmacologic triggering of effector cells and the release of hydrogen peroxide. J Exp Med 149:84–99

Nathan CF, Silverstein SG, Brukner LH, Cohn ZA (1979b) Extracellular cytolysis by activated macrophages and granulocytes II. Hydrogen peroxide as a mediator of cytotoxicity. J Exp Med 149:100–113

Nazy B (1980) Vaccination of cows with a K99 extract to protect newborn calves against experimental enterotoxic collibacillus. Infect Immun 27:21–24

Neal AL, Strong FM (1943) Microbiological determination of pantothenic acid. Further studies. Ind Eng Chem Anal Ed 15:654–657

Neilands JB (1957) Some aspects of microbial iron metabolism. Bacteriol Rev 21:101–111

Nieman C (1954) Influences of trace amounts of fatty acids on the growth of microorganisms. Bacteriol Rev 18:147–161

Nikaido H (1962) Studies on the biosynthesis of cell wall polysaccharides in mutant strain Salmonella. Proc Natl Acad Sci USA 48:1337–1341 & 1542–1548

Nikaido H (1976) Outer membrane of *Salmonella typhimurium*. Transmembrane diffusion of some hydrophobic substances. Biochem Biophys Acta 433:118–132

Nishioka K (1980) Complement and defense mechanisms. In: Mizuno D, Takeya K, Ishida N (eds) The defense mechanisms of living organisms. Tokyodaigaku-Shutspankai, Tokyo, pp 287–299

Noda M, Kato I, Matsuda F, Hirayama T (1980) Fixation and inactivation of staphylococcal leukocidin by phosphatidylcholine and ganglioside GM1 in rabbit polymorphonuclear leukocytes. Infect Immun 29:678–684

Noda M, Kato I, Matsuda F, Hirayama T (1981a) Studies on the leukocytolytic action of staphylococcal leukocidin. Jpn J Bacteriol 36:89

Noda M, Kato I, Matsuda F, Hirayama T (1981b) Mode of action of staphylococcal leukocidin: relationship between binding of ^{125}I-labeled S and F components of leukocidin to rabbit polymorphonuclear leukocytes and leukocidin activity. Infect Immun 34:362–367

Noll H, Bloch H, Asselineau J, Lederer E (1956) The chemical structure of the cord factor of *Mycobacterium tuberculosis*. Biochim Biophys Acta 20:299–309

North RJ (1973) Importance of thymus-derived lymphocytes in cell-mediated immunity to infection. Cell Immunol 7:166–176

North RJ, Bereche PA (1982) Non-H-2 restriction of T-cell-mediated macrophage activation in vivo. In: Mizuno D, Cohn ZA, Takeya K, Ishida N (eds) Self-defense mechanisms: role of macrophages. Tokyodaigaku-Shutsupankai (Univ Tokyo Press), Tokyo, pp 265–268

O'Brien IG, Gibson F (1970) The structure of enterochelin and related 2,3-dihydroxy-N-benzoylserine conjugate from *Escherichia coli*. Biochim Biophys Acta 215:393–402

O'Brien AD, Kapral FA (1976) Increased cyclic adenosine 3′5′-monophosphate content in guinea pig ileum after exposure to *Staphylococcus aureus* delta toxin. Infect Immun 13:152–162

Ogawa H, Yamazaki Y (1978) Components of complement, immunoglobulins and lysozyme in the tonsil. Keio J Med 27:99–109

Ogawa H, Nakamura A (1969) Experimental bacillary dysentery in rabbits. Jpn J Med Sci Biol 22:383–389

Ogawa H, Honjo S, Takasaka M, Fujiwara T, Imaizumi K (1966) Shigellosis in cynomolgus monkeys. (Macaca irus) IV. Bacteriological and histopathological observations on the earlier stage of experimental infection with *Shigella flexneri* 2a. Jpn J Med Sci Biol 19:23–32

Ogawa H, Yoshikura H, Nakamura A, Nakaya R (1967) Susceptibility of cell cultures from various mammalian tissues to the infection by Shigella. Jpn J Med Sci Biol 20:329–339

Ogawa H, Nakamura A, Nakaya R (1968) Cinematographic study of tissue cell cultures infected with *Shigella flexneri*. Jpn J Med Sci Biol 22:259–273

Ohashi M, Zen'Yoji H (1977) *Escherichia coli* enterotoxin (in Japanese). Jpn J Bacteriol 32:455–468

Ohashi M, Shimada T, Fukumi H (1972) In vitro production of enterotoxin and hemorrhagic principle by *Vibrio cholerae* NAG. Jpn J Med Sci Biol 25:179–194

Okanishi M (1977) The production of the second metabolites and plasmid (in Japanese). Amino Acid Nucleic Acid 35:15–30

Olds GR, Ellner JJ, Kearse LA Jr, Kazura JW, Mahmoud AAF (1980) Role of arginase in killing of Schistosomula of *Schistosoma mansoni*. J Exp Med 151:1557–1562

Oppenheim JJ, Lugar T, Sztein MB, Steeg PS (1982) Circuit of cytokine-cell interactions that regulate immunological and inflammatory reactions. In: Mizuno D, Cohn ZA, Takeya K, Ishida N (eds) Self-defense mechanisms: role of macrophages. Tokyodaigaku-Shutspankai (Univ Press of Tokyo), Tokyo, pp 127–136

Oren R, Farham AE, Saito K, Milofsky E, Karnovsky ML (1963) Metabolic patterns in three types of phagocytizing cells. J Cell Biol 17:487–501

Ørskov I, Ørskov F (1966) Episome-carried surface antigen K88 of *Escherichia coli*. I. Transmission of the determinant of the K88 antigen and the influence on the transfer of chromosomal markers. J Bacteriol 91:69–75

Ørskov F, Ørskov H, Smith W, Sojka WJ (1975) The establishment of K99, a thermolabile *Escherichia coli* Kanyigen, previously called "kco", possessed by calf and lamb enteropathogenic strains. Acta Path Microbiol Scand B 83:31–36

Osebold JW, Sawyer MT (1957) Immunization studies on listeriosis in mice. J Immunol 78:262–268

Owen JJT (1980) B cell development. In: Fougereau M, Dausset J (eds) Immunology 80. Academic Press, London, pp 303–314

Packer L, Rust JH, Ajl S (1959) Action of plague murine toxin on mammalian mitochondrial respiration. J Bacteriol 78:658–664

Panton PM, Valentine FCO (1932) Staphylococcal toxin. Lancet 222:506–508

Payne SM, Finkelstein RA (1977) Detection and differentiation of iron-responsive avirulent mutants on Congo red agar. Infect Immun 18:94–98

Payne SM, Finkelstein RA (1978) Siderophore production by *Vibrio cholerae*. Infect Immun 20:310–311

Payne SM, Finkelstein RA (1978) The critical role of iron in host-bacterial interactions. J Clin Invest, pp 1428–1440

Pekarek RS, Bostian KA, Bartelloni PJ, Calia FM, Beisel WR (1969) The effects of *Francisella tularensis* infection on iron metabolism in man. Am J Med Sci 258:14–25

Perry RD, Brubaker RR (1979) Accumulation of iron by *Yersiniae*. J Bacteriol 137:1290–1298

Pfeiffer R, Issaeff B (1894) cited from p 912 Kawakita Y (1977) The historical basis of the modern medicine, The last volume. Iwanamishoten, Tokyo, p 1206 (Z Hyg Infektionskr 18:1–14)
Pirquet C von (1910) Allergie. Springer, Berlin
Pirquet C von (1906) Allergie. Münch Med Wochenschr 53:1457–1458
Podack ER, Esser AF, Biesecker G, Müller-Eberhard HJ (1980) Membrane attack complex of complement. A structural analysis of its assembly. J Exper Med 151:301–313
Pollack JR, Neilands JB (1970) Enterobactin, an iron transport compound from *Salmonella typhimurium*. Biochem Biophys Res Commun 38:989–992
Portnoy DA, Falkow S (1981) Virulence-associated plasmids from *Yersinia enterocolitica* and *Yersinia pestis*. J Bacteriol 148:877–883
Pollitzer R (1954) Plague. WHO Monograph series No. 22, Geneva Portnoy DA, Moseley SL, Falkow S (1981) Characterization of plasmids and plasmid-associated determinants of *Yersinia enterocolitica* pathogenesis. Infect Immun 31:775–782
Prausnitz C, Kustner H (1921) Studien über Überempfindlichkeit. Zentralbl Bakteriol 86:160–169
Prusiner SB (1982) Novel proteinaceous infectious particles cause scrapie. Science 216:136–138
Puhvel SM (1966) *Corynebacterium acnes*. Arch Dermatol 93:364–366
Puschmann M, Ganzoni AM (1977) Increased resistance of iron-deficient mice to Salmonella infection. Infect Immun 17:663–664
Raffel S (1955) The mechanism involved in acquired immunity to tuberculosis. CIBA foundation symposium and host, with an addendum on leprosy. Little, Brown, Boston
Reggiardo Z, Middlebrook G (1974) Failure of passive serum transfer of immunity against aerogenic tuberculosis in rabbits. Proc Soc Exp Biol Med 145:173–175
Rich AR (1951) The pathogenesis of tuberculosis, 2nd edn. Charles C. Thomas, Springfield
Richét PC (1902) De l'action anaphylactique de certains venins. C R Soc Biol 54:170–172
Ritter von Rittershain G (1878) Die exfoliative Dermatitis jüngerer Säuglinge. Z Kinderheilkd 2:2–23
Roberts RJ (1976) Restriction endonucleases. CRC Crit Rev Biochem 4:123–164
Robins-Browne RM, Still CS, Isaacson M, Koornhof HJ, Appelbaum PC, Scragg JN (1977) Pathogenic mechanisms of a non-agglutinable *Vibrio cholerae* strain: Demonstration of invasive and enterotoxigenic properties. Infect Immun 18:542–545
Rogers DE (1956) Studies on bacteriemia 1. Mechanisms relating to the persistence of bacteriemia in rabbits following the intravenous injection of staphylococci. J Exp Med 103:713–742
Rogolsky M (1979) Nonenteric toxins of *Staphylococcus aureus*. Microbiol Rev 43:320–360
Rogolsky M, Warren R, Wiley BB, Nakamura HT, Glasgow LA (1974) Nature of the genetic determinant controlling exfoliative toxin production in *Staphylococcus aureus*. J Bacteriol 117:157–165
Rosen H, Klebanoff SJ (1979) Hydroxyl radical generation by polymorphonuclear leucocytes measured by electron spin resonance spectroscopy. J Clin Invest 64:1725–1729
Rosenblum ED, Tyrone S (1976) Chromosomal determinants for exfoliative toxin production in two strains of staphylococci. Infect Immun 14:1259–1260
Rowley D, Turner KJ, Jenkin CR (1964) The basis for immunity to mouse typhoid 3. cell-bound antibody. Aust J Exp Biol Med Sci 42:237–248
Rubin RH, Weinstein L (1977) Salmonellosis, microbiologic, pathologic and clinical features. Stratton, New York
Ryder RW, Wachsmuth IK, Buxton AE, Evans DG, Dupont HL, Mason E, Barrett FF (1976) Infantile diarrhea produced by heat-stable enterotoxigenic *Escherichia coli*. N Engl J Med 295:849–853
Ryder LP, Andersen E, Svejgaard A (1979) HLA and disease registry. Third report. Munksgaard, Copenhagen
Sack DA, Sack RB (1975) Test for enterotoxigenic *Escherichia coli* using Y1 adrenal cells in miniculture. Infect Immun 11:334–336

Sack RB, Gorbach SL, Banwell JD, Jacobs B, Chatterjee BD, Mitra RC (1971) Enterotoxigenic *Escherichia coli* isolated from patients with severe cholera-like disease. J Infect Dis 123:378–385

Sack DA, Merson MH, Well JD, Sack RB, Morris GK (1975) Diarrhoea associated with heat-stable enterotoxin-producing strains of *Escherichia coli*. Lancet 2:238[7928]–241

Saito H, Kranz DM, Takagaki Y, Hayday AC, Eisen HN, Tonegawa S (1984) Complete primary structure of a heterodimeric T-cell receptor deduced from cDNA sequences. Nature 309:757–762

Saito K, Akiyama T, Nakano M, Ushiba D (1960) Interaction between *Salmonella enteriditidis* and tissue cultured macrophages derived from immunized animals. Jpn J Microbiol 4:395–407

Sakurai S, Kondo I (1978) Characterization of exfoliatin A as a metallotoxin with special reference to determination of the contained metal by radioactivation analysis. Jpn J Med Sci Biol 31:208–211

Samuelsson B (1983) Leukotrienes: Mediators of immediate hypersensitivity reactions and inflammation. Science 220:568–575

Samuelsson K, Lindberg B, Brubaker RR (1974) Structure of O-specific side chains of lipopolysaccharides from *Yersinia pseudotuberculosis*. J Bacteriol 117:1010–1016

Sansonetti PJ, Kopecko DJ, Formal SB (1981) *Shigella sonnei* plasmids: evidence that a large plasmid is necessary for virulence. Infect Immun 34:75–83

Sansonetti PJ, Kopecko DJ, Formal SB (1982) Involvement of a plasmid in the invasive ability of *Shigella flexneri*. Infect Immun 35:852–860

Sbarra AJ, Karvovsky ML (1959) The biochemical basis of phagocytosis. I. Metabolic changes during the ingestion of particles by polymorphonuclear leukocytes. J Biol Chem 234:1355–1362

Schade AL, Caroline L (1944) Raw hen egg white and the role of iron in growth inhibition of *Shigella dysenteriae, Staphylococcus aureus, Escherichia coli* and *Saccharomyces cerevisiae*. Science 100:14–15

Schade AL, Caroline L (1946) An iron binding component in human blood plasms. Science 104:340–341

Schiemann DA, Devenish JA (1980) Virulence of *Yersinia enterocolitica* determined by lethality in mongolian gerbils and by the Serény test. Infect Immun 29:500–506

Schlifstein JI, Coleman MB (1939) An unidentified microorganism resembling *B. lignieri* and *Past. pseudotuberculosis,* and pathogenic for man. NY State J Med 39:1749–1753

Schrank GD, Verwey WF (1976) Distribution of cholera organisms in experimental *Vibrio cholerae* infections. Proposed mechanisms of pathogenesis and antibacterial immunity. Infect Immun 13:195–203

Schreiber RD, Morrison DC, Podack ER, Müller-Eberhard HJ (1979) Bactericidal activity of the alternative complement pathway generated from 11 isolated plasma proteins. J Exp Med 149:870–882

Schütze H (1929) A system of bacteriology. Vol 4. His Majesty's Stationary Office

Schütze H (1939) Studies on *B. pestis* antigen of prophylactic agents. Br J Exp Pathol 20:235–244

Segal W, Bloch H (1956) Biochemical differentiation of *Mycobacterium tuberculosis* grown in vivo and in vitro. J Bacteriol 72:132–141

Serény B (1957) Experimental keratoconjunctivitis Shigellosa. Acta Microbiol Hung 4:367–376

Shen CW, Freeze E (1972) Effects of fatty acids on growth and envelope proteins of *Bacillus subtilis*. J Bacteriol 111:516–524

Shiga K (1898) Über den Erreger der Dysenterie in Japan. Zentralbl Bakteriol Abt 1 23:599–600

Shipley PL, Dallas WS, Dougan G, Falkow S (1979) Expression of plasmid genes in pathogenic bacteria. In: Schlesinger D (ed) Microbiology – 79, AMS. Washington DC, pp 176–180

Siegel I, Gleicher N (1983) The red army. In the war against disease, white cells may not stand alone. Science 23:30–34

Skamene E, Gros P, Forget A, Konghavn PAL, St. Charles CS, Tayler BA (1982) Genetic regulation of resistance to intracellular pathogens. Nature 297:506–511

Smith DA, Burrows TW (1962) Phage and bacteriocin studies with *Pasteurella pestis* and other bacteria. Nature 193:397–398

Smith H (1972) The little known determinants of microbial pathogenicity. Symposia of the Society for General Microbiology. Number XXII. Microbial pathogenicity in man and animals, pp 1–24

Smith HW, Gyles CL (1970) The relationship between two apparently different enterotoxins produced by enteropathogenic strains of *Escherichia coli* of porcine origin. J Med Microbiol 3:387–401

Smith KO, Gehle WD (1977) Magnetic transfer devices for the use in solid phase radioimmunoassay and enzyme-linked immunosorbent assays. J Infect Dis 136 suppl:s329–336

Snow GA, White AJ (1969) Chemical and biological properties of mycobactins isolated from various mycobacteria. Biochem J 115:1031–1045

Snyder DS, Beller DI, Unanue ER (1982) Prostaglandins modulate macrophage Ia expression. Nature 299:163–165

So M, Falkow S (1977) Molecular cloning as a tool in the study of pathogenic *E. coli*. In: Beers RF Jr, Bassett EG (eds) Recombinant molecules. Impact on science and society. Raven, New York, pp 107–122

So M, Boyer HW, Betlach M, Falkow S (1976) Molecular cloning of an *Escherichia coli* plasmid determinant that encodes for the production of heat-stable enterotoxin. J Bacteriol 128:463–472

Southern EM (1975) Detection of specific sequences among DNA fragments separated by gel electrophoresis. J Mol Biol 98:503–517

Spika JS, Verbrough HA, Verhoef J (1981) Protein A effect on alternative pathway complement activation and opsonization of *Staphylococcus aureus*. Infect Immun 34:455–460

Spivack ML, Karler A (1958) Purification of the toxin of *Pasteurella pestis* by contineous-flow paper electrophoresis. J Immunol 81:441–445

Steinman RM, Cohn ZA (1973) Identification of a novel cell type in peripheral lymphoid organs of mice. I. Morphology, quantitation, tissue distribution. J Exp Med 137:1142–1162

Steinman RM, Cohn ZA (1974) Identification of a novel cell type in peripheral lymphoid organs of mice. II. Functional properties in vitro. J Exp Med 139:380–397

Steinman RM, Nussenzweig MC, Austyn JM, Witmer MD, Gutchinov B, Unkeless JC, Cohn ZA (1982) Monoclonal antibody studies of mouse dendritic cells and macrophages. In: Mizuno D, Cohn ZA, Takeya K, Ishida N (eds) Self-defense mechanisms: role of macrophages. Tokyodaigaku-Shutspankai (Univ Tokyo Press), Tokyo

Stocker BAD (1955) Bacteriophage and bacterial classification. J Gen Microbiol 12:375–381

Stocker BAD, Wilkinson RC, Makela PH (1966) Genetic aspects of biosynthesis and structure of *Salmonella* somatic polysaccharide. Ann NY Acad Sci 133:334–343

Straley SC, Brubaker RR (1981) Cytoplasmic and membrane proteins of yersiniae cultivated under conditions simulating mammalian intracellular environment. Proc Natl Acad Sci USA 78:(2)1224–1228

Straley SC, Harmon PA (1984) Growth in mouse peritoneal macrophage of *Yersinia* pestis lacking established virulence determinants. Infect Immun 45:649–654

Strominger JL (1980) Structure of products of the major histocompatibility complex in man and mice. In: Fougereau M, Dausset J (eds) Immunology 80. Academic Press, London, pp 541–554 (Progress in immunology IV.)

Surgalla MJ, Beesley ED (1971) Infectivity and virulence of nonpesticinogenic *Pasteurella pestis*. Infect Immun 4:416–418

Surgalla MJ, Andrews AW, Cavanaugh DC (1968) Studies on virulence factors of *Pasteurella pestis* and *Pasteurella pseudotuberculosis*. Symp Ser Immunol Stand 9:293–302

Surgalla MJ, Bessley ED, Albizo JN (1970) Practical applications of new laboratory methods for plague investigations. Bull WHO 42:991–997
Sutoh Y, Wake A (1979) The transfer of R-factor into *Yersinia pestis* and its relationship to virulence VI. Jpn J Bacteriol 34:156
Svejgaard A, Morling N, Platz P, Ryder LP, Thomsen M (1980) HLA and disease. In: Fougereau M, Dausset J (eds) Immunology 80. Academic Press, London (Progress in Immunology IV)
Takasaka M, Honjo S, Imaizumi H, Ogawa H (1970) Dysentery produced by oral administration of the colonic contents of Cynomolgus monkeys infected with *Shigella flexneri* 2a. Jpn J Med Sci Biol 23:249–254
Takeda Y, Honda T, Taga S, Miwatani T (1981) In vitro formation of hybrid toxins between subunits of *Escherichia coli* heat-labile enterotoxin and those of cholera enterotoxin. Infect Immun 34:341–346
Tauber H, Russell H (1961) General method for the isolation of exotoxins. Study of lipid and polysaccharide moieties. Exp Med Surg 19:161–170
Taverne J, Dockrell HM, Playfair JHL (1981) Enterotoxin-induced serum factor kills malarial parasites in vitro. Infect Immun 33:83–89
Tiselius A (1937) Electrophoresis of purified antibody preparations. J Exp Med 65:646
Tsubokura M, Itagaki K, Kawamura K, Sasaki T, Nagai T (1971) Studies on *Yersinia (Pasteurella pseudotuberculosis)*. II. A new type of *Y. pseudotuberculosis,* type VI, and subdivision of type V strains. Jpn J Vet Sci 33:137–144
Tsukano H, Wake A (1982) Virulence determinants of *Yersinia pestis* and plasmid. Jpn J Bacteriol 37:236
Tsukano H, Sutoh Y, Wake A (1978) The transfer of R-factor to *Yersinia pestis* and its relationship to virulence V. Jpn J Bacteriol 33:135
Uchida T, Gill DM, Pappenheimer AM Jr (1971) Mutation in the structural gene for diphtheria toxin carried by temperate phage β. Nature New Biol 233:8–11
Ushiba D, Saito K, Akiyama T, Nakano M, Sigiyama T, Shirono S (1959) Studies on experimental typhoid: Bacterial multiplication and host cell response after infection with *Salmonella enteritidis* in mice immunized with live and killed vaccines. Jpn J Microbiol 3:231–242
Ushiba D, Saito K, Nakano M (1962) Selective inbreeding of mouse strains uniformly susceptible to mouse typhoid and experimental infections of these strains. II. Susceptibility of DK1 strain of mice to various infection with *S. enteritidis* between DK1 and other uniform strains. Jpn J Exp Med 32:519–529
van Furth R (1970) Origin and kinetics of monocytes and macrophages. Semin Hematol 7:125
van Furth R, van der Meer JWM, van Oud Alblas Blussé, Sluiter W (1982) Development of mononuclear phagocytes. In: Mizuno D, Cohn ZA, Takeya K, Ishida N (eds) Self-defense mechanisms. Role of macrophages. A Naito Foundation Symposium. Univ Tokyo Press. Elsevier Biomedical, Tokyo, pp 25–43
Vasil ML, Holmes RK, Finkelstein RA (1975) Conjugal transfer of a chromosomal gene determining production of enterotoxin in *Vibrio cholerae*. Science 187:849–850
Vaughn LK, Kluger MJ (1977) Fever and survival in bacterially infected rabbits. Proc Fed Am Soc Exp Biol 36:511
Viscari G, Olitzki AL, Olitzki Z (1960) The action of the thermolabile toxin of *Shigella dysenteriae* on cells cultivated in vitro. Br J Exp Pathol 41:179–189
Vogel F, Pettenkoffer HJ, Helmbold W (1960) Über die Populationsgenetik der AB0-Blutgruppen 2. Mitteilung. Genhäufigkeit und epidemische Erkrankungen. Acta Genet Stat Med 10:267–294
Voino-Yasenetsky MV, Khavkin TN (1964) A study of intraepithelial localization of dysentery causative agents with the aid of fluorescent antibodies. Zh Microbiol (Moscow) 12:98–100
Wadstrom T (1974) Biological properties of extracellular protein from staphylococcus. Ann NY Acad Sci 236:343–361

Waite M, Scherphof GL, Boshouweres FMG, van Deenen LLM (1969) Differentiation of phospholipase A in mitochondria and lysosomes of rat liver. J Lipid Res 10:411–420
Wake A (1976) The transfer of a R-factor to *Yersinia pestis* and its relationship to virulence I. Jpn J Bacteriol 31:228
Wake A (1978) Plague bacteria (in Japanese). Rinshoukensa (Clin Invest) 17:710–711
Wake A (1980 a) Prerequisite for intracellular destruction of virulent *Yersinia pestis*. 11.3.25, Abstr, 4th Int Congr Immunol, Paris
Wake A (1980 b) Siderophores. In: The target of microbiology. Eikensosho, Eikenkagaku, Tokyo, 13:172–180
Wake A (1984) Virulence determinants of *Yersiniae*. In: Yoshikawa M (ed) The pathogenicity of bacteria – its molecular genetics. Maruzen, Tokyo, pp 133–153
Wake A (1985) Higher resistance of DBA/2N mouse strain to *Yersinia pestis* infection. In: Skamene E (ed) Genetic control of host resistance to infection and malignancy. Alan R Liss, New York, pp 361–366
Wake A, Sutoh Y (1983) Mechanisms of protection against virulent Yersinia pestis infection without participation of humoral antibody: H-2 restriction in athymic mouse model. Curr Microbiol 8:79–84
Wake A, Yamamoto M, Fukumi H (1954) A taxonomic study of Vibrio cholerae and cholera-like vibrios by the use of Dutta test. Jpn J Bacteriol 19:338–339
Wake A, Morita H, Yamamoto M (1970) *Yersinia pestis*, animal body and iron: a discussion on the bacterial pathogenicity. Seibutsukagaku (Biol Sci) 22:113–129
Wake A, Misawa M, Morita H (1972) Studies on the immunity against proliferative antigens, using an iron drug and *Yersinia pestis*, *Yersinia pseudotuberculosis*. Proc IInd Annu Meet Jpn Soc Immun 2:112–114
Wake A, Takeuchi S, Morita H (1973) Mechanisms of antiplague cellular immunity induction by *Yersinia pseudotuberculosis* suspended in an iron-drug. Results from the applications of fluorescent antibody technic and of viable-bacteria counting within cellular fraction of the spleen. Proc IIIrd Annu Meet Jpn Soc Immunol 3:219–221
Wake A, Misawa M, Matsui A (1975) Siderochrome production by *Yersinia pestis* and its relation to virulence. Infect Immun 12:1211–1213
Wake A, Tsukano H, Ikeda N, Ichiki S (1977) The transfer of R-factor to *Yersinia pestis* and its relationship to virulence III. Jpn J Bacteriol 32:120
Wake A, Morita H, Wake M (1978) Mechanisms of long and short term immunity to plague. Immunology 34:1045–1052
Wake A, Maruyama T, Akiyama K, Yamamoto M (1983) The role of virulence antigens (VW) in the protection of mice against *Yersinia pestis* infection. Curr Microbiol 8:73–77
Walker RV, Barnes MG, Higgins ED (1966) Composition and physiopathology produced by plague endotoxin. Nature 209:1246
Warren R, Rogolsky M, Wiley BB, Glasgow LA (1975) Isolation of extrachromosomal desoxyribonucleic acid for exfoliative toxin production from phage group II *Staphylococcus aureus*. J Bacteriol 122:99–105
Wassom DL, David CS, Gleich J (1980) MHC-linked genetic control of the immune response to parasites: *Trichinella spiralis* in the mouse. In: Skamene E, Konghavn PAL, Landy M (eds) Genetic control of natural resistance to infection and malignancy. Academic Press, New York, pp 75–82
Watanabe H, Timmis KN (1984) A small plasmid in *Shigella dysenteriae* 1 specifies one or more functions essential for O antigen production and bacterial virulence. Infect Immun 43:391–396
Watanabe H, Nakamura A, Timmis KN (1984) Small virulence plasmid of *Shigella dysenteriae* 1 strain W30864 encodes a 41,000 dalton protein involved in formation of specific lipopolysaccharide side chains of sertype 1 isolates. Infect Immun 46:55–63
Wauters G (1981) Antigens of *Yersinia enterocolitica*. In: Botton EJ (ed) *Yersinia enterocolitica*. CRC, Florida, pp 41–53

Weinberg ED (1966) Roles of metallic ions in host-parasite interactions. Bacteriol Rev 30:135–151
Weinberg ED (1971) Roles of iron in host-parasite interactions. J Infect Dis 124:401–410
Weinberg ED (1974) Role of temperature in trace metal metabolism in host-pathogen interaction. In: Hoekstra WG, Bremner I, Davies NT (eds) Trace element metabolism in animals – 2. Univ Park Press, Baltimore, pp 241–258
Wetzler TF, Hubbert WT (1968) *Pasteurella pseudotuberculosis* in North America. Symp Ser Immunobiol Stand 9:33–44
Wiley BB, Rogolsky M (1977) Molecular and serological differentiation of staphylococcal exfoliative toxin synthesized under chromosomal and plasmid control. Infect Immun 18:487–494
Wiley BB, Glasgow LA, Rogolsky M (1976) Staphylococcal scalded skin syndrome: development of a primary binding assay for human antibody to the exfoliative toxin. Infect Immun 13:512–520
Williams VR, Fieger EA (1946) Oleic acid as a growth stimulant for *Lactobacillus casei*. J Biol Chem 166:335–343
Williams WL, Broquist HP, Snell EE (1947) Oleic acid and related compounds as growth factors for lactic acid bacteria. J Biol Chem 170:619–630
Wilson MR, Hohman AW (1974) Immunity to *Escherichia coli* in pigs; adhesion of enteropathogenic *Escherichia coli* to isolated intestinal epithelial cells. Infect Immun 10:776–782
Wiseman GM (1975) The hemolysins of *Staphylococcus aureus*. Bacteriol Rev 39:317–344
Witte W (1976) Control of alpha-hemolysin formation by plasmid in distinct strains of *Staphylococcus aureus:* influence of erythromycin, rifampin and streptomycin. In: Jeliaszewicz J (ed) Staphylococci and staphylococcal diseases. Gustav Fischer, Stuttgart, pp 298–304
Wittig W (1965) Zum Vorkommen des K-antigens 88(L) bei *Escherichia coli* – Stamm von Schweinen. Zentralbl Bakteriol Parasitenkd Infektionskr Hyg Abt I Orig 197:487–499
Woodin AM, Wieneke AA (1966a) The modification of the cytotoxic effect of leucocidin by N-ethylmaleimide, flavine mononucleotide and menadione. Biochem J 99:469–478
Woodin AM, Wieneke AA (1966b) The interaction of leucocidin with the cell membrane of the polymorphonuclear leukocyte. Biochem J 99:479–492
Woodin AM, Wieneke AA (1967) The participation of phospholipids in the interaction of leucocidin and the cell membrane of the polymorphonuclear leucocyte. Biochem J 105:1029–1038
Woodin AM, Wieneke AA (1969) The action of phosphonates on the leucocyte in relation to the mode of action of leucocidin, the properties of the potassium pump and the inhibition of chemotaxis. Br J Exp Pathol 50:295–308
Woodin AM, Wieneke AA (1970) Leucocidin, TEA and membrane acyl phosphatase in relation to the leucocyte potassium pump. J Gen Physiol 56:16–32
Woodin AM, Wieneke AA (1971) Action of phospholipids and leucocidin on the p-nitrophenyl phosphatase of the leucocyte membrane. Biochem Biophys Acta 233:702–715
World Health Organization (1969) Outbreak of gastroenteritis by non-agglutinable (NAG) vibrios. Wkly Epid Rec 44:10
Wright AC, Douglass SR (1903) Proc R Soc (Lond) 72:375
Wright J (1936) Staphylococcal leucocidin (Neisserwechsberg type) and antileucocidin. Lancet 230:1002–1004
Yoshinaga M, Yoshida K, Tashiro A, Hayashi H (1971) The natural mediator for PMN emigration in Arthus lesions. I. Purification and characterization of leucoegresin from Arthus skin site. Immunology 21:281–298
Zahorchak RJ, Brubaker RR (1982) Effect of exogeneous nucleotides on Ca^{2+} dependence and V antigen synthesis in *Yersinia pestis*. Infect Immun 38:953–959
Zahorchak RJ, Charnetzky WT, Little RV, Brubaker RR (1979) Consequences of Ca^{2+} deficiency on macromolecular synthesis and adenylate energy charge in *Yersinia pestis*. J Bacteriol 139:792–799

Zen-Yoji H (1975) Bacteriology, ecology and infections of *Yersinia enterocolitica* and *Yersinia pseudotuberculosis* (in Japanese). Jpn J Bacteriol 30:571–581

Zen-Yoji H (1980) The mechanism of infection, pathogenesis and complications of Yersiniosis. In: Fujino T et al. (eds) The target of microbiology (in Japanese). Eiken-sosho, Tokyo 13:209–217

Zen-Yoji H, Maruyama T, Sakai S, Kimura S, Mizuno T, Momose T (1973) An outbreak of enteritis due to *Yersinia enterocolitica* occuring at a junior high school. Jpn J Microbiol 17:220–222

Zink DL, Feeley JC, Well JG, Vanderzant C, Vickery JC, Roof WD, O'Donovan GA (1980) Plasmid-mediated tissue invasiveness in *Yersinia enterocolitica.* Nature 251:230–233

Zinkernagel RM (1974) Restriction by the H-2 gene complex of the transfer of cell-mediated immunity to *Listeria monocytogenes*. Nature 251:230–233

Zinkernagel RM (1979) Associations between major histocompatibility antigens and susceptibility to disease. Annu Rev Microbiol 33:201–213

Zinkernagel RM, Althage A, Adler B, Blanden RV, Davidson WF, Kees U, Dunlop MBC, Schreiffler DC (1977) H-2 restriction of cell-mediated immunity to an intracellular bacterium. Effector T cells are specific for Listeria antigen in association with H-2I region-coded self-markers. J Exp Med 145:1353–1367

Zinnaka Y (1971) Characterization of vibrio isolated from the cases of cholera-like diarrheal disease in the Sudan. J Egypt Public Health Assoc 46:161–174

Zinnaka Y, Carpenter CCJ (1972) An enterotoxin produced by noncholera vibrios. Johns Hopkins Med J 131:403–411

Zinsser H (1921) On the essential identity of the antibodies. J Immunol 6:289–299

Zsigray RM, Lawton WD, Surgalla MJ (1983) Repression of the virulence of *Yersinia pestis* by an F'plasmid. Infect Immun 39:974–976

Subject Index

MIX
Papier aus verantwortungsvollen Quellen
Paper from responsible sources
FSC® C105338

If you have any concerns about our products,
you can contact us on
ProductSafety@springernature.com

In case Publisher is established outside the EU,
the EU authorized representative is:
Springer Nature Customer Service Center GmbH
Europaplatz 3, 69115 Heidelberg, Germany

Printed by Libri Plureos GmbH
in Hamburg, Germany